An Introduction to Numerical Computations

An Introduction to Numerical Computations

Sidney Yakowitz

Systems and Industrial Engineering Department
University of Arizona

Ferenc Szidarovszky

Department of Computer Science
University of Agriculture, Budapest

Macmillan Publishing Company
New York
Collier Macmillan Publishers
London

Copyright © 1986, Macmillan Publishing Company, a division of Macmillan, Inc.

Printed in the United States of America

Macmillan Publishing Company
866 Third Avenue, New York, New York 10022

Collier Macmillan Canada, Inc.

Library of Congress Cataloging in Publication Data

Yakowitz, Sidney.
  An introduction to numerical computations.

  Includes index.
  1. Numerical analysis.   I. Szidarovszky, Ferenc.
II. Title.
QA297.Y35   1986   511   85-3114
ISBN 0-02-430810-2

Printing: 1 2 3 4 5 6 7 8   Year: 6 7 8 9 0 1 2 3 4 5

ISBN 0-02-430810-2

Preface ▬▬▬▬▬▬▬▬▬▬

This book is an outgrowth of class notes used for a sophomore/junior-level course required of most engineering students at the University of Arizona. The prerequisites are a FORTRAN course and a calculus sequence. Previous or concurrent registration in an introductory course on differential equations and matrices is recommended. This computer methods course constitutes the only systematic numerical methods training most of our engineering graduates will receive, and the text is written with that fact in mind. On the other hand, a portion of the students do take the senior/graduate-level course in numerical analysis. *An Introduction to Numerical Computations* provides a solid foundation for such further study. It is rife with FORTRAN subroutines and computational examples and "experiments," but leaves rigorous development of theoretical superstructure and error analysis to future course work.

The subject matter here is traditional. The divergence of the present book from the many other numerical methods textbooks is by way of orientation. With some exceptions, we find other introductory textbooks more authoritarian and pedantic than we would like. Here we build intuition and self-reliance through an experimentalist viewpoint more akin to physics than to mathematics. Typically, we motivate the common techniques for a given class of problems through the following pattern of inquiry:

1. Devise a crude "commonsense" algorithm from first principles.
2. Solve various problems and find that trouble sometimes arises: The computed answer may not be sufficiently accurate to warrant the processing time or the method may even fail to provide an answer.
3. Understand the nature of the failure and that approximation polynomials or other analytic devices point the way to more sophisticated algorithms.
4. Try them out and find that they are effective.
5. Look for checks, error bounds, and so on, to gain assurance that methods thus devised are in fact working well.

Following this pattern, we provide incentive for seeking higher-order methods by pushing simplistic procedures beyond their limits, showing through concrete examples that in some reasonable circumstances, they cannot deliver a required accuracy. We do not settle for a more-sophisticated rule without evidence that it does indeed outperform the more primitive ones. This book is unique in its ambitious effort to compare features of various methods for a given problem area. Our aim in such analysis and computational experimentation is not just to provide this taxonomy but to encourage and guide the student in developing a critical eye for computational matters.

Although we aim to cultivate the reader's computational intuition, mathematical principles are at the forefront of this exposition. The reader will find this to be an honest and demanding work: The theoretical bases and fundamental properties of various methods are carefully derived. This book gives a solid foundation for numerical analysis but sensibly leaves completion of the edifice to future study.

PROGRAMMING ASPECTS

Developments in computer technology have assisted us in our dialectic and experimental pedagogic orientation. The students now use an interactive computer. Because of various capabilities and rapid throughput times on these machines, the students avoid much of the drudgery of earlier days, and concentrate their efforts on performing and interpreting their computational experiments. For example, they are able to copy the subroutines of this book directly into their file areas; the only coding required, therefore, is the (usually brief) problem-specific calling program. To assist the student in concentrating on essentials, this book supplies simple but adequate VAX-11 FORTRAN 77 subroutines for virtually all the methods described. (These subroutines are available on VAX/VMS tapes in either FORTRAN or Pascal from the first author. Also, they are offered in MS FORTRAN or TURBO Pascal on IBM-PC DOS-readable disks.) In composing these subroutines, we have held clarity and simplicity above all else, including efficiency, elegance, and safeguards. Most of the subroutines are accompanied by calling programs, output, and discussion of sample numerical experiments that show, by example, the experimental and pragmatic view toward computation which we believe to be most healthy and valuable. In addition to providing a direct link from theory to action, our approach to programming and use of subroutines exemplifies the modular top-down philosophy which has proven so effective over the past decade.

PLANS OF STUDY

The "unstarred" sections of this book constitute our basic numerical computations course. We have taken great pains to make these sections, which include the traditional fare of introductory numerical methods courses, readily accessible to our students. The "starred" sections contain material we regard as important but not essential in a first course.

We remark on a few procedural issues. In discussing splines (Chapter 2) and interpolatory quadrature (Chapter 3), we introduce linear equations before giving appropriate methodology (Chapter 4). Our justification for this inverse ordering is that we want to introduce students to livelier computational ideas early and to motivate the relatively staid linear equation techniques before describing them. But, with respect to unstarred sections, the instructor can order the material in any fashion within the prerequisite chapter constraints shown in the following figure.

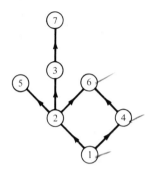

ACKNOWLEDGMENTS

We are deeply grateful for encouragement and support for this effort from the Systems and Industrial Engineering Department and the Engineering College of the University of Arizona. Also, we are indebted to innumerable students, secretaries, and graduate assistants, who provided help beyond the call of duty in refining early versions of the book. Tommy Hassett, Karen Markey, and Brian Rutherford particularly deserve our deep gratitude. Finally, we mention that it was through various National Science Foundation grants that the authors met and enjoyed a fruitful and happy decade of collaboration.

Sidney Yakowitz   Ferenc Szidarovszky
Tucson, Arizona   Budapest, Hungary

Contents

3 Numerical Differentiation and Integration   79

4 The Solution of Simultaneous Linear Equations   135

5   Nonlinear Equations   **191**

6   Function Approximation and Data Fitting   **229**

7   The Solution of Ordinary Differential Equations   285

CHAPTER 1

Computer Number Representation and Roundoff

1.1
FUNDAMENTAL CAPABILITIES OF A COMPUTER

To appreciate the concerns of "numerical computations," one must understand what a computer can and cannot do with numbers. Within limits to be discussed in this chapter, the computer has two fundamental capabilities:

Capability 1. The computer can store a finite set of numbers.
Capability 2. The computer can perform arithmetic (addition, subtraction, multiplication, and division), and it can find the order of any two stored numbers x and y. That is, it can decide whether x is greater than, equal to, or less than y.

Every computational solution to a numerical problem must ultimately be built up of operations in which stored numbers are compared or operated on arithmetically and the resultant stored. Most problems of interest cannot be solved by a finite sequence of such operations. We must be content with some program that yields an approximation to the solution and accept that there will typically be approximation error. Such error resulting from replacing a desired mathematical operation by a realizable computation will be referred to as *truncation error*. Techniques for bounding this error or assuring a specified accuracy should accompany a computational method.

For reasons to be disclosed in this chapter, neither capability 1 nor capability 2 can be achieved exactly. Computer words are equivalent to strings of 0's and 1's of uniform length. There are consequently only a fixed finite number of distinct computer words available for approximation of numbers. Thus regardless of what scheme is used to map the infinite set of real numbers into computer numbers, error will usually occur. Such error is referred to as *roundoff error*. In several sections of this chapter, origins, bounds, illustrations, and remedies for roundoff error will be presented. Later chapters offering numerical methods for various

1

classes of computational problems will give us insight into the cause and magnitude of truncation error.

Examples 1.1 and 1.2 demonstrate roundoff and truncation error.

EXAMPLE 1.1

It is possible, in fact customary, for a computer to store a number different from that presented as input, and to commit this error without warning. We perform the simple experiment of reading the number $x = 0.1234567890123$ into a DEC 10 computer, and then printing out the number actually stored. The program for doing this is shown in Table 1.1. The first of the numbers listed in Table 1.2 is the number stored in the input file, and the second is the representation received in the output file. Note that the format field length is adequate and is not the cause of the discrepancy. The difference between these numbers, which is about 2×10^{-9} is roundoff error.

TABLE 1.1   Program to Illustrate Roundoff Error

```
C       PROGRAM STORE
C
C       *****************************************************
C       THIS PROGRAM WILL READ A NUMBER FROM ONE FILE AND
C       WRITE IT TO ANOTHER
C       INPUT: X
C       OUTPUT: X
C       *****************************************************
C
        READ(11,1) X
        WRITE(10,1) X
      1 FORMAT(2X,F15.13)
        STOP
        END
```

TABLE 1.2   Input and Output

Input:	0.1234567890123
Output:	0.1234567910433

∎

EXAMPLE 1.2

Recall from calculus that the derivative $f'(x)$ of a function $f(x)$ at x_0 is defined to be the limit of

$$D(h) = \frac{f(x_0 + h) - f(x_0)}{h}$$

as h converges to 0. This formula for $D(h)$, which we have illustrated in Figure 1.1, is representative of finite difference methods used in Chapter 3 to approximate derivatives numerically. The true derivative is the slope of the tangent line, in Figure 1.1, and the approximation is the slope of the line connecting $f(x_0)$ and $f(x_0 + h)$.

If the magnitude of h is "too large," then $D(h)$ is inaccurate because h is not sufficiently close to the limit. This error is the truncation error associated with

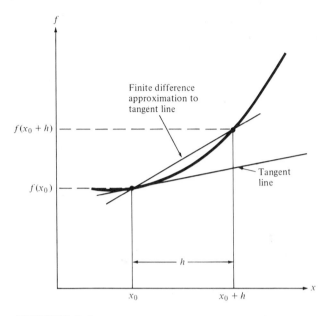

FIGURE 1.1 Finite Difference Approximation

using the "realizable" arithmetic formula $D(h)$ for approximating an unrealizable limiting operation, the derivative. Such error would occur even if $f(x)$ and $D(h)$ could be evaluated exactly. For reasons discussed in Section 1.4.2, as h becomes small, inaccuracies due to roundoff error dominate. The truncation and roundoff error in computing $f(x_0 + h) - f(x_0)$ is large relative to the actual value of this difference. Roundoff error would not be present if $D(h)$ could somehow be computed perfectly.

The program and printout displayed in Tables 1.3 and 1.4 give clear evidence

TABLE 1.3 Program for Derivative Approximation

```
C      PROGRAM NUMAPP
C
C      *********************************************************
C      THIS PROGRAM DEMONSTRATES ERRORS IN NUMERICAL
C      APPROXIMATION OF THE DERIVATIVE OF THE
C      EXPONENTIAL FUNCTION
C      OUTPUT: H=THE MAGNITUDE OF THE CHANGE IN INDEPENDENT
C                  VARIABLE USED IN THE FINITE DIFFERENCE
C                  APPROXIMATION
C              ERR=THE DIFFERENCE BETWEEN THE TRUE VALUE
C                  OF THE DERIVATIVE AS COMPUTED BY THE
C                  LIBRARY FUNCTION EXP AND THE APPROXIMATION
C      *********************************************************
C
       X=1.
       TRUE=EXP(1.)
       H=0.5
       B=EXP(X)
       DO WHILE(H.GT.1.E-8)
         APROX=(EXP(X+H)-B)/H
         ERR=TRUE-APROX
         WRITE(10,*)H,ERR
         H=H/2.
       END DO
       STOP
       END
```

NOT ANSI 77.

TABLE 1.4 Printout Showing Errors in Numerical Approximation of a Derivative

h	exp(1) - D(h)	
0.5000000	-0.8085327	
0.2500000	-0.3699627	
0.1250000	-0.1771994	
6.2500000E-02	-8.6745262E-02	
3.1250000E-02	-4.2918205E-02	Truncation
1.5625000E-02	-2.1357536E-02	Error
7.8125000E-03	-1.0661125E-02	Dominates
3.9062500E-03	-5.3510666E-03	
1.9531250E-03	-2.6655197E-03	
9.7656250E-04	-1.4448166E-03	
4.8828125E-04	-9.5653534E-04	
2.4414063E-04	-4.6825409E-04	
1.2207031E-04	-4.6825409E-04	
6.1035156E-05	-4.6825409E-04	
3.0517578E-05	-4.6825409E-04	
1.5258789E-05	-4.6825409E-04	
7.6293945E-06	-4.6825409E-04	
3.8146973E-06	-3.1718254E-02	Roundoff
1.9073486E-06	-3.1718254E-02	Error
9.5367432E-07	-3.1718254E-02	Dominates
4.7683716E-07	-0.2817183	
2.3841858E-07	-0.2817183	
1.1920929E-07	-1.281718	
5.9604645E-08	-5.281718	
2.9802322E-08	2.718282	
1.4901161E-08	2.718282	

that there is a limit to how well the derivative of exp (x_0) at $x_0 = 1$ can be approximated by the finite difference formula $D(h)$. The approximation error is graphed against step size h in Figure 1.2. (The scales are logarithmic.)

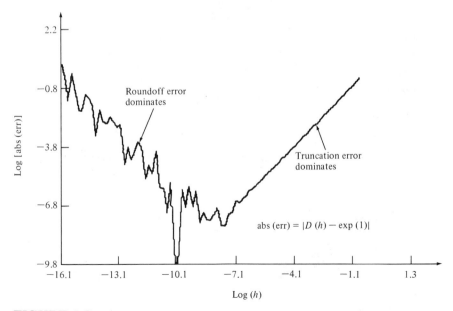

FIGURE 1.2 Error Versus Step Size

In view of the foregoing remarks, computers might seem to be unexpectedly limited instruments, but one should note that compared to human beings, computers can do those operations in capability 2 very quickly. Therein lies the beauty and power of digital computation. But exploitation of this power ultimately requires that someone approximate more complicated operations [such as evaluating sin (x)] by a sequence of arithmetical operations and comparisons.

1.2
CONCEPTS OF COMPUTATION ERROR

Toward surveying the ground to be covered in this book, let us sharpen some notions introduced informally in the preceding section. We will view a *program* as a finite sequence of instructions involving only comparisons and arithmetic operations on stored numbers, and conditional branches from one place to another in the sequence of instructions. For a concrete example, think of FORTRAN. We disallow library functions such as sin (x) and exp (x) until Chapter 2, where their rationale is revealed. Also, input and output facilities are of little concern to us.

With this notion of "program" in mind, we define:

Roundoff error: The error introduced in approximating a given number by a computer number.
Truncation error: The error introduced by approximating some desired mathematical operation by computations directed by a program. It is presumed that computations are done without roundoff error.

The term *truncation error* stems from recognition that many numerical methods are constructed by first finding a Taylor's series representation $\sum_{j=0}^{\infty} \alpha_j x^j$ of the mathematical operation, and then computing a truncation [i.e., an initial (polynomial) segment $\sum_{j=0}^{n} \alpha_j x^j$] of this series. We will find that polynomial approximations and power series expansions lie behind a large proportion of the computational methods to be encountered. Analysis of truncation error is strongly dependent on the application area and computational methodology under consideration. Meaningful discussion must therefore await the special topics of later chapters.

Roundoff error has its origins in computer operations regardless of problem area. In this chapter we describe these origins and examine some frequently encountered settings in which its effects can be distressing if not confronted intelligently.

Roundoff error tends to become a bothersome and perhaps a limiting factor when at some point in a computation, small changes of an input parameter to a calculation result in relatively large deviations of the calculated output.

Linear equations have the form

$$a_{i1}x_1 + a_{i2}x_2 + \cdots + a_{in}x_n = b_i, \qquad 1 \le i \le m,$$

and arise quite naturally from engineering considerations. In many cases linear equations have solutions that depend sharply on the coefficient values a_{ij}. For such equations, roundoff error is a prime factor.

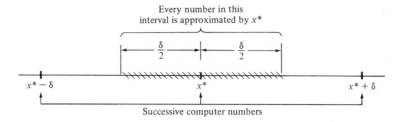

FIGURE 1.3 Computer Number Approximation

Regardless of problem area, roundoff error limits the accuracy to which a solution can be computed. Assume that in the neighborhood of a solution x, the gap between successive computer numbers is δ. If the computer approximates x by the closest computer number, clearly we cannot count on being able to locate x more closely than $\delta/2$. For if near the computer number x^*, the gap between successive computer numbers is δ, any number x between $x^* - \delta/2$ and $x^* + \delta/2$ is stored as x^*. We have illustrated this in Figure 1.3. Any number x in the crosshatched region must be approximated by x^*. Some computers use schemes other than the nearest-machine-number rule, and even larger error can result. We discuss approximation strategies later in this chapter.

Toward understanding the source and magnitude of roundoff error, we need to have a clearer notion of how computers store numbers. This topic is considered next.

1.3
COMPUTER REPRESENTATION OF NUMBERS

With some exceptions, a computer number is stored in one, two, or four computer words. For purposes of this discussion, one may imagine a *computer word* to be a binary string (i.e., a string of 0's and 1's) with t coordinates, where t is dependent on machine design. That is, a computer word may be visualized as the array

$$\boxed{b_0} \; \boxed{b_1} \; \boxed{\cdots} \; \boxed{b_{t-1}}$$

TABLE 1.5 Word Length of Some Common Computers

Computer	Word Length, t
Control Data Corporation	
CDC 6000, 7000, and CYBER series	60
IBM Corporation	
IBM 360/370 series and 303X, 308X series	32
IBM PC, AT, and XT	16
Digital Equipment Corporation	
DEC 10	36
PDP 11 series	16
VAX 11 series	32
Prime computers	32

where the b_j's are either 0 or 1. These terms b_j are referred to as binary coefficients, or *bits*, in the sense that decimal coefficients are known as ''digits.'' We have tabulated the word lengths of some popular computer families in Table 1.5.

Physically, the bits of a computer word are stored by means of two-state electronic circuits that are embedded in integrated circuit chips or alternatively by the direction of flux in magnetic elements. Other storage devices appear to be on the horizon.

Everything—machine instructions, alphanumeric strings, memory pointers, and so on—not just computer numbers, must be stored as machine words. The problem of computer number representation is the problem of finding a sensible coding scheme for mapping real numbers into machine words.

1.3.1 Number Systems

Before proceeding to the details of machine numbers, we remind the reader of a few facts about number representation. Our conventional number system is decimal. The symbol 546.3 in the decimal number system designates the sum

$$\underline{5} \times 10^2 + \underline{4} \times 10^1 + \underline{6} \times 10^0 + \underline{3} \times 10^{-1}.$$

Thus 546.3 can be regarded as a member of a code for designating nonnegative numbers. The technique of the decimal representation can be generalized to any integer *base* $N > 1$. Assuming for each j that a_j is in the set $\{0, 1, \ldots, N - 1\}$, define the string $(a_k a_{k-1} \cdots a_0)_N$ to be the code word for the integer

$$M = a_k N^k + a_{k-1} N^{k-1} + \cdots + a_0 N^0. \tag{1.1}$$

For reasons mentioned at the outset of this section, the *binary* ($N = 2$ and $a_j = 0$ or 1) system is the natural base for machine number representation.

EXAMPLE 1.3

Here we decode the string $(10110)_2$. In this specific case, (1.1) takes the form

$$M = \underline{1} \times 2^4 + \underline{0} \times 2^3 + \underline{1} \times 2^2 + \underline{1} \times 2^1 + \underline{0} \times 2^0 = 2^4 + 2^2 + 2 = (22)_{10}.$$

In brief, $(10110)_2 = (22)_{10}$.

■

Extending these ideas, and confining attention to the binary system, we find that any positive number x can be represented as

$$x = a_k 2^k + \cdots + a_0 2^0 + a_{-1} 2^{-1} + a_{-2} 2^{-2} + \cdots. \tag{1.2}$$

We may unambiguously write $(a_k \cdots a_0.a_{-1}a_{-2} \cdots)_2$ for x, with the understanding that (1.2) gives the conversion. The radix point is positioned between a_0 and a_{-1}. It partitions the number into its integral and fractional parts.

───── **EXAMPLE 1.4** ──────────────────────────────

By evaluating the right side of the equation below according to (1.2), the reader will see that $(1.8)_{10}$ has the nonterminating binary representation

$$(1.8)_{10} = (1.110011001100...)_2,$$

in which the pattern 1100 repeats ad infinitum.   ■

───

The binary representation just discussed is fundamental to all three principal computer representation systems, which are

1. Integer representation.
2. Floating-point representation.
3. Multiple-precision representation.

These systems are described below in this order.

1.3.2 Integer Representation

Figure 1.4 shows by vertical lines the locations of computer integers on a segment of the real number line. Note that the integers form a uniform grid: The distance between neighboring integers is always 1, regardless of the magnitudes of these numbers. The implication is that the roundoff error encountered in approximating a given positive number x by an integer can be just as large for small values of x as for large values.

The most obvious encryption of integers into computer words is the *signed magnitude* method. Here the first (b_0) bit indicates the sign of the number, and the remaining bits b_j, $1 \le j \le t - 1$, store the coefficients a_i, where, as in (1.1),

$$|M| = a_k 2^k + a_{k-1} 2^{k-1} + \cdots + a_0. \tag{1.3}$$

Here, if $2^{t-1} > M \ge 0$, the bits a_i in the representation above are stored directly with the understanding that $a_j = 0$, $t - 1 \ge j > k$:

a_{t-1}	$\cdots$	a_3	a_2	a_1	a_0

In comparing the format above with the bit designation at the beginning of Section 1.3, we see that

$$b_j = a_{t-j-1}, \qquad 1 \le j \le t - 1.$$

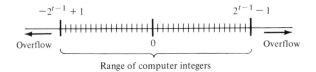

FIGURE 1.4 Distribution of Computer Integers

For reasons of speed and convenience in hardware design, negative integers are often represented by the *two's-complement* technique. Under this regime, for M negative, the bits of the positive integer representation of $2^t - |M|$ are stored. It turns out that the coefficients of $2^t - |M|$ can be obtained in the following way. First, replace all the 0's in the binary representation (1.3) of $|M|$ by 1's and all the 1's by 0, and then add 1 to the result.

EXAMPLE 1.5

We give the DEC 10 representation of 155 and -155. Recall from Table 1.5 that the word length of this computer is $t = 36$. One may check that

$$155 = (10011011)_2.$$

If the integer number is represented in the binary system, the binary coefficients may be stored directly, with a_0 placed in the rightmost box, b_{35}; a_1 stored in the next-to-last box, b_{34}; and so on. Consequently, the DEC 10 integer representation of 155 is

all 0's

j:	0	1		26	27	28	29	30	31	32	33	34	35
b_j:	0	0	$\cdots$	0	0	1	0	0	1	1	0	1	1

The two's complement representation for -155 is the reverse ordering of the coefficients of $2^{36} - 155$, which is

all 1's

j:	0	1		26	27	28	29	30	31	32	33	34	35
b_j:	1	1	$\cdots$	1	1	0	1	1	0	0	1	0	1

This representation can be obtained, as stated earlier, by taking the complements of the bits in the representation for $+155$ and adding 1, in binary, to the resulting string.

EXAMPLE 1.6

If some arithmetic operation results in an integer too large to store, some computers "remember" only the least significant t bits of the binary representation. The program and output in Tables 1.6 and 1.7, from a PDP 11, illustrates this point. In this case, $t = 16$, so the magnitude of the largest integer representation is bounded by $2^{15} - 1 = 32767$. Note what happens when the number accumulated by the variable KSUM exceeds this bound.

TABLE 1.6   Program for Integer Overflow

```
C       PROGRAM OVRFLO
C
C       ****************************************************
C       THIS PROGRAM ILLUSTRATES THE CONSEQUENCES OF
C       INTEGER OVERFLOW ON THE PDP 11
C       OUTPUT: KSUM=SUM OF 20 INTEGERS
C       ****************************************************
C
        KSUM=0
        NUM=5000
        DO 1 I=1,20
           KSUM=KSUM+NUM
           WRITE(10,*) KSUM
      1 CONTINUE
        STOP
        END
```

TABLE 1.7   Printout for Example 1.6

Sum
5000
10000
15000
20000
25000
30000
-30536
-25536
-20536
-15536
-10536
-5536
-536
4464
9464
14464
19464
24464
29464
-31072

1.3.3   Floating-Point Representation

The *floating-point representation,* and its close cousin, the multiple-precision representation, are the most important machine number systems for numerical computations.

The floating-point representation is conveniently described in terms of what is known as the *normal-form representation,* which we define next. If x is a positive number, then for some unique integer E and real number f such that $\frac{1}{2} \le f < 1$, we may write

$$x = f \cdot 2^E. \tag{1.4}$$

For if x is written in the form (1.2), then

$$x = a_k 2^k + a_{k-1} 2^{k-1} + \cdots + a_0 + a_{-1} 2^{-1} + a_{-2} 2^{-2} + \cdots$$
$$= (a_k 2^{-1} + a_{k-1} 2^{-2} + \cdots) \cdot 2^{k+1}.$$

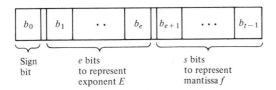

FIGURE 1.5 Segmentation of Computer Word for Floating Point Representation

With k so determined, we see that

$$f = a_k 2^{-1} + a_{k-1} 2^{-2} + \cdots \qquad \text{and} \qquad E = k + 1.$$

For instance,

$$x = (1001.01)_2 = 2^3 + 2^0 + 2^{-2} = (2^{-1} + 2^{-4} + 2^{-6}) \cdot 2^4.$$

In this example, $f = 2^{-1} + 2^{-4} + 2^{-6}$ and $E = 4$. The integer E as described above is called the *exponent* of the representation, and the term f, $\frac{1}{2} \le f < 1$, is the *mantissa*. Assume that x is represented in normal form (which is a purely mathematical construct):

$$x = f \cdot 2^E, \qquad \tfrac{1}{2} \le f < 1, \qquad E \text{ integer}.$$

The number x is stored in a computer word as a floating-point number as follows:

    1 bit stores the sign of x.
    s bits store the mantissa f.
    e bits store the exponent E.

The storage lengths s and e are usually fixed by machine design. If only one computer word, with the length t, is used, we would expect that

$$t = 1 + s + e.$$

If need be, the computer will round or chop the mantissa to the most significant s bits. (By observing that the most significant bit of f must be 1, some computers (e.g., VAXs) have effectively $s + 1$ bits of storage). The bit allocation of the floating-point number is illustrated in Figure 1.5.

───── **EXAMPLE 1.7** ─────────────────────────────

We now describe the floating-point-number representation employed by the DEC 10 computer. In this representation, the first bit (b_0) of the binary string of length $t = 36$ gives the sign of the number (here 0 means "positive" and 1 means "negative"). The exponent is represented by bits 1 to 8 and the mantissa is given by the binary coordinates 9 to 35, as shown in Figure 1.6. Let b_j denote the bit stored in location j, $0 \le j \le 35$, in Figure 1.6. For positive numbers the mantissa is determined according to the rule

$$f = (0.b_9 b_{10} \cdots b_{35})_2 = b_9 \cdot 2^{-1} + b_{10} \cdot 2^{-2} + \cdots + b_{35} \cdot 2^{-27}.$$

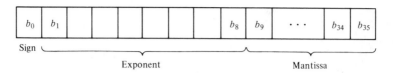

FIGURE 1.6 DEC 10 Floating-Point Number Representation

The conversion between the exponent E and the stored bits b_j $(1 \leq j \leq 8)$ is given by

$$E = (b_1 b_2 \cdots b_8)_2 - 128 = -(1 - b_1) \cdot 128 + (b_2 b_3, \ldots, b_8)_2.$$

From these considerations, one may conclude that a number may be represented as a floating-point number on a DEC 10 computer only if the magnitude of its exponent E lies between $(-128)_{10}$ and $(127)_{10}$. If, additionally, the mantissa f in the normal-form representation (1.4) is an integral multiple of 2^{-27}, then x is exactly equal to its DEC 10 floating-point representation. Otherwise, the mantissa is obtained by rounding f to 27 binary places.

Consider the specific DEC 10 floating-point number

all 0's

0	0	1	1	0	1	0	0	1	1	1	0	1	0	$\cdots$	0	0
b_0	b_1	b_2	b_3	b_4	b_5	b_6	b_7	b_8	b_9	b_{10}	b_{11}	b_{12}	b_{13}	$\cdots$	b_{34}	b_{35}

We calculate that

$$E = (01101001)_2 - 128 = 2^6 + 2^5 + 2^3 + 2^0 - 128$$

$$= 64 + 32 + 8 + 1 - 128 = -23$$

and

$$f = (2^{-1} + 2^{-2} + 2^{-4}) = (0.8125)_{10},$$

so

$$x = (0.8125)_{10} \times 2^{-23} \approx (9.7 \times 10^{-8})_{10}.$$

Figure 1.7 illustrates the distribution of floating-point numbers on a segment of the real number line. As in Figure 1.5, e and s are the numbers of bits used for exponent and mantissa representations. In contrast to the integer representation, floating-point numbers are more dense near the origin and more sparse at the extremes. It is this feature that gives the floating-point representation the capacity to preserve significant digits, as discussed later in this chapter. If the number x to be approximated has small magnitude, the real representation error will be correspondingly small. In fact, the gap between a real number x and its floating-point representation x^* is essentially proportional to the magnitude of x itself. In the next section we will see how to find the constant of proportionality.

In Figure 1.7 we have also illustrated a gap between zero and the smallest

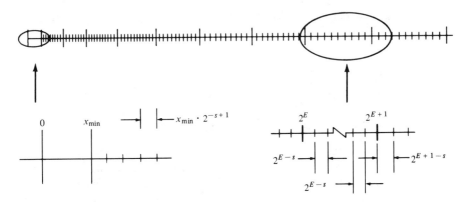

FIGURE 1.7 Distribution of Floating-Point Numbers

positive floating-point number, $x_{\min}$. We give a detailed derivation of $x_{\min}$. If e bits store the exponent E, then $E_{\min}$, the minimum value of E, is

$$E_{\min} = -2^{e-1}.$$

Also, $f \geq \frac{1}{2}$. So

$$x_{\min} = 2^{-1} \cdot 2^{E_{\min}} = 2^{-1} \cdot 2^{-2^{e-1}}.$$

The next larger floating-point number x is obtained by adding 1 to the least significant bit of f, so the new mantissa equals $2^{-1} + 2^{-s}$. Thus

$$x = (2^{-1} + 2^{-s}) \cdot 2^{E_{\min}} = x_{\min}(1 + 2^{-s+1}).$$

In the *multiple-precision representation,* two or more computer words are used to encode a given number x. The rationale of multiple-precision representation is exactly that of floating-point representation; the only difference is that the number of available storage bits is increased.

In Table 1.8 we have tabulated the number of bits assigned to the mantissas

TABLE 1.8   Floating-Point Characteristics of Some Computers*

Computer	Single Precision		Multiple Precision	
	s	e	s	e
Control Data Corporation	48	11	96	11
CDC 6000, 7000, and CYBER				
series				
IBM Corporation				
IBM 360/370 series and 303X and	24	7	56	7
308X series				
IBM PC†, AT, and XT	24	7	56	7
Digital Equipment Corporation				
DEC 10	27	8	63	8
PDP 11 series†	23	8	55	8
VAX 11 series	23	8	55	8
Prime computers	23	8	47	15

*s, number of mantissa bits; e, number of exponent bits.
†Uses two computer words for a single-precision floating-point number.

and the exponents of floating-point and double-precision numbers in various computers.

1.4
ROUNDOFF ERROR

1.4.1   Roundoff Error Bounds and Machine Epsilon

Recall the normal-form representation $x = f \cdot 2^E$, with $\frac{1}{2} \leq f < 1$. By adding 1 to the least significant place of the floating-point mantissa, we conclude that if for a given computer, s is the number of floating-point mantissa coefficients, then the difference between x^* and the next larger floating-point number, say y^*, is $2^{-s} \cdot 2^E$. As in Figure 1.3, let δ denote the separation of computer numbers. Then $\delta = 2^{E-s}$. The roundoff error in approximating a number x between x^* and y^* is bounded by 2^{E-s-1}, assuming that the computer finds the nearest floating-point number to x. This procedure is called *rounding*. Define *machine epsilon,* or simply ε, to be 2^{-s}. Note that since $f \geq \frac{1}{2}$, $|x| \geq 2^{E-1}$. We combine this with our earlier (rounding) bound to write

$$|x - x^*| \leq \frac{\delta}{2} = 2^{-s+E-1} = 2^{-s} \cdot 2^{E-1} \leq 2^{-s}|x| = \varepsilon|x|.$$

In summary,

$$\boxed{|\text{roundoff error } (x)| = |x - x^*| \leq \varepsilon|x|.} \tag{1.5}$$

Intuitively, $1 + \varepsilon$ is the smallest number greater than 1 that the computer in question will distinguish from 1.

It is good practice to design programs that are "portable" in the sense that their performance does not depend on which computer is being used. We saw in

TABLE 1.9   Subroutine EPS for Finding Machine Epsilon

```
      SUBROUTINE EPS(E)
C
C   *****************************************************************
C   *   FUNCTION: THIS SUBROUTINE COMPUTES THE MACHINE EPSILON   *
C   *   USAGE:                                                   *
C   *        CALL SEQUENCE: CALL EPS(E)                          *
C   *   PARAMETERS:                                              *
C   *        INPUT:                                              *
C   *             NONE                                           *
C   *        OUTPUT:                                             *
C   *             E=MACHINE EPSILON                              *
C   *****************************************************************
C
      E=1.0
      DO WHILE(E+1.0.GT.1.0)
         E=E/2.0
      END DO
      E=2.0*E
      RETURN
      END
```

Table 1.8 that the number s of mantissa bits, and hence machine epsilon, can vary widely. Therefore, it is often advisable to include in a program a code that automatically estimates machine epsilon. In Table 1.9, subroutine EPS for computing machine epsilon is offered. Certain subroutines require the user to supply a "stopping rule" threshold. The choice of such a value should be based on an estimate of machine epsilon.

───── **EXAMPLE 1.8** ─────────────────────────────────────

Machine epsilon of the DEC 10 computer was obtained by means of subroutine EPS. To three significant decimal places, the output was 7.45×10^{-9}. This number was actually 2^{-27}, which accords with the entry $s = 27$ in Table 1.8 for this machine.

With regard to the experiment described in Example 1.1, examination of the output Table 1.2 reveals that the observed roundoff error in the number actually stored for input $x = 0.1234567890123$ was 1.7×10^{-10}. By the considerations in this section, this error should be less than $\varepsilon \cdot |x| \approx 9.2 \times 10^{-10}$. Thus machine epsilon did lead to a reasonably close upper bound for the observed roundoff error. ■

1.4.2 Roundoff Error in Action

Here we present two examples illustrating common situations in which a computed solution is needlessly inaccurate as a result of roundoff error. These examples show further that mathematical and computational reasoning can be beneficial. For in each instance, once we see the origin of the problem, we are able to devise a more cunning computational strategy and thereby achieve an accurate answer. The ultimate source of difficulty in these examples is clearly roundoff error. Were computers able to store and operate on real numbers, rather than floating-point approximations, these difficulties would not arise.

───── **EXAMPLE 1.9** ─────────────────────────────────────

We here examine a common "trapdoor" known as *subtractive cancellation*. It arises when two nearly equal floating-point numbers are subtracted from one another.

For convenience of notation, assume that we have a machine with decimal words of mantissa length $s = 4$. Suppose that we are to estimate the difference of square roots of two given integers.

$$y = \sqrt{1985} - \sqrt{1984}.$$

and we are given the rounded approximations

$$\sqrt{1985} = 44.55$$
$$\sqrt{1984} = 44.54.$$

Then blind subtraction would have us estimate y by

$$\hat{y} = 44.55 - 44.54 = 0.01.$$

The correct answer to the accuracy shown is 0.011224. Could we have done better with the given approximations? Yes! The basic trouble is that the definition of y has us subtract two very nearly equal numbers, and the only information about their difference resides in their unequal digits, which in this case is only the last digit. There are several paths that bypass this subtractive cancellation problem. Recall from algebra that

$$\sqrt{A} - \sqrt{B} = \frac{A - B}{\sqrt{A} + \sqrt{B}}. \qquad (1.6)$$

Here we presume that A and B are exact, so no cancellation occurs in the numerator of (1.6), and the denominator retains nearly four significant digits of accuracy. Specifically, by use of the right side of (1.6), we compute the new estimate

$$\tilde{y} = \frac{1}{44.55 + 44.54} = 0.011225.$$

Here the error resulting from using the rounded values is only about 10^{-6}. ∎

EXAMPLE 1.10

From Taylor's series developments in calculus books (and as we learn in Chapter 2), it is known that

$$e^x \approx 1 + \frac{x}{1!} + \frac{x^2}{2!} + \cdots + \frac{x^n}{n!}. \qquad (1.7)$$

The error of this approximation does not exceed

$$\frac{|x|^{n+1}}{(n+1)!} \max\{1, e^x\}.$$

Recall that for a positive integer j,

$$j! = j \times (j - 1) \times (j - 2) \times \cdots \times 1.$$

A sensible computational procedure would seem to be to sum terms $x^j/j!$ until the magnitude of the addends is less than machine epsilon. The program listed in Table 1.10 follows this strategy and on a VAX, computes an approximation, which we denote by $S(-8)$, of e^{-8} to be 0.319328×10^{-3}. The correct value, to the accuracy shown, is 0.3354626×10^{-3}.

The error of the estimate $S(-8)$ is 1.6×10^{-5}, and the answer is correct only in the first significant decimal. The fundamental trouble is that whereas the magnitude of the final sum is relatively small, as shown in Table 1.11, several of the terms $(-8)^j/j!$ are large, and these large terms, while eventually canceling, determine the number of significant places. Henrici (1982) has termed this phenomenon *smearing*. Smearing can be anticipated whenever magnitudes of individual terms in a summation are considerably larger than the sum itself. Since the error in a stored number x is in the neighborhood of $\varepsilon |x|$, in the notation of (1.5), the

TABLE 1.10   Program for Approximation of Exponential Function

```
C       PROGRAM TAYEXP
C
C       *************************************************************
C       COMPUTE THE TAYLOR SERIES APPROXIMATION OF EXP(X), (X=-8.)
C       OUTPUT: S=THE TAYLOR SERIES APPROXIMATION
C               TRUE=THE VALUE OF EXP(X) AS COMPUTED BY THE
C                    LIBRARY FUNCTION EXP
C               ERR=THE DIFFERENCE BETWEEN THE TRUE VALUE
C                    AND THE TAYLOR SERIES APPROXIMATION
C       *************************************************************
C
        S=0.
        A=1.
        X=-8.
        J=0
C
C       *** THIS LOOP WILL COMPUTE THE TAYLOR SERIES APPROX.   ***
C       *** UNTIL ADDEND IS NEAR MACHINE EPS                   ***
C
        DO WHILE(ABS(A).GT.1.E-7)
           WRITE(10,*)J,A
           S=S+A
           A=A*(X/(J+1))
           J=J+1
        END DO
        TRUE=EXP(X)
        ERR=TRUE-S
        WRITE(10,*)S,TRUE,ERR
        STOP
        END
```

TABLE 1.11   Addends in Series for exp (-8)

j	$\dfrac{(-8)^j}{j!}$
0	1.000000
1	-8.000000
2	32.00000
3	-85.33334
4	170.6667
5	-273.0667
6	364.0889
7	-416.1017
8	416.1017
9	-369.8681
10	295.8945
11	-215.1960
12	143.4640
13	-88.28555
14	50.44889
15	-26.90608
16	13.45304
17	-6.330842
18	2.813707
19	-1.184719
20	0.4738875
21	-0.1805286

error in computing a sum is typically at least as large as ε times the magnitude of the largest addend. If the sum itself is much smaller than the largest addend, smearing will almost surely occur, as in this example. The moral: Beware of summing a series with mixed signs.

For the particular case of evaluating exp (-8), smearing can be sidestepped by observing that for $x = -1$, for instance, the largest addends have the same order of magnitude as the sum. Thus the procedure in Table 1.10 should be accurate for finding an approximation $S(-1)$ of e^{-1}. In fact, with $x = -1.0$ the program in Table 1.10 computes the estimate 0.36787945, which is correct in all but the last decimal place. Then, since $e^{-8} = (e^{-1})^8$, one may offer $S(-1)^8$ as an estimate. The error in following this tactic is only about 2×10^{-9}, whereas direct use of the algorithm resulted in an error of 1.6×10^{-5}. ∎

1.5
ABSOLUTE AND RELATIVE ERROR BOUNDS

The error analysis of computations is characterized according to several different criteria. Let x denote an exact value and let variable x^* denote an approximation of x. One may think of x^* as the result of a computation. Then the absolute difference $|x - x^*|$ gives the *absolute error*. Any nonnegative number $\delta(x^*)$ satisfying the inequality $|x - x^*| \leq \delta(x^*)$ is called an *absolute bound* or *upper bound* for the error of x^* as an approximation of x. Note that all values larger than $\delta(x^*)$ also serve as upper bounds.

An absolute error bound $\delta(x^*)$ for a floating-point approximation x^* that rounds a nonzero input x to the closest value is, in view of developments in Section 1.4, given by $\delta = \delta(x^*) = \varepsilon|x^*|$.

The *relative error* of approximating x by x^* is defined to be $|x - x^*|/|x|$. We summarize these definitions as follows:

$$
\begin{array}{l}
\text{absolute error} = |\text{true value} - \text{approximation}| \\[2mm]
\text{relative error} = \dfrac{\text{absolute error}}{|\text{true value}|}.
\end{array}
\qquad (1.8)
$$

We will use the notation $\delta(x^*)$ to denote an absolute bound for the error in approximating x by x^*, and Rel (x^*), a relative error bound.

A feature of integer representation is that the absolute roundoff error bound is always $\frac{1}{2}$ (assuming that rounding to the nearest integer is done). The relative error depends on what the number x to be approximated happens to be. By contrast, in floating-point representation, the absolute error depends on x, but in view of (1.5), Rel $(x^*) = \varepsilon$, ε being machine epsilon. In summary, integer representation assures that the absolute error bound is constant in x, and floating-point representation preserves the relative error.

In accordance with usage in our description of floating-point numbers, we say that $x = f \cdot 10^E$ is a normal-form representation of x if $0.1 \leq f < 1$ and E is an integer. Let x^* be an approximation of x. We say that x^* has k *significant digits* if

$$|x - x^*| \le \tfrac{1}{2} \cdot 10^{-k+E}$$

[There is some divergence on the definition of "significant"; we follow Rice (1983), for example, in ours.]

─────── **EXAMPLE 1.11** ───────────────────────────────────

If $x^* = 12.765$ and $x = 12.8111$, then

$$x = 0.128111 \times 10^2$$

and $|x^* - x| < \tfrac{1}{2} \times 10^{-1}$. Since $E = 2$ in the normal-form representation of x^*, one concludes that $k = 3$, and therefore x^* has three significant digits. ■

1.6
ERROR PROPAGATION

1.6.1 Propagation of Error in a Single Computation

We now derive upper bounds for the errors that result from computing functional values using incorrect data. Assume that x^* is an approximation of x. If the functional value $f(x)$ is desired, and only the approximating value x^* of x is known, we approximate $f(x)$ by $f(x^*)$. Assume that $\delta(x^*)$ is an absolute error bound for $|x - x^*|$ and that the function $f(x)$ is differentiable. Then the mean-value theorem of calculus (Appendix B, item 19) implies that

$$f(x) - f(x^*) = f'(\zeta)(x - x^*), \tag{1.9}$$

where ζ is some number in the interval with endpoints x and x^* and $f'(\zeta)$ is the derivative of $f(x)$ at ζ. Since the value of ζ is unknown, equation (1.9) cannot be used directly. To obtain practical formulas, we have to make further assumptions about function f.

Assume first that for arbitrary t between x^* and x, and for some fixed number D,

$$|f'(t)| \le D.$$

Then (1.9) implies that

$$|f(x) - f(x^*)| = |f'(\zeta)| \cdot |x - x^*| \le D\,\delta(x^*);$$

consequently, the quantity $D\,\delta(x^*)$ can be accepted as an error bound for $f(x^*)$. That is, for the interval I containing all points between x^* and x,

$$\delta(f(x^*)) = \max_{t \in I} |f'(t)|\,\delta(x^*). \tag{1.10}$$

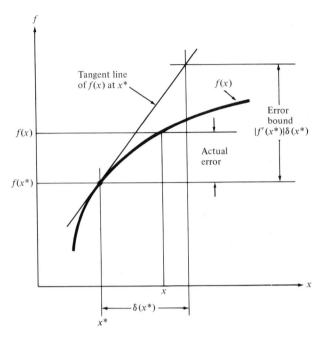

FIGURE 1.8   Error Bound for Operation of Inexact Data

The application of this error bound has the disadvantage that an upper bound for the derivative f' of $f(x)$ must be determined.

Assume next that $\delta(x^*)$ is small, $f'(x^*) \neq 0$ and $f'(t)$ is nearly constant near x^*. Then (1.10) implies that $D \approx |f'(x^*)|$. Consequently,

$$\delta(f(x^*)) \approx |f'(x^*)|\ \delta(x^*). \qquad (1.11)$$

Figure 1.8 illustrates the error-bound construct (1.11).

———— EXAMPLE 1.12 ————

Let $f(x) = \sqrt{x}$ and assume that x^* is an approximation of x with an error bounded by $\delta(x^*)$. Then (1.11) implies that

$$\delta(\sqrt{x^*}) = \frac{1}{2\sqrt{x^*}}\ \delta(x^*).$$

If, for example, $x^* = 4.00$ and $\delta(x^*) = 0.005$, then

$$\delta(\sqrt{4.00}) = \frac{1}{2\sqrt{4.00}} \times 0.005 = 0.00125.$$

In fact, if $x = x^* - \delta(x^*) = 3.995$, then

$$\sqrt{x} - \sqrt{x^*} = -0.0012504.$$

In this case, the approximation (1.11) does give an accurate error estimate. ■

Error bounds for functions of two or more variables can also be obtained by the approach described above. Consider a real-valued function $f(x, y)$ of two variables x and y. Assume that the "true" values x and y are unknowns but their approximations, x^* and y^* are given. Then the desired value $f(x, y)$ is approximated by $f(x^*, y^*)$. Assume again that f is differentiable, the error bounds $\delta(x^*)$ and $\delta(y^*)$ are small, and the first partial derivatives are not both zero and change slowly. Then

$$f(x, y) - f(x^*, y^*) = f(x, y) - f(x^*, y) + f(x^*, y) - f(x^*, y^*).$$

Consequently,

$$\begin{aligned}
|f(x, y) - f(x^*, y^*)| &\leq |f(x, y) - f(x^*, y)| + |f(x^*, y) - f(x^*, y^*)| \\
&\approx \left|\frac{\partial f}{\partial x}(x^*, y)\right| \cdot |x - x^*| + \left|\frac{\partial f}{\partial y}(x^*, y^*)\right| \cdot |y - y^*| \\
&\approx \left|\frac{\partial f}{\partial x}(x^*, y^*)\right| \cdot |x - x^*| + \left|\frac{\partial f}{\partial y}(x^*, y^*)\right| \cdot |y - y^*| \\
&\leq \left|\frac{\partial f}{\partial x}(x^*, y^*)\right| \cdot \delta(x^*) + \left|\frac{\partial f}{\partial y}(x^*, y^*)\right| \cdot \delta(y^*).
\end{aligned}$$

Hence analogously to (1.11),

$$\delta(f(x^*, y^*)) \approx \left|\frac{\partial f}{\partial x}(x^*, y^*)\right| \delta(x^*) + \left|\frac{\partial f}{\partial y}(x^*, y^*)\right| \delta(y^*). \qquad (1.12)$$

Generalization of this approach to more variables should be evident. If x_i^* is an estimate of x_i, $1 \leq i \leq n$, and $\delta(x_i^*)$ denotes a bound for $|x_i - x_i^*|$, $1 \leq i \leq n$, then with $x^* = (x_1^*, \ldots, x_n^*)$, we have

$$\begin{aligned}
\delta(f(x_1^*, \ldots, x_n^*)) \approx & \left|\frac{\partial f(x^*)}{\partial x_1}\right| \delta(x_1^*) + \left|\frac{\partial f(x^*)}{\partial x_2}\right| \delta(x_2^*) + \cdots \\
& + \left|\frac{\partial f(x^*)}{\partial x_n}\right| \delta(x_n^*).
\end{aligned} \qquad (1.13)$$

───── **ENGINEERING EXAMPLE** ─────

We given an upper bound for the error of the computation

$$t = 2\pi \sqrt{\frac{l}{g}}$$

if π, l, and g are not known precisely. The relation above gives the time period of a "linearized" pendulum (see Figure 1.9) with g the acceleration of gravity and l the pendulum length, which we take as being nominally 3 m, with an error bound $\delta(l)$ of 1 cm. As usual, π is the ratio of circle circumference to diameter.

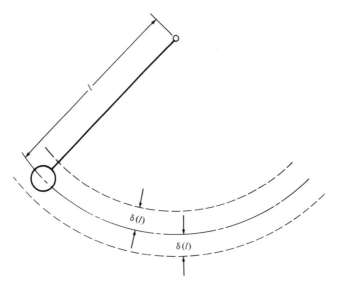

FIGURE 1.9   Uncertainty in Pendulum Length

In evaluating π by $4.0 \times \tan^{-1}$, one can approximate π with an error bound of $\delta(\pi) = \pi \cdot$ (machine epsilon). This term will give a negligible contribution to the error. Gravitational acceleration at $45°$ latitude is nominally 9.80 m/s^2 and varies from location to location. Unless site measurements are made, $\delta(g) \approx 10^{-2}$ m/s^2. With these error bounds in hand, we are ready to proceed with our computation. The nominal period will be

$$t = 2\pi \frac{\sqrt{3}}{\sqrt{9.80}} \approx 3.476.$$

By the use of the three-variable version of (1.13), we conclude that

$$\delta(t) \approx \left| \frac{\partial t}{\partial \pi} \right| \delta(\pi) + \left| \frac{\partial t}{\partial l} \right| \delta(l) + \left| \frac{\partial t}{\partial g} \right| \delta(g)$$

$$= 2\sqrt{\frac{l}{g}} \, \delta(\pi) + 2\pi \frac{1}{2\sqrt{l/g}} \frac{1}{g} \delta(l) + 2\pi \frac{1}{2\sqrt{l/g}} \frac{l}{g^2} \delta(g).$$

As stated, the $\delta(\pi)$ term is negligible compared to the other uncertainties, so

$$\delta(t) \approx \frac{\pi}{\sqrt{lg}} \delta(l) + \frac{\pi\sqrt{l}}{\sqrt{g^3}} \delta(g)$$

$$= 0.58 \times 0.01 + 0.18 \times 0.01$$

$$= 0.0076 \text{ s}.$$

In summary,

$$t = 3.48 \pm 0.01 \text{ s}.$$

*1.6.2 Error Propagation in Arithmetic

Bounds for error in arithmetic can be derived from the general rule (1.12), as the following example illustrates.

───── **EXAMPLE 1.13** ─────

Assume that $f(x, y) = x + y$. Then

$$\frac{\partial}{\partial x} f(x^*, y^*) = \frac{\partial}{\partial y} f(x^*, y^*) = 1,$$

so, by (1.12),

$$\delta(f(x^*, y^*)) = \delta(x^* + y^*) = 1 \times \delta(x^*) + 1 \times \delta(y^*) = \delta(x^*) + \delta(y^*).$$

In addition, then, the error bounds are added.
  Assume next that $f(x, y) = x - y$. Then

$$\delta(f(x^*, y^*)) = \delta(x^* - y^*) = 1 \times \delta(x^*) + 1 \times \delta(y^*) = \delta(x^*) + \delta(y^*).$$

Hence in subtraction the error bounds also add.
  If we define $f(x, y) = xy$, then

$$\delta(f(x^*, y^*)) = \delta(x^* y^*) \approx |y^*| \, \delta(x^*) + |x^*| \, \delta(y^*),$$

and if $f(x, y) = x/y$, then

$$\delta(f(x^*, y^*)) = \delta\left(\frac{x^*}{y^*}\right) \approx \frac{1}{|y^*|} \delta(x^*) + \frac{|x^*|}{|y^*|^2} \delta(y^*) = \frac{|y^*| \, \delta(x^*) + |x^*| \, \delta(y^*)}{|y^*|^2}.$$

■

───── **EXAMPLE 1.14** ─────

Consider the function

$$f(x, y, z) = xy + z,$$

where

$$x \approx x^* = 2.20, \qquad \delta(x^*) = 0.005$$
$$y \approx y^* = 1.15, \qquad \delta(y^*) = 0.005$$
$$z \approx z^* = 3.05, \qquad \delta(z^*) = 0.005.$$

Then repeated application of the relation above gives that

$$\delta(x^* y^*) = |1.15| \times 0.005 + |2.20| \times 0.005 = 0.01675$$

and therefore

$$\delta(x^* y^* + z^*) = \delta(x^* y^*) + \delta(z^*) = 0.02175.$$

■

TABLE 1.12   Absolute Error Bounds for Arithmetic Operations on Inexact Data

$$\delta(x^* + y^*) = \delta(x^*) + \delta(y^*)$$

$$\delta(x^* - y^*) = \delta(x^*) + \delta(y^*)$$

$$\delta(x^*y^*) \approx |x^*|\,\delta(y^*) + |y^*|\,\delta(x^*)$$

$$\delta\left(\frac{x^*}{y^*}\right) \approx \frac{|x^*|\,\delta(y^*) + |y^*|\,\delta(x^*)}{|y^*|^2}$$

TABLE 1.13   Relative Error Bounds for Arithmetic Operations on Inexact Data

$$\text{Rel}\,(x^* + y^*) = \max\,\{(\text{Rel}\,(x^*),\,\text{Rel}\,(y^*)\},\ \text{for } x,\,y \text{ having same sign}$$

$$\text{Rel}\,(x^* - y^*) = \frac{\delta(x^*) + \delta(y^*)}{|x - y|},\ \text{any } x,\,y$$

$$\text{Rel}\,(x^* \cdot y^*) \approx \text{Rel}\,(x^*) + \text{Rel}\,(y^*),\ \text{any } x,\,y$$

$$\text{Rel}\,(x^*/y^*) \approx \text{Rel}\,(x^*) + \text{Rel}\,(y^*),\ \text{any } x,\ \text{and any } y \neq 0$$

In Tables 1.12 and 1.13 we collect the absolute and relative error bounds induced by performing arithmetic on approximating data.

EXAMPLE 1.15

We derive the entry in Table 1.13 for Rel $(x^* + y^*)$, with x and y having the same sign:

$$\text{Rel}\,(x^* + y^*) = \frac{\delta(x^*) + \delta(y^*)}{|x + y|}.$$

Suppose, by exchanging labels if necessary, that

$$\frac{\delta(x^*)}{|x|} = \text{Rel}\,(x^*) \geq \text{Rel}\,(y^*) = \frac{\delta(y^*)}{|y|}.$$

Then $\delta(x^*)|y| \geq \delta(y^*)|x|$ and

$$\delta(x^*)|x + y| = \delta(x^*)(|x| + |y|) \geq |x|(\delta(x^*) + \delta(y^*)),$$

which gives the relation

$$\frac{\delta(x^*)}{|x|} \geq \frac{\delta(x^*) + \delta(y^*)}{|x + y|}$$

or

$$\text{Rel}\,(x^*) \geq \text{Rel}\,(x^* + y^*).$$

*1.6.3 Propagation of Errors Through a Computation

In our discussion of bounds for propagated error, we assumed that all operations and computation of function values are done exactly. But in fact, at each stage of a computation, errors are introduced through truncation and rounding. Thus typically, each stage of a lengthy program operates on faulty data and is itself a source of new errors.

Figure 1.10 illustrates the distinction between an ideal sequence of computations and its computer-realizable counterpart. In this figure the exact input to the first stage is represented by x_1. After the first stage of computation, ideally the output should be $y_1 = F_1(x_1)$. This output serves as input to the next computational stage or module. In the realizable computation, the input x_1^* is an approximation of the desired input x_1. Whereas ideally, we wished the output of the first stage to be $F_1(x_1)$, in the actual computation, it will be $y_1^* = F_1(x_1^*)$. Added to this faulty output is an undesired error e_1, which can represent effects of inexact calculation of the operator F_1 as well as roundoff error accumulated in the first stage of computation. The sum of y_1^* and e_1 serves as input to the next computational stage or module, and this process continues, with error propagating and accumulating at each segment of the computation.

Armed with an understanding from the preceding section about how error propagates through an arithmetical operation, and assuming that the relative error introduced by roundoff after each such operation F_j is e_j, we are in a position to analyze the overall accumulated error in any program-directed computation. Such analysis is seldom undertaken for lengthy programs because, while conceptually straightforward, it involves excruciating detail due to all the many operations and branches in a typical computation. Moreover, such "worst-case" analysis tends to lead to pessimistic bounds if carried out over many stages. However, for relatively small computations that occur repeatedly within a larger program, error bounding is frequently useful.

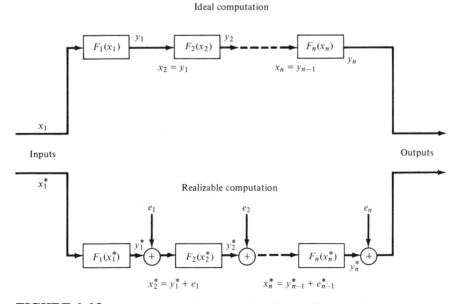

FIGURE 1.10 Comparison of Ideal and Realizable Computations

───── **EXAMPLE 1.16** ───────────────────────────────

As illustration of the principle involved, let us bound the relative error in computing $xy + z$ using a machine with epsilon value ε. Further, presume the floating-point numbers x^*, y^*, and z^*, which approximate x, y, and z, are positive. Then, according to Table 1.13,

$$\text{Rel}\,(x^*y^*) \approx \text{Rel}\,(x^*) + \text{Rel}\,(y^*). \tag{1.14}$$

Now the number $(x^*y^*)^*$, stored as the computed x^*y^*, is subject to roundoff error with relative error bound ε. This error is added directly to the propagated error. We then have that

$$\text{Rel}\,((x^*y^*)^*) = \frac{\delta(x^*y^*) + \varepsilon|xy|}{|xy|} = \text{Rel}\,(x^*y^*) + \varepsilon$$

$$\approx \text{Rel}\,(x^*) + \text{Rel}\,(y^*) + \varepsilon.$$

Now again by Table 1.13,

$$\text{Rel}\,((x^*y^*)^* + z^*) = \max\,\{\text{Rel}\,((x^*y^*)^*),\ \text{Rel}\,(z^*)\},$$

and after approximating $(x^*y^*) + z^*$ by the floating-point number $((x^*y^*)^* + z^*)^*$ and again accounting for roundoff, we obtain the relative error bound of the output approximation $((x^*y^*)^* + z^*)^*$ for $xy + z$, as

$$\text{Rel}\,(((x^*y^*)^* + z^*)^*) = \max\,\{\text{Rel}\,(x^*) + \text{Rel}\,(y^*) + \varepsilon,\ \text{Rel}\,(z^*)\} + \varepsilon. \qquad ■$$

───── **EXAMPLE 1.17** ───────────────────────────────

Suppose for the sake of simplicity, that a binary computer allows only $s = 4$ binary coefficients to represent the mantissa of a number, and that in normal form representation, $x = (0.1101)_2 \times 2^3$ and $y = (0.1100)_2 \times 2^{-1}$. Then

$$x + y = (110.1)_2 + (0.011)_2 = (110.111)_2.$$

If the surplus mantissa coefficients are simply ignored (or "chopped"), this sum is stored as $(0.1101)_2 \times 2^3$. Note that y was not large enough to change the floating-point representation after addition. ■

Although the error resulting from a single arithmetic operation is usually insignificant in itself, over the course of a long computation, the cumulative effect of many such errors can make itself felt.

───── **EXAMPLE 1.18** ───────────────────────────────

We perform the experiment of forming a sum by adding the number $A = 10^{-6}$ one million times. The program and outcome on a VAX are given in Tables 1.14 and 1.15. Note that the sum is correct only to two significant decimals, whereas from Table 1.8, or direct computation of machine epsilon, we are assured that the number A itself was stored with an accuracy of seven or eight significant decimals.

TABLE 1.14   Program for Accumulating a Sum

```
C       PROGRAM ADDALOT
C
C       ******************************************************
C       THIS PROGRAM DEMONSTRATES THE CUMULATIVE EFFECT OF
C       ADDING NUMBERS OF DIFFERENT MAGNITUDES MANY TIMES
C       OUTPUT: SUM=SUM OF THE VARIABLE A 1000000 TIMES
C       ******************************************************
C
        A=1.E-6
        SUM=0.
        DO 1 I=1,1000000
           SUM=SUM+A
      1 CONTINUE
        WRITE(10,2) SUM
      2 FORMAT(33X,'SUM=',F10.7)
        STOP
        END
```

TABLE 1.15   Output for Accumulating a Sum

```
        SUM = 1.0090389
```

The major cause of this loss of precision is repeated loss of significance in adding a relatively small number to the much larger sum, as the number of terms increases. This is the phenomenon illustrated in Example 1.10. To some extent this loss of significance can be sidestepped by accumulating the sums in a great many smaller partial sums, and then adding these partial sums together. The program shown in Table 1.16 follows this idea, accumulating the sums of the small number A in a thousand different partial sums S(J), which are later added to one another to achieve an answer (Table 1.17) with five significant decimal places.

TABLE 1.16   Program for Accumulating a Sum

```
C       PROGRAM ADD
C
C       ******************************************************
C       THIS PROGRAM ILLUSTRATES HOW MORE ACCURACY IS OBTAINED
C       WHEN CARE IS TAKEN TO ADD NUMBERS OF SIMILAR MAGNITUDE
C       OUTPUT: SUM=SUM OF THE VARIABLE A COMPUTED AS THE SUM
C                   OF 1000 SUMS OF 1000 A'S
C       ******************************************************
C
        DIMENSION S(1000)
        A=1.E-6
        SUM=0.
        DO 1 J=1,1000
           S(J)=0.
           DO 1 I=1,1000
              S(J)=S(J)+A
      1    CONTINUE
        DO 2 J=1,1000
           SUM=SUM+S(J)
      2 CONTINUE
        WRITE(10,3) SUM
      3 FORMAT(33X,'SUM=',F10.7)
        STOP
        END
```

TABLE 1.17   Output for Accumulating a Sum

SUM = 0.9999985

1.7
SUPPLEMENTARY NOTES AND DISCUSSIONS

During the period from 1940 to 1955, ideas and methodology for computer number representation underwent lively and innovative development, as indeed this was the period of the emergence of the electronic digital computer. Prior to this period, computers were mechanical and, with the exception of experimental devices, strictly decimal. The arithmetic mode was equivalent to integer representation. Metropolis, et al. (1980) give ample documentation of the evolution of floating-point representations. The advantage of this representation was clearly foreseen by the pioneers of modern computers, but in the first machines there was a tendency to force the programmer to encode the number representation. In early computer designs, essentially all user conveniences were sacrificed for the sake of computational speed. Since the advent of FORTRAN in the late 1950s, automatic floating-point representation seems to have become firmly established as the medium for numerical calculations. Many implementations of BASIC, for example, allow only such representation. The evolution will undoubtedly continue. For instance, some experimental computers have variable-length floating-point mantissas.

For details of number representation and the actual operations involved in computer arithmetic, the assembly language manual of the computer in question should be consulted. Knuth (1969, Chap. 4) provides many more details on this issue than we have given and illustrates the principles of computer number representation and arithmetic with a hypothetical assembly language.

We have described situations such as subtractive cancellation and smearing in which numerical error leads to large or intolerable "loss of significance." As illustrated in our examples, in many such situations these difficulties can be circumvented through more careful computational and analytical thinking. Stegun and Abramowitz (1956), Henrici (1982), and Rice (1983) provide further examples and insights along these lines. Our plan is to delay until later chapters (especially Chapter 4 on linear equations) a discussion of "ill-conditioned" problems in which slight roundoff or measurement error on input parameters can lead to devastating error in the computed output.

At the close of this chapter we sketched principles involved in finding the absolute bound of propagated error. For full-scale computations, such analysis, in addition to being excruciatingly tedious, tends to result in pessimistic answers. Early in the computer age, scientists occasionally concluded from such calculations that a given problem could not possibly be solved on present-day computers, only to find that somebody else had indeed found and verified the solution. At present, absolute bound analysis finds use in small-scale tasks such as bounding the effect of computer word roundoff in a library function routine. For large-scale computations, the tendency is to push ahead with the calculations but provide accuracy checks at key points along the way. For example, it may be possible to substitute a proposed solution back into an equation to be solved and to check that the equation is indeed satisfied. Alternatively, a method may be checked on prototypical problems having known closed-form solutions.

At the outset of this chapter we described the goals of numerical analysis as including construction of computer-realizable approximations for unrealizable mathematical operations, and derivation of error bounds for these constructs. A further theme that will come to the fore in subsequent chapters is that methods should be efficient. If one method is essentially as accurate as another but requires only half the computation time, it is clearly the method of choice.

PROBLEMS

Section 1.3

1. Convert the following binary numbers to decimal.
   (a) $(111001)_2$.
   (b) $(110.1101)_2$.

2. Suppose, for our convenience, that we have a computer with decimal (instead of binary) words, and that the mantissa length s is only three digits, but in this problem the exponent length e is of no concern. Since $s = 3$, and assuming that the computer rounds a given number to the nearest computer number, it will store 416.8, for example, as 417., and store 0.04164 as 0.0416. Assuming that the computer does arithmetical operations perfectly and the only error is that incurred in storing the resultant as a computer number find the stored sum of the numbers in parts (a) to (d).
   (a) 30. + 4.12.
   (b) 300. + 1.312.
   (c) 922. + 106.73.
   (d) 12345. + (−12344.).
   (e) Suppose that each of the numbers in parts (a) to (d) must first be stored and then the stored representations added. What results form these operations?

3. Find the decimal value of the following floating-point representation by the DEC 10 computer.

0	1	2	3	4	5	6	7	8	9	10	11	12		33	34	35
0	1	0	0	1	1	1	1	1	1	1	0	0	$\cdots$	0	0	0

4. How many distinct machine words of length t are there? (**HINT:** Examine the situation for $t = 1, 2, \ldots$ and generalize. Establish your result rigorously by finite induction.) Note that this provides an upper bound to the number of distinct single-precision integers or floating-point numbers.

5. Write a subroutine which, when presented with positive integers N and K and integer array $A = (A(1),\ldots,A(K+1))$, returns the integer

$$M = A(1) + A(2)N + \cdots + A(K+1)N^K.$$

Check your subroutine by verifying that $(35411)_6 = (5119)_{10}$.

6. Write a subroutine that does the converse of Problem 5. That is, this subroutine, when presented with the input variables M and N, should successively compute integers A(1), A(2), . . ., A(K + 1), satisfying the equation in Problem 5 as well as the condition that $0 \le A(j) < N$, $1 \le j \le K + 1$. [**HINT:** Let "Rem" denote "remainder." Then A(1) = Rem(M/N). Let I = ((M − A(1))/N). Then A(2) = Rem(I/N). Figure out why these statements are so, and then continue the algorithm in a recursive loop.] Check your method by finding the representation of $(5119)_{10}$ in the base $N = 6$. Note that in FORTRAN, Rem (I/N) = MOD(I, N).

Section 1.4

7. Write a program that subtracts 50,000,000 from 50,000,001 in both integer and floating-point arithmetic. Explain the output.

8. Write a program and find a set of N numbers X(1), X(2), . . ., X(N) illustrating that the computed sum

$$S = X(1) + X(2) + \cdots + X(N)$$

can depend on the order in which the X(j)'s are added. Explain this effect.

9. Use Subroutine EPS (Table 1.9) to find machine epsilon on your computer. Check to see that if β is the smallest positive floating-point number such that

$$A + \beta \ne A,$$

then β is "not far" from $|A|\varepsilon$, ε being your computer machine epsilon. Take $A = 10$, $\frac{1}{30}$, and 104.

10. Write a program to approximate sin (x) by means of the expansion

$$\sin (x) = x - \frac{x^3}{3!} + \cdots + (-1)^j \frac{x^{2j+1}}{(2j + 1)!} + \cdots.$$

Compute sin (10) and compare it with the library function estimate. Devise a "wise" procedure for approximating sin (10). Try $j = 1, 5, 10$, and 15.

11. The quadratic formula

$$x_1, x_2 = (-b \pm \frac{\sqrt{b^2 - 4ac)}}{(2a)}$$

for the roots of $p(x) = ax^2 + bx + c$ is prone to subtractive cancellation error when ($|b| \gg |a|$ and $|c|$). Illustrate this fact by showing that $p(x_j)$ is not very close to 0. If you have mathematical background, a cure might occur to you.

12. One must distinguish between machine epsilon EPS and x_{min}, the smallest positive floating-point number, which is far smaller than EPS. Find a theoretical number value of this latter number in terms of the word length t and the mantissa length s of your computer. Experimentally approximate the latter number on your computer. A number having magnitude less than this quantity cannot reliably be distinguished from zero by the computer.

Section 1.6

13. Presume that numbers $x^* = 11.33$ and $y^* = 2.15$ are obtained by rounding x and y, respectively, to the number of significant digits shown. That is, $\delta(x^*) = \delta(y^*) = 0.005$. Give error bounds for the following quantities.
    (a) $x + xy$.
    (b) $\sqrt{xy}$.
    (c) $\sin(2x - y) + \cos(2x - y)$.
    (d) $\dfrac{x}{y} - \dfrac{y}{x}$.

14. With x and y as in Problem 13, estimate the relative errors for each of the following.
    (a) $x + xy$.
    (b) $\sqrt{xy}$.
    (c) $\sin(2x - y) + \cos(2x - y)$.
    (d) $\dfrac{x}{y} - \dfrac{y}{x}$.

15. In theory, after performing the recursion SUM = SUM + A, N times, we should have SUM = $N * A$. However, because of roundoff error, the branching statement "IF (SUM .EQ. $N * A$)" may fail to branch. Give a computational illustration of this phenomenon.

16. Write a program to compute the sum

$$x = 1 - \frac{1}{a} + \frac{1}{a^2} - \frac{1}{a^3} + \cdots + \frac{1}{a^{10,000}},$$

where $a = 1.001$. Then print the sum and recalculate the sum backward from the right and print the result. Compare the two values, stating which you believe to be more accurate, and why.

17. A real number x is approximated by 135.12 and we are told that the relative error bound is 0.01. What can be said about the true value of x?

18. The developments in this section have a bearing on the illustration of subtractive cancellation in Example 1.9. Show how to use the mean-value theorem to estimate $f(a) - f(b)$, for a and b given numbers. Use this technique to estimate $f(1985) - f(1984)$, where $f(x) = $ SQRT(x). Apply your formula to the numbers given in Example 1.9.

19. The area of a right triangle is completely determined by the length x of a side and the angle θ which that given side makes with the hypotenuse. Let $x^* = 5$ and the angle $\theta = 45°$. Suppose that $\delta(x^*) = 0.1$ and $\delta(\theta) = 5°$. Estimate bounds for the absolute and relative errors.

CHAPTER 2

Interpolation

2.1
ORIENTATION AND PREVIEW

The two main practical tasks that motivate the methods of this chapter are (1) given some function $f(x)$ and an interval I, find a program that gives an adequate approximation to $f(x)$ for any x in I; and (2) given a collection of data points $(x_k, f(x_k))$, $k = 1, \ldots, n$, estimate the value of $f(x)$ at some domain point x not among the x_k's.

These problems arise in many areas of engineering and economics. As an illustration, consider the following circumstances.

1. Suppose that we are writing software for a new computer. It is necessary that certain library functions [e.g., sin (x), cos (x), exp (x), . . .] be available.

2. Suppose that the sales rate of a certain product as a function of price has been established at a certain finite set of prices, through previous sales history. We would like to have a curve relating sales rate to price, as an aid to the management problem of optimal price selection. In this regard there is currently great commercial interest in developing software for graphical display of business and engineering data.

In addition to immediate business and engineering tasks, the subject of interpolation plays a crucial role within the realm of numerical methods itself. For example, prominent methods for other numerical tasks such as integration (Chapter 3) and solution of differential equations (Chapter 7) are motivated by polynomial approximation. In particular, if we are assured that the polynomial $p(x)$ is close to a given function $f(x)$, it might rightly be suspected that for some interval $[a, b]$ of moderate length, $\int_a^b p(x)\,dx$ ought to be close to $\int_a^b f(x)\,dx$. The value of this insight is that polynomial integration is very simple.

We use polynomials as function approximators and as curves for data fitting, since they and their ratios are the only functions that can be reduced to elementary arithmetic. However, as we suggest in Section 6.4, other function classes, notably sinusoids, are useful alternatives in function approximation. The gist of the situation is that once computer methodology has been developed for approximation

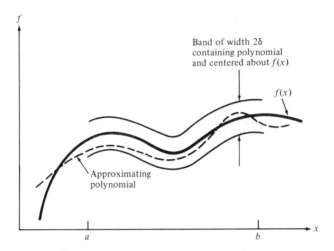

FIGURE 2.1 Illustration of the Weierstrass Theorem

of members of some class (such as sinusoids) of functions, that class itself becomes available as a basis for approximation of yet more general function classes.

It turns out, however, that the class of polynomials is surprisingly powerful; one can feasibly approximate a great many common functions reasonably accurately over a limited domain. Examples supporting this claim will be offered, but so will exceptions. There are theoretical reasons why we might hope for success with polynomials, the most noteworthy being the following statement.

> ***Theorem (Weierstrass).*** *Let $f(x)$ be any continuous function defined on a closed bounded interval $[a, b]$, and let δ be any positive number. Then there is a polynomial $p(x)$ such that for all $x \in [a, b]$,*
>
> $$|f(x) - p(x)| < \delta.$$

The proof of this theorem is found in many numerical analysis texts [e.g., Szidarovszky and Yakowitz (1978), pp. 26–28]. Figure 2.1 illustrates this theorem. No matter how small a band you draw about the continuous function, there is some polynomial lying entirely inside this band.

As polynomials are fundamental to the methods of this chapter, discussion begins with an efficient scheme for their computer evaluation. Then use of polynomials as function approximation devices commences with a description of Taylor's series methods, which the reader has presumably encountered in calculus. Following that we describe polynomial and spline interpolation techniques.

2.2
HORNER'S RULE

A recurring task in this chapter and elsewhere is that of evaluating a given polynomial

$$p(x) = a_k x^k + a_{k-1} x^{k-1} + \cdots + a_0 \tag{2.1}$$

at some specified number x. The coefficient a_k of the highest power in k is the *leading* coefficient, and k itself is the polynomial *degree*.

It is in the tradition of numerical analysis to seek efficient ways to evaluate frequently used formulas. If we thoughtlessly programmed the right side of (2.1) as it stands, with $x \times x$, $x \times x \times x$, . . ., replacing x^2, x^3, and so on, then evaluation of $p(x)$ would require $k + (k - 1) + (k - 2) + \cdots + 1 = k(k + 1)/2$ multiplications and k additions. On the other hand, polynomial evaluation by Horner's rule requires only k additions and multiplications, k being the degree of the polynomial.

The idea behind Horner's rule is that of rewriting (2.1) as a "nested" formula,

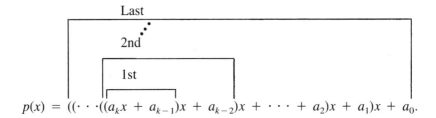

$$p(x) = ((\cdots((a_k x + a_{k-1})x + a_{k-2})x + \cdots + a_2)x + a_1)x + a_0.$$

The convention is that the innermost expression $a_k x + a_{k-1}$ is to be calculated first. The resulting number constitutes a multiplicand for the expression at the next level. In this fashion, successive terms are computed as values for the lower levels (or inner parenthetical terms) become available. The number of levels equals the polynomial degree, k.

By this procedure, only one addition and one multiplication is needed at each level. Although some improvement over Horner's rule is possible, it is known (Knuth, 1969, Sec. 4.6) that any general polynomial algorithm requires at least $k/2$ multiplications.

Horner's rule is implemented by construction of the sequence b_k, b_{k-1}, . . ., b_0 from (2.1) according to

$$
\begin{aligned}
b_k &= a_k \\
b_{k-1} &= b_k x + a_{k-1} \\
&\vdots \\
b_j &= b_{j+1} x + a_j \\
&\vdots \\
b_1 &= b_2 x + a_1 \\
p(x) = b_0 &= b_1 x + a_0.
\end{aligned}
\tag{2.2}
$$

By this construct, $p(x) = b_0$.

EXAMPLE 2.1

Let us apply Horner's rule to evaluate the quadratic (second-degree polynomial)

$$p(x) = 3x^2 + x + 5$$

at $x = 6$. In the notation of (2.1), $k = 2$ and $a_2 = 3$, $a_1 = 1$, and $a_0 = 5$. So, according to (2.2),

$$b_2 = 3$$

$$b_1 = 3 \times 6 + 1 = 19$$

$$b_0 = 19 \times 6 + 5 = 119 = p(6).$$

Horner's rule entails performing the arithmetic operations as indicated by the following nested representation:

$$p(x) = (3x + 1)x + 5 = \underbrace{(\overbrace{b_2 x + 1}^{b_1})x + 5}_{b_0}.$$

Subroutine HORNER for Horner's rule is given in Table 2.1. The arguments j for the coefficients $A(j)$ run from 1 to $k + 1$, instead of 0 to k because FORTRAN array variables must have positive indices. The role of b_j, in (2.2) is assumed by the variable P.

TABLE 2.1   Subroutine for the Horner's Rule

```
      SUBROUTINE HORNER(K,A,X,P)
C
C     ****************************************************************
C     *  FUNCTION: THIS SUBROUTINE COMPUTES THE VALUE OF A K-TH     *
C     *            DEGREE POLYNOMIAL P(X) AT A GIVEN INDEPENDENT     *
C     *            VALUE X USING THE HORNER'S RULE. THE K+1 BY 1     *
C     *            INPUT ARRAY A REPRESENTS THE POLYNOMIAL           *
C     *            ACCORDING TO:                                     *
C     *            P(X)=A(K+1)*X**K+A(K)*X**(K-1)+...+A(2)*X+A(1)    *
C     *  USAGE:                                                      *
C     *      CALL SEQUENCE: CALL HORNER(K,A,X,P)                     *
C     *  PARAMETERS:                                                 *
C     *      INPUT:                                                  *
C     *          K=POLYNOMIAL DEGREE                                 *
C     *          A=(K+1) BY 1 ARRAY OF POLYNOMIAL COEFFICIENTS       *
C     *          X=GIVEN INDEPENDENT VALUE                           *
C     *      OUTPUT:                                                 *
C     *          P=POLYNOMIAL VALUE P(X)                             *
C     ****************************************************************
C
      DIMENSION A(K+1)
C     *** INITIALIZATION ***
      P=A(K+1)
C     *** COMPUTE K-TH DEGREE POLYNOMIAL VALUE P(X) ***
      DO 1 I=1,K
        P=P*X+A(K+1-I)
    1 CONTINUE
      RETURN
      END
```

2.3
TAYLOR'S POLYNOMIALS

Toward unifying the methodology of this chapter, Taylor's series ideas are now viewed from an unorthodox vantage point. Suppose that at some domain point x_0, we are given the values $f'(x_0), f''(x_0), f^{(2)}(x_0), \ldots, f^{(n-1)}(x_0)$ of $f(x)$ and its first $n - 1$ derivatives. On the basis of these n numbers, we are to provide a computer-implementable approximation of $f(x)$. In keeping with our stated resolve to work with polynomials, it is natural to seek a minimal-degree polynomial that is consistent with these data. Our objective then is to find the polynomial $p_{n-1}(x)$ of degree less than n satisfying the n conditions

$$p_{n-1}(x_0) = f(x_0), \quad p'_{n-1}(x_0) = f'(x_0), \ldots, p_{n-1}^{(n-1)}(x_0) = f^{(n-1)}(x_0). \quad (2.3)$$

Suppose that we could construct a system of polynomials $t_0(x), t_1(x), \ldots, t_j(x), \ldots$ with $t_j(x)$ having degree j, and such that

$$t_j^{(k)}(x_0) = \begin{cases} 0 & \text{if } k \neq j \\ 1 & \text{if } k = j \end{cases}.$$

Then the polynomial

$$p_{n-1}(x) = f(x_0)t_0(x) + f^{(1)}(x_0)t_1(x) + \cdots + f^{(n-1)}(x_0)t_{n-1}(x) \quad (2.4)$$

would satisfy (2.3). But construction of the requisite polynomials is easy. Note that if we set

$$\tilde{t}_j(x) = (x - x_0)^j,$$

then

$$\tilde{t}_j^{(k)}(x_0) = \begin{cases} 0, & k \neq j \\ j!, & k = j, \end{cases}$$

where $j! = j \times (j - 1) \times \cdots \times 2 \times 1$. We obtain the polynomial $t_j(x)$ by simply normalizing $\tilde{t}_j(x)$. That is,

$$t_j(x) = \frac{1}{j!} \tilde{t}_j(x) = \frac{(x - x_0)^j}{j!}. \quad (2.5)$$

It turns out that the polynomial (2.4) is the only polynomial of degree less than n that satisfies (2.3) (see Problem 8).

The polynomial $p_{n-1}(x)$ obtained through the constructs (2.4) and (2.5) is referred to as the *Taylor's polynomial*. It coincides with the first n terms of the

Taylor's series expansion of $f(x)$ about x_0. One can combine these equations to write the $(n-1)$st-degree Taylor's polynomial for $f(x)$ about x_0 as

$$p_{n-1}(x) = f(x_0) + \frac{f'(x_0)}{1!}(x - x_0) + \cdots + \frac{f^{(n-1)}(x_0)}{(n-1)!}(x - x_0)^{n-1}.$$

(2.6)

We will have occasion to use some standard notation that allows more tidy representation of formulas such as (2.6). Let a_0, a_1, a_2, . . . be any numbers and m a positive integer. Then $\sum_{j=0}^{m} a_j$ denotes the sum $a_0 + a_1 + a_2 + \cdots + a_m$. In this notation (2.6) may be written

$$p_{n-1}(x) = \sum_{k=0}^{n-1} \frac{f^{(k)}(x_0)}{k!}(x - x_0)^k.$$

Subroutine TAYLOR (Table 2.2), when presented with derivative values (in array DF), computes (2.6). The subroutine and the next example make use of the fact that

$$p_j(x) = p_{j-1}(x) + \frac{f^{(j)}(x_0)}{j!}(x - x_0)^j.$$

TABLE 2.2 Subroutine TAYLOR for Taylor's Polynomials

```
      SUBROUTINE  TAYLOR(N,DF,X0,X,P)
C
C
C     ****************************************************************
C     * FUNCTION: THIS SUBROUTINE COMPUTES THE VALUE OF THE N-TH *
C     *           DEGREE TAYLOR POLYNOMIAL P(X)                  *
C     *           GIVEN THE VALUE OF THE FUNCTION AND ITS FIRST  *
C     *           N DERIVATIVES AT A POINT X0                    *
C     * USAGE:                                                   *
C     *      CALL SEQUENCE:CALL TAYLOR(N,DF,X0,X,P)              *
C     * PARAMETERS:                                              *
C     *    INPUT:                                                *
C     *           N=DEGREE OF TAYLOR POLYNOMIAL                  *
C     *           DF=N+1 BY 1 ARRAY DF(J) IS (J-1)ST DERIVATIVE  *
C     *           OF F EVALUATED AT X0                           *
C     *           X=GIVEN INDEPENDENT VALUE                      *
C     *           X0=EXPANSION POINT                             *
C     *    OUTPUT:                                               *
C     *           P=TAYLOR POLYNOMIAL VALUE AT X                 *
C     ****************************************************************
C
      DIMENSION DF(N+1)
C     *** PROD=(X-X0)**(I-1) ***
      PROD=1.
      DIFF=X-X0
C     *** FACT= I FACTORIAL ***
      FACT=1
C     *** P ACCUMULATES THE TAYLOR POLYNOMIAL VALUE ***
      P=0.
      DO 10 I=1,N+1
         P=P+PROD*DF(I)/FACT
         PROD=PROD*DIFF
         FACT=FACT*I
   10 CONTINUE
      RETURN
      END
```

──── **EXAMPLE 2.2** ────────────────────────────────────

Let $f(x) = \sin(x)$ and $x_0 = 1$. Define $p_n(x)$ to be the nth-degree Taylor's polynomial for $f(x)$. Then

$$p_0(x) = f(x_0) = \sin(1)$$

$$p_1(x) = p_0(x_0) + f'(x_0)(x - x_0) = \sin(1) + \cos(1)(x - 1)$$

$$p_2(x) = p_1(x) + \tfrac{1}{2}f''(x_0)(x - x_0)^2$$

$$= \sin(1) + \cos(1)(x - 1) - \tfrac{1}{2}\sin(1)(x - 1)^2$$

and

$$p_3(x) = p_2(x) + \tfrac{1}{6}f^{(3)}(x_0)(x - x_0)^3$$

$$= \sin(1) + \cos(1)(x - 1) - \tfrac{1}{2}\sin(1)(x - 1)^2$$

$$- \tfrac{1}{6}\cos(1)(x - 1)^3.$$

In Table 2.3 we compare the formulas above and their error at 10 evenly spaced points on the interval $[0, 2]$. The program calling subroutine TAYPOL by which the table was computed is displayed in Table 2.4. Figure 2.2 plots $p_1(x)$, $p_2(x)$, $p_3(x)$, and $\sin(x)$ on the interval $[0, 2]$. Note that $p_1(x)$ is the tangent line at $x_0 = 1$.

Note that $p_3(x)$ gives a credible approximation to $\sin(x)$ over the interval $[0, 2]$. To reflect on what we have accomplished, observe that from just two numbers, $\sin(1)$ and $\cos(1)$, we can compute a Taylor's polynomial of arbitrarily high degree. This is a peculiarity of the trigonometric functions, but in general, $p_{n-1}(x)$ depends on only n numbers.

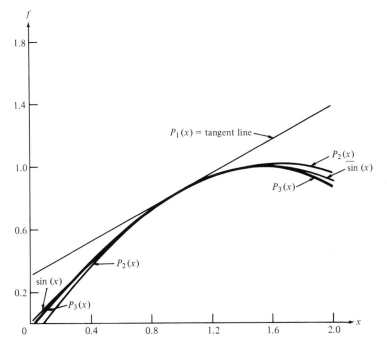

FIGURE 2.2   Taylor's Polynomial Approximation of $\sin(x)$

TABLE 2.3   **Application of Taylor's Formula**

x	$\sin(x)$	$p_1(x)$	$\sin(x) - p_1(x)$	$p_2(x)$	$\sin(x) - p_2(x)$	$p_3(x)$	$\sin(x) - p_3(x)$
0.2	0.198669	0.409229	-0.211E+00	0.139958	0.587E-01	0.186064	0.126E-01
0.4	0.389418	0.517290	-0.128E+00	0.365825	0.236E-01	0.385276	0.414E-02
0.6	0.564642	0.625350	-0.607E-01	0.558032	0.661E-02	0.563796	0.847E-03
0.8	0.717356	0.733411	-0.161E-01	0.716581	0.775E-03	0.717301	0.546E-04
1.0	0.841471	0.841471	0.000E+00	0.841471	0.000E+00	0.841471	0.000E+00
1.2	0.932039	0.949531	-0.175E-01	0.932702	-0.663E-03	0.931982	0.575E-04
1.4	0.985450	1.057592	-0.721E-01	0.990274	-0.482E-02	0.984511	0.939E-03
1.6	0.999574	1.165652	-0.166E+00	1.014187	-0.146E-01	0.994737	0.484E-02
1.8	0.973848	1.273713	-0.300E+00	1.004442	-0.306E-01	0.958336	0.155E-01
2.0	0.909297	1.381773	-0.472E+00	0.961038	-0.517E-01	0.870987	0.383E-01

TABLE 2.4   **Program for Taylor's Polynomial Example**

```
C       PROGRAM TAYPOL
C
C       ****************************************************************
C       COMPUTES TAYLOR POLYNOMIALS OF DEGREE 1 TO 3, FOR
C       SIN(X), ABOUT X0=1
C       CALLS:  TAYLOR
C       OUTPUT: X=THE VALUE OF THE INDEPENDENT VARIABLE
C               TRUE=THE VALUE OF SIN(X) AS COMPUTED BY THE
C                    FORTRAN LIBRARY FUNCTION SIN
C               P(K)=THE POLYNOMIAL APPROXIMATION AT X FOR DEGREE K
C               E(K)=THE ERROR IN APPROXIMATION AT X FOR EACH
C                    POLYNOMIAL
C       ****************************************************************
C
        DIMENSION DF(4),P(3),E(3)
        PI=4*ATAN(1.)
        N=3
        X0=1.
C
C       *** COMPUTE THE DERIVATIVES                               ***
C
        DO 10 I=1,N+1
            DF(I)=SIN(PI*(I-1)/2.+X0)
     10 CONTINUE
C
C       *** COMPUTES THE TRUE VALUES AT 10 EVENLY SPACED          ***
C       *** POINTS                                                ***
C
        DO 20 J=1,10
            X=J*.2
            TRUE=SIN(X)
C
C       *** SUBROUTINE TAYLOR COMPUTES THE K-TH DEGREE            ***
C       *** TAYLOR POLYNOMIAL THE ERROR IS THEN                   ***
C       *** COMPUTED AND ALL VALUES PRINTED                       ***
C
            DO 30 K=1,3
                CALL TAYLOR(K,DF,X0,X,F)
                P(K)=F
                E(K)=TRUE-P(K)
     30     CONTINUE
            WRITE(10,2)X,TRUE,P(1),E(1),P(2),E(2),P(3),E(3)
      2     FORMAT(1X,F3.1,1X,F10.6,3(1X,F10.6,1X,E12.3))
     20 CONTINUE
        STOP
        END
```

Incidentally, nowadays, FORTRAN library routines can be relied on to compute intrinsic functions such as sin (x) to an accuracy commensurate with the computer roundoff error. In Problem 14 we suggest ways to use trigonometric identities to check the accuracy of sin (x) and other functions.

The remainder formula for partial Taylor's series expansions is often useful for finding absolute bounds for the truncation error of the Taylor's polynomial. If $f(x)$ is n times differentiable in any interval $[a, b]$ that contains x and x_0 and $p_{n-1}(x)$ is the $(n - 1)$st-degree Taylor's polynomial (2.6), then

$$f(x) - p_{n-1}(x) = \frac{f^{(n)}(\zeta)}{n!} (x - x_0)^n, \qquad (2.7)$$

where ζ is between x_0 and x. The term on the right side of (2.7) is called the *remainder term*.

A drawback to the Taylor's polynomial approach is that it is not immediately applicable to the data-fitting problem (2) described in Section 2.1. Even for function approximation under the best of circumstances [i.e., when $f(x)$ is a known, sufficiently differentiable function], the Taylor's polynomial is typically awkward to implement because the expressions for higher derivatives are cumbersome.

The interpolation polynomial approach described in the following section does not require that the derivatives of $f(x)$ be known, or even exist. Moreover, its computer implementation is typically simple. Yet for "well-behaved" functions, the interpolation approach yields polynomial approximations of accuracy quite competitive with those of Taylor's series methods. A characteristic of the Taylor's polynomial is that it tends to be very accurate near x_0, but less accurate at a distance from x_0, whereas interpolation polynomials tend to "distribute" the error. The Taylor's series approach will be more important to us as an analytic device than as a computational technique.

2.4
POLYNOMIAL INTERPOLATION

2.4.1 Definition of Interpolation

Let $x_1, \ldots, x_n$ denote distinct real numbers and let $f_1, \ldots, f_n$ be arbitrary real numbers. The points (x_k, f_k), $k = 1, \ldots, n$, can be imagined to be data values to be connected by a curve. Any function $p(x)$ satisfying the conditions

$$p(x_k) = f_k \qquad (k = 1, 2, \ldots, n) \qquad (2.8)$$

is called an *interpolation function*. An interpolation function, then, is a curve that passes through the data points, as shown in Figure 2.3. In view of our computer orientation, we will be concerned with the case that $p(x)$ is a polynomial. Assume, then, that $p(x)$ is the mth-degree polynomial

$$p(x) = a_m x^m + a_{m-1} x^{m-1} + \cdots + a_1 x + a_0.$$

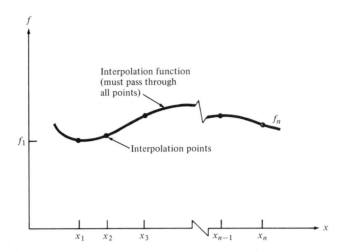

FIGURE 2.3 An Interpolation Function

Then conditions (2.8) imply the relation

$$a_m x_k^m + a_{m-1} x_k^{m-1} + \cdots + a_1 x_k + a_0 = f_k \qquad (k = 1, 2, \ldots, n),$$

$$(2.9)$$

where the unknowns are the $m + 1$ polynomial coefficients a_m, a_{m-1}, . . ., a_1, a_0. Since the number of equations is n, typically we have no solutions for $m + 1 < n$, infinitely many solutions exist for $m + 1 > n$, and the solution is unique for $n = m + 1$.

Notice that the Taylor's polynomial was constructed from the consecutive derivatives $f(x_0)$, $f'(x_0)$, . . ., $f^{(n-1)}(x_0)$ of $f(x)$ at a specified point x_0, whereas the interpolation polynomial is determined by the function values $f(x_1)$, $f(x_2)$, . . ., $f(x_n)$ of $f(x)$ at specified domain points x_1, . . ., x_n. In both cases, n data values determine unique polynomials of degree less than n.

We state at the outset that in many situations, data interpolation is not recommended. Such situations include erratic functions and data with random components. Procedures in Chapter 6 should be considered in these contexts.

2.4.2 Lagrange Interpolation

Whereas (2.9) can be shown to uniquely determine the $(n - 1)$st-degree interpolation polynomial, for computational and expository reasons, other constructs are preferred.

Toward motivating the Lagrange representation, suppose that for each j, $1 \leq j \leq n$, we can find an $(n - 1)$st-degree polynomial $l_j(x)$ such that for any k, $1 \leq k \leq n$,

$$l_j(x_k) = \begin{cases} 0 & \text{if } j \neq k \\ 1 & \text{if } j = k. \end{cases} \qquad (2.10)$$

Then since the sum of polynomials of degree less than n is itself a polynomial of degree less than n,

$$p_{n-1}(x) = f_1 l_1(x) + f_2 l_2(x) + \cdots + f_n l_n(x)$$

is an interpolation polynomial of requisite degree. For observe that

$$p_{n-1}(x_j) = f_1 \cdot 0 + f_2 \cdot 0 + \cdots + f_{j-1} \cdot 0$$
$$+ f_j \cdot 1 + f_{j+1} \cdot 0 + \cdots + f_n \cdot 0 = f_j.$$

But the requisite polynomial $l_j(x)$ satisfying (2.10) is easy enough to construct. As an example we give a detailed description of $l_1(x)$. In light of (2.10), the distinct points $x_2, x_3, \ldots, x_n$ must be roots of $l_1(x)$. But this condition determines $l_1(x)$ up to a scalar multiple. For a standard factorization theorem of polynomial algebra states that if x_k is a root of any polynomial $g(x)$, then $(x - x_k)$ is a factor of $g(x)$. From this and (2.10) we conclude that $l_1(x)$ must have the factors $(x - x_2), (x - x_3), \ldots, (x - x_n)$. Since the degree must be less than n, these must be the only factors. Thus

$$l_1(x) = C(x - x_2) \cdots (x - x_n).$$

The remaining job is to find a scalar C so that $l_1(x)$ satisfies the remaining condition of (2.10), that $l_1(x_1) = 1$. But this is clearly achieved by setting

$$C = \frac{1}{(x_1 - x_2)(x_1 - x_3) \cdots (x_1 - x_n)}.$$

The other terms $l_j(x)$, $j > 1$, can be obtained by the same procedure. We have in this fashion constructed the *Lagrange interpolation polynomial* for the data (x_k, f_k), $k = 1, \ldots, n$. We summarize this construction as

$$p_{n-1}(x) = f_1 l_1(x) + f_2 l_2(x) + \cdots + f_n l_n(x),$$

where for $j = 1, 2, \ldots, n$,

$$l_j(x) = \frac{(x - x_1)(x - x_2) \cdots (x - x_{j-1})(x - x_{j+1}) \cdots (x - x_n)}{(x_j - x_1)(x_j - x_2) \cdots (x_j - x_{j-1})(x_j - x_{j+1}) \cdots (x_j - x_n)}.$$

$$(2.11)$$

Analogously to the Σ notation introduced in connection with Taylor's polynomial, the Π symbol allows concise representation of multiplication operations. For any numbers $y_1, \ldots, y_n$,

$$\prod_{k=1}^{n} y_k$$

denotes the product $y_1 \times y_2 \times \cdots \times y_n$, and the symbol

$$\prod_{\substack{k=1 \\ k \neq j}}^{n} y_k$$

denotes the product $y_1 \times y_2 \times \cdots \times y_{j-1} \times y_{j+1} \times \cdots \times y_n$. Thus we may now write the Lagrange factors in (2.11) as

$$l_j(x) = \left[\prod_{\substack{k=1 \\ k \neq j}}^{n} (x - x_k) \right] \Big/ \left[\prod_{\substack{k=1 \\ k \neq j}}^{n} (x_j - x_k) \right].$$

or even

$$l_j(x) = \prod_{\substack{k=1 \\ k \neq j}}^{n} \frac{x - x_k}{x_j - x_k}.$$

Subroutine LAGR for computing values of the Lagrange interpolation polynomial according to (2.11) is given in Table 2.5. One inputs the number N of points, arrays $X = (X(1), \ldots, X(N))$ and $F = (F(1), \ldots, F(N))$ of data values, and domain point T. The subroutine output variable P is the value $p(t)$ of the interpolation polynomial at $t = T$. The DO loop with label 2 computes the values $l_k(t)$, and they are stored as the variable FACTOR. The DO loop with label 1 multiplies these values by the functional values $F(k)$ and adds these products. The sum is returned as the variable P.

TABLE 2.5   Subroutine LAGR for Polynomial Interpolation

```
      SUBROUTINE LAGR(N,X,F,T,P)
C
C     ****************************************************************
C     *   FUNCTION: THIS SUBROUTINE COMPUTES THE LAGRANGE INTER-    *
C     *             POLATION POLYNOMIAL VALUE, GIVEN A SET OF N     *
C     *             DATA POINTS, AT A SPECIFIED VALUE T             *
C     *   USAGE:                                                    *
C     *       CALL SEQUENCE: CALL LAGR(N,X,F,T,P)                   *
C     *   PARAMETERS:                                               *
C     *       INPUT:                                                *
C     *           N=NUMBER OF DATA POINTS                           *
C     *           X=N BY 1 ARRAY OF INDEPENDENT DATA POINTS         *
C     *           F=N BY 1 ARRAY OF FUNCTIONAL VALUES               *
C     *           T=DESIRED POINT FOR INTERPOLATION                 *
C     *       OUTPUT:                                               *
C     *           P=INTERPOLATION POLYNOMIAL VALUE AT T             *
C     ****************************************************************
C
      DIMENSION X(N),F(N)
C     *** INITIALIZATION ***
      P=0.0
C     *** COMPUTE LAGRANGIAN POLYNOMIAL VALUE P AT T ***
      DO 1 K=1,N
         FACTOR=1.0
C     *** COMPUTE LAGRANGIAN INTERPOLATION FACTORS ***
         DO 2 I=1,N
            IF (I.NE.K) THEN
               FACTOR=FACTOR*(T-X(I))/(X(K)-X(I))
            END IF
    2    CONTINUE
         P=P+FACTOR*F(K)
    1 CONTINUE
      RETURN
      END
```

━━━━ **EXAMPLE 2.3** ━━━━

A three-point Lagrange interpolation polynomial for $f(x) = \sin(x)$ is calculated in detail. We take the data given in Table 2.6.

TABLE 2.6 Data for Three-Point Interpolation Example

j	1	2	3
x_j	0	1	2
f_j	sin (0)	sin (1)	sin (2)

Then by (2.11),

$$l_1(x) = \frac{(x - 1)(x - 2)}{(0 - 1)(0 - 2)} = \frac{(x - 1)(x - 2)}{2}$$

$$l_2(x) = \frac{(x - 0)(x - 2)}{(1 - 0)(1 - 2)} = -x(x - 2)$$

$$l_3(x) = \frac{(x - 0)(x - 1)}{(2 - 0)(2 - 1)} = \frac{x(x - 1)}{2}.$$

From (2.11) and the preceding equations,

$$p_2(x) = \sin(0)\frac{(x - 1)(x - 2)}{2} - \sin(1)x(x - 2) + \sin(2)\frac{x(x - 1)}{2}.$$

For $x = 1.5$, $p_2(1.5) \approx 0.9721$, and the correct value is sin (1.5) ≈ 0.9975. The relative error is about 2.6%.

In Table 2.7 we have provided a calling program that utilizes subroutine LAGR (Table 2.5) to compute first-, second-, and third-degree Lagrange approximations to sin (x). These are denoted, respectively, by p_1, p_2, and p_3 in Table 2.8, which gives the results of evaluating these polynomials at 10 evenly spaced points on [0, 2].

In Figure 2.4, the first-, second-, and third-degree Lagrange interpolation polynomials have been plotted for comparison with sin (x) over the interval $I = [0, 2]$. For the two-point formula, only the endpoints are used, and this of course resulted in the secant line connecting the endpoints. The points in Table 2.6 served as data for the three-point formula, and the interpolation points x_1, x_2, x_3, and x_4 were 0, $\frac{2}{3}$, $\frac{4}{3}$, and 2, for the four-point (cubic) interpolation polynomial. In comparing Figure 2.4 to the corresponding Figure 2.2 of the Taylor's polynomial example, note that the latter tends to be more accurate near the middle of the interval, but the interpolation polynomial maintained good approximation toward the extremes. This illustrates developments discussed in the next section.

TABLE 2.7   Program for Lagrange Polynomial Example

```
C        PROGRAM LAGRANGE
C
C        ******************************************************************
C        COMPUTES  LAGRANGE  INTERPOLATION POLYNOMIALS FOR THE
C        FUNCTION F(X)=SIN(X) AT 10 EQUALLY SPACED POINTS ON
C        THE INTERVAL [0,2]
C        CALLS: LAGR
C        OUTPUT: XD=DOMAIN POINTS FOR EVALUATION
C                TRUE=SIN(XD) AS COMPUTED BY THE LIBRARY FUNCTION
C                INTER(I)=VALUE OF THE INTERPOLATION POLYNOMIAL OF
C                         DEGREE (I)
C                ERR(I)= THE CORRESPONDING ERROR I=1,2,3
C        ******************************************************************
C
         REAL X(10),F(10),INTER(10),ERR(10)
         DO 10 N=1,10
            XD=N*0.2
            TRUE=SIN(XD)
            DO 20 I=1,3
               DO 30 J=1,I+1
C
C        *** COMPUTE THE DATA VALUES                                   ***
C
                  X(J)=(J-1)*2./I
                  F(J)=SIN(X(J))
30             CONTINUE
C
C        *** SUBROUTINE LAGR COMPUTES THE INTERPOLATION VALUES         ***
C        *** AND ERR GIVES THE CORRESPONDING ERROR                     ***
C
               CALL LAGR(I+1,X,F,XD,P)
               ERR(I)=TRUE-P
               INTER(I)=P
20          CONTINUE
            WRITE(10,2)XD,TRUE,(INTER(I),ERR(I),I=1,3)
2           FORMAT(3X,F4.2,F10.6,3(3X,F10.6,3X,E12.3))
10       CONTINUE
         STOP
         END
```

TABLE 2.8   Numerical Results of Example 2.3

x	sin (x)	$p_1(x)$	$sin(x) - p_1(x)$	$p_2(x)$	$sin(x) - p_2(x)$	$p_3(x)$	$sin(x) - p_3(x)$
0.2	0.198669	0.090930	0.108E+00	0.230186	-0.315E-01	0.204306	-0.564E-02
0.4	0.389418	0.181859	0.208E+00	0.429426	-0.400E-01	0.394319	-0.490E-02
0.6	0.564642	0.272789	0.292E+00	0.597720	-0.331E-01	0.565951	-0.131E-02
0.8	0.717356	0.363719	0.354E+00	0.735068	-0.177E-01	0.715113	0.224E-02
1.0	0.841471	0.454649	0.387E+00	0.841471	0.000E+00	0.837717	0.375E-02
1.2	0.932039	0.545578	0.386E+00	0.916928	0.151E-01	0.929676	0.236E-02
1.4	0.985450	0.636508	0.349E+00	0.961439	0.240E-01	0.986901	-0.145E-02
1.6	0.999574	0.727438	0.272E+00	0.975004	0.246E-01	1.005306	-0.573E-02
1.8	0.973848	0.818368	0.155E+00	0.957624	0.162E-01	0.980800	-0.695E-02
2.0	0.909297	0.909297	0.000E+00	0.909297	0.000E+00	0.909297	0.000E+00

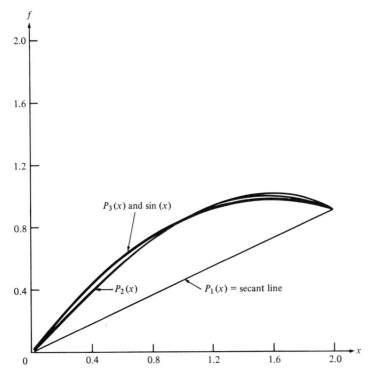

FIGURE 2.4 Interpolation Polynomial Approximation of sin (x)

Because of various constructions such as (2.9), (2.11), and more to come, of "the" interpolation polynomial, it is reassuring to know that they are all different expressions for precisely the same function. That is, not only do the polynomials constructed by these various schemes all satisfy the requisite interpolation condition (2.8), but they also coincide for all other arguments x. To demonstrate this assertion, suppose that $p(x)$ and $q(x)$ are both interpolation polynomials of degree less than n for the same n data points. Then the polynomial

$$r(x) = p(x) - q(x)$$

must satisfy

$$r(x_k) = p(x_k) - q(x_k) = 0, \qquad 1 \le k \le n.$$

That is, $r(x)$ must have n distinct roots, although its degree is at most $n - 1$. But by the factorization property of polynomials cited earlier, this is impossible, unless $r(x)$ is the zero polynomial, in which case $p(x) = q(x)$, as we claimed.

If a Lagrange interpolation is to be evaluated at many different arguments x, it is worthwhile noting that we can calculate, once and for all, the terms

$$y_j = \frac{f_j}{\prod_{\substack{k=1 \\ k \neq j}}^{n} (x_j - x_k)}, \qquad 1 \leq j \leq n. \tag{2.12}$$

At each desired argument x, we compute

$$w_n(x) = \prod_{k=1}^{n} (x - x_k).$$

The Lagrange interpolation polynomial value $p_{n-1}(x)$ is then given by

$$p_{n-1}(x) = w_n(x)\left(\frac{y_1}{x - x_1} + \frac{y_2}{x - x_2} + \cdots + \frac{y_n}{x - x_n}\right), \tag{2.13}$$

which, as the reader may confirm, is algebraically identical to the defining condition (2.11). By this procedure, the main cost is the computation of the y_j terms, which together require $n(n - 2)$ multiplications and $n(n - 1)$ additions. But this need be done only once for each polynomial. Thereafter, (2.13) is evaluated at the cost of n multiplications and $2n - 1$ additions and n divisions. For purposes of simplicity, our subroutine LAGR did not follow this path, but evaluates (2.11) directly, at the cost of approximately n^2 multiplications per call. Our view is that for various reasons such as suggested in Section 2.4.4, the proper place of polynomial interpolation is instances in which the degree is 15 or less, and not a great many evaluations are to be made of any single polynomial. Techniques such as to be presented in Chapter 6 have more appeal for oft-repeated tasks such as library functions.

2.4.3 Truncation Error of Interpolation Polynomials

In the spirit of (2.7) for Taylor's polynomial truncation error, the truncation error of the interpolation polynomial now occupies our attention. For this discussion it is presumed that $p_{n-1}(x)$ is the interpolation polynomial determined by the data (x_k, f_k), $k = 1, \ldots, n$, with $f_k = f(x_k)$ for some function $f(x)$.

Let $f(x)$ be n times differentiable in an interval $[a, b]$ that contains the points $x_1, \ldots, x_n$ and x. Then there exists a point $\zeta = \zeta(x) \in [a, b]$ such that

$$\boxed{f(x) - p_{n-1}(x) = \frac{1}{n!} f^{(n)}(\zeta)w_n(x),} \tag{2.14}$$

where $f^{(n)}(x)$ denotes the nth derivative of $f(x)$ and $w_n(x)$ is defined by

$$\boxed{w_n(x) = (x - x_1)(x - x_2) \cdots (x - x_n).} \tag{2.15}$$

Derivation of this error bound is given by Szidarovszky and Yakowitz (1978, pp. 31–32).

It is instructive to compare this truncation error formula with the remainder formula (2.7) for the error of the Taylor's polynomial approximation. It is clear that for x near x_0, Taylor's formula ought to be more accurate. But if $f^{(n)}(x)$ does not change appreciably on the interval of interest, and if the domain points x_k, $k = 1, \ldots, n$, are "evenly distributed," one would anticipate that for x closer to the extremes of $[a, b]$,

$$|w_n(x)| < |(x - x_0)^n|$$

and that Lagrange interpolation might well be more accurate. For example, if $f(x)$ happened to be an nth-degree polynomial, then $f^{(n)}(x)$ is a constant and our discussion would apply forcefully. The next section gives a plan for choosing the points x_i, $1 \le i \le n$, in such a fashion as to make the maximum of $|w_n(x)|$ as small as possible.

──────── **EXAMPLE 2.4** ──

In Examples 2.2 and 2.3, the Taylor's and the interpolation polynomials, respectively, were applied to the task of approximating sin (x) over the interval $[0, 2]$. Figure 2.5 shows the actual error, as a function of argument x, associated with the third-degree Taylor's and interpolation polynomial approximations. In keeping with our theoretical anticipations, the Taylor's polynomial error is relatively small near the expansion point $x_0 = 1$ and relatively large near the extremes of the interval.

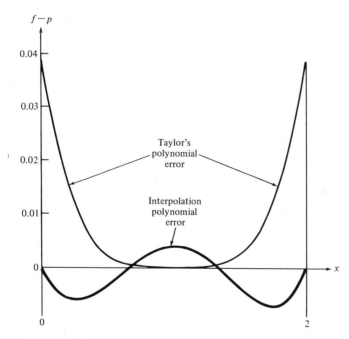

FIGURE 2.5 Comparison of Actual Error of Cubic Taylor's and Interpolation Polynomials of $f(x) = \sin(x)$

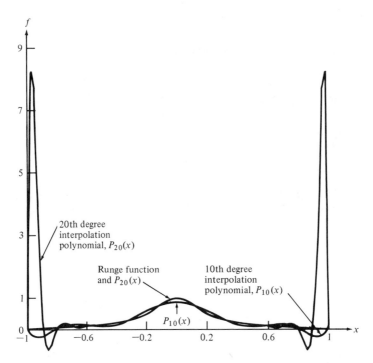

FIGURE 2.6   Interpolation Polynomial Approximation of the Runge Function

2.4.4   Refinements and Limitations of Polynomial Interpolation

One might hope that if a function $f(x)$ is continuous and "well behaved" on a given finite interval $[a, b]$, then as the number of interpolation points x_k increases and becomes dense on that interval, the associated polynomial sequence $\{p_{n-1}(x)\}$ converges to $f(x)$ at each point x in $[a, b]$. In Example 2.5, with the x_k's evenly spaced on $[-1, 1]$ and $f(x)$ not only continuous but differentiable, we see an approximation actually get worse.

EXAMPLE 2.5

Let $f(x) = 1/(1 + 25x^2)$, where $x \in [-1, 1]$. This function is called the *Runge function*. Figure 2.6 shows this function and its interpolation polynomials of degrees 10 and 20. It is known (Isaacson and Keller, 1966, pp. 275–278) that the sequence $\{p_{n-1}(x)\}$ of these interpolation polynomials with equally spaced interpolation points does not converge to $f(x)$ at any x in the intervals $[-1, -0.727]$ or $[0.727, 1]$. Note that the Runge function is fairly well represented by interpolation polynomials over the central 72.7% of the interval, but the approximation is very bad near the extremes. ∎

The power of mathematical reasoning is most astonishing in cases in which it directs us to unintuitive conclusions. It would seem "reasonable" that equally spaced interpolation points x_k (i.e., $x_{k+1} - x_k$ the same for all k) ought to be best in the sense that they provide the most representative $f(x_k)$ samples. Thus it might

FIGURE 2.7 Distribution of Chebyshev Points

be expected that these ought to be the most logical points for constructing an interpolation polynomial. This line of reasoning is incorrect.

Let us recall from Section 2.4.3 that the truncation error for approximating an n-times differentiable function by an interpolation polynomial $p_{n-1}(x)$ with interpolation points $x_1, \ldots, x_n$ is

$$f(x) - p_{n-1}(x) = \frac{f^{(n)}(\zeta)w_n(x)}{n!}. \tag{2.16}$$

In (2.16), ζ is some number in the interval containing all interpolation points as well as x, and

$$w_n(x) = (x - x_1)(x - x_2) \cdots (x - x_n).$$

The term $f^{(n)}(\zeta)$ above is beyond our control. However, the term $w_n(x)$ offers possibilities if we are free to choose the interpolation points. A sensible ambition might then be to choose these points so that the maximum of $|w_n(x)|$ is as small as possible over the interval $[a, b]$ of interpolation. It turns out (Szidarovszky and Yakowitz, 1978, Sec. 2.2) that this ambition is realized by choosing the so-called Chebyshev points as the interpolation points. For the interval $[a, b]$, the *Chebyshev points* are defined to be

$$x_k = \frac{a + b}{2} + \frac{a - b}{2} \cos\left(\frac{2k - 1}{2n} \pi\right) \qquad (k = 1, \ldots, n). \tag{2.17}$$

In Figure 2.7 we have located the Chebyshev points for $n = 10$ on the interval $[-1, 1]$. Figure 2.8 compares $w_{10}(x)$ when x_i's are evenly spaced against $w_{10}(x)$ for x_i's the Chebyshev points.

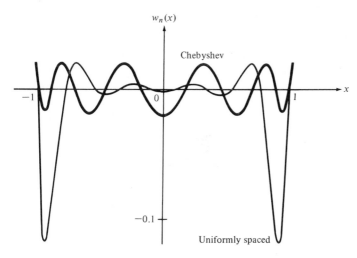

FIGURE 2.8 Comparison of $w_n(x)$ Functions

**TABLE 2.9   Interpolation of Runge Function
Using Chebyshev Points**

```
C       PROGRAM RUNGE
C
C       ************************************************************
C       COMPUTES 30TH DEGREE INTERPOLATION POLYNOMIAL USING
C       CHEBYSHEV POINTS FOR THE RUNGE FUNCTION
C       CALLS:  LAGR
C       OUTPUT: T=THE VALUE OF THE INDEPENDENT VARIABLE
C               TRUE=THE FUNCTIONAL VALUE AT THE POINT T AS
C                    DETERMINED BY THE STATEMENT FUNCTION FF
C               P=THE VALUE OF THE INTERPOLATION POLYNOMIAL AT T
C               E=THE ERROR OF THE APPROXIMATION AT T
C       ************************************************************
C
        DIMENSION X(40),F(40)
C
C       *** RUNGE FUNCTION STATEMENT                           ***
C
        FF(X)=1.0/(1.0+(5.0*X)**2)
C
C       *** COMPUTE VALUES AT CHEBYSHEV POINTS                 ***
C
        N=31
        PI=4.0*ATAN(1.0)
        DO 10 I=1,N
          X(I)=COS((2.0*I-1)*PI/(2.0*N))
          F(I)=FF(X(I))
     10 CONTINUE
C
C       *** SUBROUTINE LAGR WILL COMPUTE VALUE OF THE          ***
C       *** INTERPOLATION POLYNOMIAL AT T AND E GIVES THE ERROR ***
C
        DO 30 I=1,20
          T=-1.0+I*0.1
          CALL LAGR(N,X,F,T,P)
          TRUE=FF(T)
          E=TRUE-P
          WRITE(10,2)T,TRUE,P,E
      2   FORMAT(2X,F3.1,2X,E14.7,2X,E14.7,2X,E14.7)
     30 CONTINUE
        STOP
        END
```

───── **EXAMPLE 2.6** ─────

By means of the simple calling program in Table 2.9 we have constructed the
31-point Chebyshev-interpolation polynomial for the Runge function (described
in Example 2.5). In the output listing (Table 2.10), the interpolation error is
tabulated at 20 evenly spaced values. We see that the Runge function is apparently
under control.

**TABLE 2.10   Output of Runge Function
Interpolation**

x	Runge Function	Approximation	Error
-.9	0.4705882E-01	0.4692032E-01	0.1385026E-03
-.8	0.5882353E-01	0.5837149E-01	0.4520416E-03
-.7	0.7547170E-01	0.7447892E-01	0.9927750E-03
-.6	0.9999998E-01	0.1011296E+00	-0.1129620E-02
-.5	0.1379310E+00	0.1372023E+00	0.7287264E-03

TABLE 2.10   (Continued)

x	Runge Function	Approximation	Error
-.4	0.2000000E+00	0.2003204E+00	-0.3204048E-03
-.3	0.3076923E+00	0.3076517E+00	0.4056096E-04
-.2	0.5000001E+00	0.4999129E+00	0.8720160E-04
-.1	0.8000001E+00	0.8000615E+00	-0.6145239E-04
0.0	0.1000000E+01	0.1000000E+01	-0.1192093E-06
0.1	0.8000000E+00	0.8000612E+00	-0.6127357E-04
0.2	0.4999999E+00	0.4999129E+00	0.8702278E-04
0.3	0.3076922E+00	0.3076518E+00	0.4041195E-04
0.4	0.2000000E+00	0.2003204E+00	-0.3203750E-03
0.5	0.1379310E+00	0.1372023E+00	0.7287711E-03
0.6	0.9999998E-01	0.1011297E+00	-0.1129672E-02
0.7	0.7547168E-01	0.7447891E-01	0.9927750E-03
0.8	0.5882351E-01	0.5837147E-01	0.4520416E-03
0.9	0.4705882E-01	0.4692034E-01	0.1384839E-03
1.0	0.3846154E-01	0.3764865E-01	0.8128881E-03

■

Have we mastered the approximation problem through interpolation using Chebyshev points? In some cases "yes" and in other cases "no." From de Boor (1978), for example, the following facts are known:

1. If (as in the case of the Runge function) $f(x)$ has a *continuous derivative* on $[a, b]$, then using Chebyshev points, for each x in $[a, b]$,

$$p_{n-1}(x) \rightarrow f(x), \qquad \text{as } n \rightarrow \infty.$$

2. No matter how interpolation points are chosen, there is some continuous function $f(x)$ for which $\{p_{n-1}(x)\}$ fails to converge, regardless of x.
3. Even when $f(x)$ has a continuous derivative, adequate approximation may require an impossibly large n.

Point 3 is illustrated by our printout in Table 2.10, where a 30th-degree polynomial is scarcely accurate to three significant decimals. In this regard we note that the degree must be over 1 million for $\sqrt{|x|}$ to be interpolated on $[-1, 1]$ to an accuracy of 10^{-3} (de Boor, 1978). Interpolation polynomials of degree 30 are on the borderline of the impractical, and degrees greater than 100 lead almost invariably to grievous roundoff effects.

*2.4.5   The Newton Representation

From developments in Section 2.4.2, we know that all interpolation polynomials having degree not exceeding $n - 1$, n being the number of points to be interpolated, are identical to the Lagrange representation (2.11). But other representations sometimes serve useful purposes. For example, the *Newton representation* of the interpolating polynomial has the feature that it is easier to update—that is, to include an additional point (x_{n+1}, f_{n+1}). Also, if coded wisely, it requires only about half as many arithmetic operations per call as the Lagrange form and is less subject to deterioration due to roundoff effects.

The Newton representation of the interpolation polynomial for the data (x_k, f_k), $k = 1, \ldots, n$, is

$$
\begin{aligned}
p_{n-1}(x) = a_0 &+ a_1(x - x_1) + \cdots \\
&+ a_k(x - x_1)(x - x_2) \cdots (x - x_k) + \cdots \\
&+ a_{n-1}(x - x_1) \cdots (x - x_{n-1}).
\end{aligned}
\tag{2.18}
$$

where the coefficients $a_0, a_1, \ldots, a_{n-1}$ are determined so as to ensure that the data are interpolated [i.e., that $p_{n-1}(x_k) = f_k$]. The interpolation condition for one and two points requires that

$$
\begin{aligned}
a_0 &= f_1 \\
a_1 &= \frac{f_2 - a_0}{x_2 - x_1},
\end{aligned}
\tag{2.19}
$$

and for $k = 2, \ldots, n - 1$,

$$
a_k = \frac{\left[f_{k+1} - a_0 - \sum_{l=1}^{k-1} a_l(x_{k+1} - x_1) \cdots (x_{k+1} - x_l) \right]}{(x_{k+1} - x_1) \cdots (x_{k+1} - x_k)}.
\tag{2.20}
$$

This formula, in conjunction with (2.19), provides a means of recursively computing the coefficients for the Newton representation (2.18). If one wishes to add an additional point (x_{n+1}, f_{n+1}) in order to obtain an nth-degree interpolation polynomial $p_n(x)$ for the pairs (x_k, f_k), $1 \leq k \leq n + 1$, then the coefficients $\bar{a}_k$ of p_n are identical to the coefficients a_k of p_{n-1}, for $k = 0, \ldots, n - 1$ and $\bar{a}_n$ is given by (2.20) with $k = n$. Thus in the Newton representation only one coefficient needs to be evaluated for updating, whereas in other representations, in order to accommodate an additional data pair, the computations have to be redone in their entirety.

We now rewrite the Newton representation (2.18) using some traditional notation. For $k \geq 1$,

$$
f[x_k] = f_k
$$

$$
f[x_k, x_{k+1}] = \frac{f[x_{k+1}] - f[x_k]}{x_{k+1} - x_k}
$$

$$
f[x_k, x_{k+1}, x_{k+2}] = \frac{f[x_{k+1}, x_{k+2}] - f[x_k, x_{k+1}]}{x_{k+2} - x_k}
$$

$$
\vdots
$$

$$
f[x_k, x_{k+1}, \ldots, x_i, x_{i+1}] = \frac{f[x_{k+1}, \ldots, x_{i+1}] - f[x_k, \ldots, x_i]}{x_{i+1} - x_k}
$$

$$
\vdots
$$

These quantities are called *divided differences* and can be computed recursively by the equations above. Finite induction shows that the coefficients of the Newton representation satisfy the relations

$$a_0 = f[x_1]$$

$$a_1 = f[x_1, x_2]$$

$$\vdots$$

$$a_{n-1} = f[x_1, x_2, \ldots, x_n].$$

(2.22)

Thus the Newton representation can be obtained through implementation of the above recursion. For $k = 1, \ldots, n$, define $f[x_k]$ to be equal to f_k. Then, for $k = 1, \ldots, n - 1$ compute the values of $f[x_k, x_{k+1}]$. Now, for $k = 1, \ldots,$ $n - 2$ determine the quantities $f[x_k, x_{k+1}, x_{k+2}]$, and so on. Until the value of $f[x_1, \ldots, x_n]$ is obtained. The *difference table* constructed from the interpolating points $x_1, \ldots, x_n$ and functional values $f_1, \ldots, f_n$ has the following form:

x_1	$f[x_1]$	$f[x_1, x_2]$	$\cdots$	$f[x_1, \ldots, x_{n-1}]$	$f[x_1, \ldots, x_n]$
x_2	$f[x_2]$	$f[x_2, x_3]$	$\cdots$	$f[x_2, \ldots, x_n]$	
x_3	$f[x_3]$	$f[x_3, x_4]$			
$\vdots$	$\vdots$	$\vdots$			
x_{n-2}	$f[x_{n-2}]$	$f[x_{n-2}, x_{n-1}]$	$f[x_{n-2}, x_{n-1}, x_n]$		
x_{n-1}	$f[x_{n-1}]$	$f[x_{n-1}, x_n]$			
x_n	$f[x_n]$				

The elements $f[x_1], f[x_1, x_2], \ldots, f[x_1, \ldots, x_n]$ of the top row are the coefficients $a_0, a_1, \ldots, a_{n-1}$ of the Newton representation (2.18).

EXAMPLE 2.7

Consider again the function $f(x) = \sin(x)$. The Newton representation of the interpolation polynomial with interpolating points $x_1 = 0$, $x_2 = 2$, and $x_3 = 1$, will be determined from the recursive relations (2.19)–(2.20). From (2.19),

$$a_0 = f_1 = f(x_1) = 0,$$

$$a_1 = \frac{f(x_2) - a_0}{x_2 - x_1} = \frac{\sin(2)}{2},$$

so the linear interpolating polynomial based on the points x_1 and x_2 is

$$p(x) = 0 + \frac{\sin(2)}{2} x.$$

To incorporate the additional interpolation point $x_3 = 1$ into the interpolation polynomial, in accordance with (2.20), we calculate that

$$a_2 = \frac{f(x_3) - a_0 - a_1(x_3 - x_1)}{(x_3 - x_1)(x_3 - x_2)}$$

$$= \frac{\sin(1) - 0 - a_1(1 - 0)}{(1 - 0)(1 - 2)}$$

$$= a_1 - \sin(1) = \frac{\sin(2)}{2} - \sin(1).$$

The reader may confirm that with these coefficients, (2.18) is a quadratic interpolation polynomial for sin (x) at x_1, x_2, and x_3, and because of the uniqueness of the interpolating polynomial, it must coincide with the quadratic Lagrange representation, the performance of which was investigated in Example 2.3.

In terms of divided differences, these interpolation polynomials can be expressed as follows. Since

$$f[x_1] = f_1 = f(0) = 0$$

$$f[x_2] = f_2 = f(2) = \sin(2)$$

$$f[x_3] = f_3 = f(1) = \sin(1),$$

the recursion of the divided differences implies that

$$f[x_1, x_2] = \frac{\sin(2) - 0}{2 - 0} = \frac{\sin(2)}{2},$$

$$f[x_2, x_3] = \frac{\sin(1) - \sin(2)}{1 - 2} = \sin(2) - \sin(1),$$

and

$$f[x_1, x_2, x_3] = \frac{\sin(2) - \sin(1) - \sin(2)/2}{1 - 0} = \frac{\sin(2)}{2} - \sin(1).$$

Thus the linear and quadratic interpolation polynomials are

$$p_1(x) = a_0 + a_1(x - x_1) = 0 + \frac{\sin(2)}{2} x$$

and

$$p_2(x) = a_0 + a_1(x - x_1) + a_2(x - x_1)(x - x_2)$$

$$= 0 + \frac{\sin(2)}{2} x + \left[\frac{\sin(2)}{2} - \sin(1)\right] x(x - 2). \quad\blacksquare$$

2.5
SPLINE FUNCTIONS

2.5.1 Motivation for Piecewise Polynomial Methods

There are situations in which the performance of interpolation polynomials is found to be inferior to alternative techniques such as those discussed in this section and in Chapters 5 and 6. Specific situations for which splines, the subject of the present section, afford advantages over interpolation polynomials are

> *Situation 1.* The number n of data points to be interpolated is moderate to large (e.g., n is greater than 20).
>
> *Situation 2.* The data points are associated with some known or unknown function $f(x)$ the derivatives of which are large, or do not exist.
>
> *Situation 3.* The data points arise from a natural or "nonmathematical" source (e.g., the profile of an airplane fuselage or the boundary between two countries).

Roundoff error is a prime cause of difficulty when n is large. Let $\tilde{f}_k$, $1 \leq k \leq n$, denote the data values actually used in computing the interpolation polynomial $\tilde{p}_{n-1}(x)$ and suppose the "correct" values to be f_k, with

$$|f_k - \tilde{f}_k| \leq \delta, \tag{2.23}$$

where δ is a given positive number. Then in the notation of (2.11) the difference of the Lagrange interpolation polynomials $p_{n-1}(x)$ and $\tilde{p}_{n-1}(x)$ based, respectively, on the data (x_k, f_k) and $(x_k, \tilde{f}_k)$ can be bounded as follows:

$$|p_{n-1}(x) - \tilde{p}_{n-1}(x)| = \left| \sum_{k=1}^{n} f_k l_k(x) - \sum_{k=1}^{n} \tilde{f}_k l_k(x) \right|$$

$$= \left| \sum_{k=1}^{n} (f_k - \tilde{f}_k) l_k(x) \right|$$

$$\leq \sum_{k=1}^{n} |f_k - \tilde{f}_k| \cdot |l_k(x)| \leq \delta \sum_{k=1}^{n} |l_k(x)|.$$

The quantity on the right can increase dramatically with the number of data points. As the next example shows, even under the most favorable circumstances, roundoff error causes intolerable deterioration of high-degree interpolation polynomials. We thereby motivate situation 1 above.

────── **EXAMPLE 2.8** ──────────────────────────

The function $f(x) = \sin(x)$ is interpolated using evenly spaced domain points in the interval $[0, 1]$. This function is exceptionally suited to approximation by interpolation because the magnitudes of derivatives are bounded by 1. From the printout (Table 2.11), which displays values of $\sin(x)$ and its Lagrange approx-

TABLE 2.11   Numerical Results of Example 2.8

	Argument x	sin (x)	Lagrange Interpolation $p_{n-1}(x)$
	0.1948200	0.1935899	0.1935901
	0.7324600	0.6687008	0.6687007
	0.6087400	0.5718342	0.5718342
	0.3225800	0.3170145	0.3170145
n = 20	0.1084500	0.1082375	0.1082376
	0.1884800	0.1873660	0.1873660
	0.6172300	0.5787785	0.5787784
	0.7132600	0.6543025	0.6543024
	0.6379100	0.5955178	0.5955178
	0.4875300	0.4684451	0.4684452
	0.1948200	0.1935899	0.1936576
	0.7324600	0.6687008	0.6687038
	0.6087400	0.5718342	0.5718339
	0.3225800	0.3170145	0.3170149
n = 50	0.1084500	0.1082375	0.1447353 ←
	0.1884800	0.1873660	0.1875511
	0.6172300	0.5787785	0.5787789
	0.7132600	0.6543025	0.6543033
	0.6379100	0.5955178	0.5955179
	0.4875300	0.4684451	0.4684451
	0.1948200	0.1935899	0.1949847
	0.7324600	0.6687008	0.6686892
	0.6087400	0.5718342	0.5718341
	0.3225800	0.3170145	0.3170152
n = 70	0.1084500	0.1082375	60.45348 ←
	0.1884800	0.1873660	0.1861881
	0.6172300	0.5787785	0.5787786
	0.7132600	0.6543025	0.6543000
	0.6379100	0.5955178	0.5955176
	0.4875300	0.4684451	0.4684451

imation at randomly chosen domain points in the unit interval, we see that the approximation is excellent for $n = 20$ but is a disaster for $n = 70$. This computational experiment will depend on the computer word length; this particular run used a VAX. ∎

De Boor (1978, p. 22) has offered a heuristic explanation as to why piecewise interpolation might provide relief with irregular curves (situations 2 and 3 at the start of this section):

We stress the fact that polynomial interpolation at appropriately chosen points (e.g., the Chebyshev points) produces an approximation which, for all practical purposes, differs very little from the best possible approximant by polynomials of the same order. This allows us to illustrate

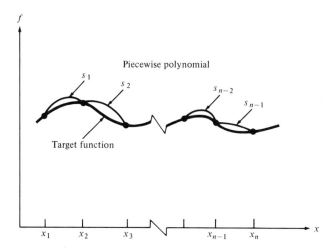

FIGURE 2.9 Piecewise Polynomial Interpolation

the essential limitations of polynomial approximation: If the function to be approximated is badly behaved anywhere in the interval of approximation, then the approximation is poor everywhere. This global dependence on local properties can be avoided when using *piecewise* polynomial approximants.

During the past decade, piecewise polynomial approximations have become prominent both in theoretical studies and in applications. Instead of trying to approximate a function over the entire interval by one polynomial of high degree, one approximates the function by a piecewise polynomial function, where the degree of the polynomial "pieces" associated with each subinterval is small. A piecewise polynomial approximation is illustrated in Figure 2.9. Polynomial $s_1(x)$ approximates $f(x)$ on the interval $[x_1, x_2]$, s_2 approximates $f(x)$ on $[x_2, x_3]$, and so on. Note in this graph that the piecewise polynomial displays jagged behavior at the interpolation points. Splines, the subject discussed next, are piecewise polynomials that prevent such erratic profiles by imposing derivative constraints at the common boundary points of the polynomial pieces.

2.5.2 Definition of Splines

Spline functions are piecewise polynomials with derivatives constrained for the purpose of making the resulting function smooth at the node points x_i. Let $[a, b]$ be a finite interval containing the points $a = x_1 < x_2 < \cdots < x_n = b$. A *spline function* of degree m with interpolating points x_k, $k = 1, \ldots, n$, is a piecewise polynomial $s(x)$ satisfying the following properties:

> 1. $s(x)$ is $m - 1$ times differentiable at each point x_i, $1 \leq i \leq n$.
> 2. On each subinterval $[x_k, x_{k+1}]$ ($1 \leq k \leq n - 1$), $s(x)$ is a polynomial of degree not exceeding m.

In words, a spline function (or, more simply, a spline) of degree m is a piecewise polynomial function whose "pieces" do not exceed degree m. Property 1

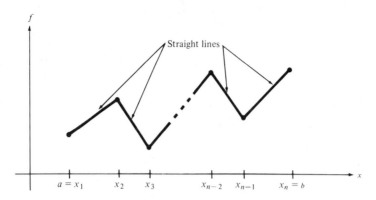

FIGURE 2.10   Spline Function of Degree 1

assures that the polynomial coefficients are constrained so that the derivatives, up to order $m - 1$, of adjacent polynomials agree at their common boundary point.

In Section 2.4 a polynomial was said to be an interpolation polynomial if it interpolates a specified data set. In the same spirit, a spline $s(x)$ is an *interpolation spline* for given points (x_i, f_i), $1 \le i \le n$, if for each i,

$$s(x_i) = f_i.$$

The spline functions of degree 1 are piecewise linear functions as shown in Figure 2.10. On each subinterval $[x_i, x_{i+1}]$, $s(x)$ has the form

$$s(x) = s_i(x) = a_i + b_i(x - x_i),$$

which is a linear polynomial of $x - x_i$. (The coefficients a_i and b_i bear no relation to the endpoints a and b of the approximation interval $[a, b]$.) The representation of splines as polynomials in $x - x_i$ rather than x will lead to simpler formulas. The coefficients a_i, b_i may vary with interval index i, but continuity (order 0 differentiability) at x_{i+1} requires that $s_{i+1}(x_{i+1}) = s_i(x_{i+1})$. That is, the coefficients must satisfy

$$a_{i+1} + b_{i+1}(x_{i+1} - x_{i+1}) = a_{i+1} = a_i + b_i(x_{i+1} - x_i), \qquad 1 \le i \le n - 2.$$

This implies that

$$b_i = \frac{a_{i+1} - a_i}{x_{i+1} - x_i}.$$

If the data values f_i are specified at each point x_i, "interpolation" requires that

$$f_i = s_i(x_i) = a_i + b_i(x_i - x_i) = a_i.$$

Thus each coefficient a_i equals the corresponding functional value f_i, and b_i equals the slope connecting the neighboring data points (x_i, f_i) and (x_{i+1}, f_{i+1}).

━━ **EXAMPLE 2.9** ━━━━━━━━━━━━━━━━━━━━━━━━━━━━━━

Here we construct a linear interpolating spline for the following data set:

i	1	2	3	4
x_i	−1	0	1	2
f_i	0	1	2	1

Since we have just observed that $a_i = f_i$, we immediately conclude that

$$a_1 = 0, \quad a_2 = 1, \quad a_3 = 2, \quad a_4 = 1.$$

Then, using the formula

$$b_i = \frac{a_{i+1} - a_i}{x_{i+1} - x_i},$$

we calculate

$$b_1 = \frac{1 - 0}{1} = 1, \quad b_2 = 1, \quad b_3 = \frac{1 - 2}{1} = -1.$$

The linear interpolating spline connects data points by broken lines as in Figure 2.10. The graph is continuous.

━━━ ■

As an application we mention that the popular numerical integration method known as the trapezoidal rule can be viewed as a method in which an interpolating spline function of degree 1 is integrated. We shall encounter this rule in Chapter 3.

The quadratic spline functions ($m = 2$) are piecewise quadratic polynomials, with continuous slopes at the interpolation points, as illustrated in Figure 2.11. On subinterval $[x_i, x_{i+1}]$, the quadratic spline $s(x)$ must have the form

$$s_i(x) = a_i + b_i(x - x_i) + c_i(x - x_i)^2.$$

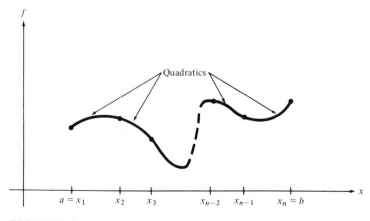

FIGURE 2.11 Spline Function of Degree 2

To be a spline, $s(x)$ must be continuous and have a first derivative at x_i and x_{i+1}. Continuity requires that $s_{i+1}(x_{i+1}) = s_i(x_{i+1})$; that is, the coefficients satisfy

$$a_{i+1} + b_{i+1}(x_{i+1} - x_{i+1}) + c_{i+1}(x_{i+1} - x_{i+1})^2 = a_{i+1}$$

$$= a_i + b_i(x_{i+1} - x_i) + c_i(x_{i+1} - x_i)^2$$

and the existence of a derivative at x_{i+1} will require that $s'_{i+1}(x_{i+1}) = s'_i(x_{i+1})$. That is,

$$b_{i+1} + 2c_{i+1}(x_{i+1} - x_{i+1})$$

$$= b_{i+1} = b_i + 2c_i(x_{i+1} - x_i), \quad i = 1, 2, \ldots, n-2.$$

Details of quadratic spline construction are omitted. The more important case, cubic splines, is investigated thoroughly in Section 2.5.3, which follows.

2.5.3 Cubic Splines

Cubic (degree 3) interpolating splines are the most common splines in the applications literature. Reasons for this popularity include their ability to interpolate data with curves that look smooth. Lower-degree splines do not "disguise" the data points well (see Figure 2.10 in this regard), and high-degree splines have the instabilities inherent in high-degree polynomials. Cubic splines appear to flow smoothly and indeed possess a minimum curvature property described in Section 2.5.4.

Let $(x_1, f_1), (x_2, f_2), \ldots, (x_n, f_n)$ be given data points. The aim of this section is to describe in detail the construction of a cubic interpolating spline for these points.

According to notation used in the preceding section, on each subinterval $[x_i, x_{i+1}]$, a cubic spline $s(x)$ has the form

$$\boxed{\begin{aligned} s_i(x) = a_i + b_i(x - x_i) + c_i(x - x_i)^2 + d_i(x - x_i)^3 \\ (i = 1, 2, \ldots, n-1), \end{aligned}} \quad (2.24)$$

where the coefficients a_i, b_i, c_i, d_i are to be determined from the definition of "cubic splines" and the interpolatory requirement.

Continuity and interpolation require that

$$s_i(x_{i+1}) = f_{i+1} = s_{i+1}(x_{i+1}),$$

so

$$a_i + b_i(x_{i+1} - x_i) + c_i(x_{i+1} - x_i)^2 + d_i(x_{i+1} - x_i)^3 = f_{i+1}$$

$$= a_{i+1} + b_{i+1}(x_{i+1} - x_{i+1}) + c_{i+1}(x_{i+1} - x_{i+1})^2 \quad (2.25)$$

$$+ d_{i+1}(x_{i+1} - x_{i+1})^3.$$

By introducing the notation

$$h_i = x_{i+1} - x_i$$

for each $i = 1, 2, \ldots, n - 1$, relation (2.25) can be rewritten as

$$a_{i+1} = f_{i+1} = a_i + b_i h_i + c_i h_i^2 + d_i h_i^3. \tag{2.26}$$

Since cubic splines are twice differentiable at the nodes x_{i+1},

$$s_i'(x_{i+1}) = s_{i+1}'(x_{i+1}) \quad \text{and} \quad s_i''(x_{i+1}) = s_{i+1}''(x_{i+1}).$$

By using the expansion (2.24), these differentiability conditions may be written

$$b_i + 2c_i h_i + 3d_i h_i^2 = b_{i+1}, \quad c_i + 3d_i h_i = c_{i+1}. \tag{2.27}$$

Solving for d_i in the second equation of (2.27) and substituting this value into (2.26) and the first equation of (2.27), we obtain the relations

$$a_{i+1} = a_i + b_i h_i + c_i h_i^2 + \frac{c_{i+1} - c_i}{3h_i} h_i^3$$

$$= a_i + b_i h_i + \frac{h_i^2}{3}(2c_i + c_{i+1}) \tag{2.28}$$

and

$$b_{i+1} = b_i + 2c_i h_i + 3\frac{c_{i+1} - c_i}{3h_i} h_i^2$$

$$= b_i + h_i(c_i + c_{i+1}).$$

The final relationship involving the coefficients is obtained by solving the first equation of (2.28) for b_i, to get

$$b_i = \frac{a_{i+1} - a_i}{h_i} - \frac{h_i}{3}(2c_i + c_{i+1}), \tag{2.29}$$

and after reducing the index by 1,

$$b_{i-1} = \frac{a_i - a_{i-1}}{h_{i-1}} - \frac{h_{i-1}}{3}(2c_{i-1} + c_i).$$

By substituting these values into the second equation of (2.28), and using the fact (2.26) that for interpolation

$$a_j = f_j, \quad 1 \le j \le n - 1,$$

we have

$$\boxed{h_{i-1}c_{i-1} + 2(h_{i-1} + h_i)c_i + h_i c_{i+1} = 3\frac{f_{i+1} - f_i}{h_i} - 3\frac{f_i - f_{i-1}}{h_{i-1}}}$$

$$(i = 2, 3, \ldots, n - 2).$$

$$\tag{2.30}$$

Once the coefficients c_i are found, the remaining coefficients are simply determined from (2.26) to (2.28) to be

$$a_i = f_i,$$

$$b_i = \frac{f_{i+1} - f_i}{h_i} - \frac{h_i}{3}(2c_i + c_{i+1}),$$

$$d_i = \frac{c_{i+1} - c_i}{3h_i},$$

In (2.30), $c_1, c_2, \ldots, c_{n-1}$ are the unknowns. The number of equations (2.30) equals $n - 3$. Consequently, two more conditions are required to uniquely determine the c_i's. The conditions $s_1''(a) = s_{n-1}''(b) = 0$ are customarily introduced, and a cubic spline, the *natural spline*, is thereby determined. Observe that

$$s_1''(a) = 2c_1 + 6d_1(x_1 - x_1) = 2c_1, \qquad s_{n-1}''(b) = 2c_{n-1} + 6d_{n-1}h_{n-1}.$$

Thus

$$c_1 = 0, \qquad c_{n-1} + 3d_{n-1}h_{n-1} = 0$$

for natural splines. The second of these conditions is most conveniently imposed by defining the (extraneous) parameter $c_n = 0$; then (2.30) holds for $i = 2, 3, \ldots, n - 1$. In detail, (2.30) becomes

$$2(h_1 + h_2)c_2 + h_2c_3 = 3\frac{f_3 - f_2}{h_2} - 3\frac{f_2 - f_1}{h_1}$$

$$h_2c_2 + 2(h_2 + h_3)c_3 + h_3c_4 = 3\frac{f_4 - f_3}{h_3} - 3\frac{f_3 - f_2}{h_2}$$

$$\cdots$$

$$h_{n-3}c_{n-3} + 2(h_{n-3} + h_{n-2})c_{n-2} + h_{n-2}c_{n-1} = 3\frac{f_{n-1} - f_{n-2}}{h_{n-2}} - 3\frac{f_{n-2} - f_{n-3}}{h_{n-3}}$$

$$h_{n-2}c_{n-2} + 2(h_{n-2} + h_{n-1})c_{n-1} = 3\frac{f_n - f_{n-1}}{h_{n-1}} - 3\frac{f_{n-1} - f_{n-2}}{h_{n-2}},$$

$$(2.31)$$

where $h_i = x_{i+1} - x_i, \qquad i = 1, 2, \ldots, n - 1.$

The computer program for solving the linear equations (2.3.1) and computing the coefficients a_i, b_i, and d_i is given by subroutine SPLN, shown in Table 2.12. Subroutine SPLE, given in Table 2.13, evaluates the spline function at any desired arguments, according to (2.24). It requires the coefficients a_i, b_i, c_i, d_i obtained

TABLE 2.12 Subroutine SPLN for Evaluating Spline Coefficients

```
        SUBROUTINE SPLN(N,X,F,A,B,C,D)
C
C       ***************************************************************
C       *    FUNCTION: THIS SUBROUTINE COMPUTES THE COEFFICIENTS FOR    *
C       *              A NATURAL CUBIC SPLINE                           *
C       *    USAGE:                                                     *
C       *         CALL SEQUENCE: CALL SPLN(N,X,F,A,B,C,D)               *
C       *         EXTERNAL FUNCTIONS/SUBROUTINES: SUBROUTINE GAUS1      *
C       *    PARAMETERS:                                                *
C       *         INPUT:                                                *
C       *              N=NUMBER OF DATA POINTS                          *
C       *              (AT LEAST 3 AT MOST 101)                         *
C       *              X=N BY 1 ARRAY OF DOMAIN POINTS                  *
C       *              F=N BY 1 ARRAY OF FUNCTIONAL VALUES              *
C       *         OUTPUT:                                               *
C       *              A,B,C,D=N BY 1 ARRAYS OF CUBIC SPLINE COEFFICIENTS *
C       ***************************************************************
C
        DIMENSION X(N),F(N),A(N),B(N),C(N),D(N),H(100),T(100,101),U(100)
C       *** INITIALIZATION ***
        DO 1 I=1,N-1
            H(I)=X(I+1)-X(I)
            U(I)=(F(I+1)-F(I))/H(I)
            A(I)=F(I)
      1 CONTINUE
C       *** COMPUTE COEFFICIENTS OF THE LINEAR EQUATIONS ***
        DO 2 I=1,N-2
            DO 2 J=1,N-2
               T(I,J)=0.0
      2 CONTINUE
        DO 3 I=1,N-2
            T(I,I)=2.0*(H(I)+H(I+1))
      3 CONTINUE
        IF(N.GT.3) THEN
            DO 4 I=2,N-2
               T(I,I-1)=H(I)
               T(I-1,I)=H(I)
      4     CONTINUE
        END IF
        DO 5 I=1,N-2
            T(I,N-1)=3.0*(U(I+1)-U(I))
      5 CONTINUE
        N2=N-2
        M=1
        ND=100
        EPS=0.0000001
C       *** COMPUTE COEFFICIENTS C(I) USING GAUSSIAN ELIMINATION ***
        CALL GAUS1(N2,M,ND,T,EPS)
        DO 6 I=2,N-1
            C(I)=T(I-1,N-1)
      6 CONTINUE
        C(1)=0.0
        C(N)=0.0
        DO 7 I=1,N-1
            B(I)=U(I)-H(I)*(2.0*C(I)+C(I+1))/3.0
            D(I)=(C(I+1)-C(I))/(H(I)*3.0)
      7 CONTINUE
        RETURN
        END
```

by SPLN as input parameters. In subroutine SPLE, Horner's rule is used for evaluating the cubic polynomials on the right-hand side of (2.24). The rationale for solving linear equations such as (2.31) is deferred to Chapter 4, where subroutine GAUS1, which is called by SPLN, is explained.

**TABLE 2.13 Subroutine SPLE for Evaluating
Natural Spline Values**

```
        SUBROUTINE SPLE(N,X,A,B,C,D,T,P)
C
C  ***************************************************************
C  *  FUNCTION: THIS SUBROUTINE COMPUTES THE VALUE OF A CUBIC    *
C  *            SPLINE P(T) AT T   USING THE COEFFICIENTS        *
C  *            A,B,C,D GENERATED BY THE SUBROUTINE SPLN         *
C  *  USAGE:                                                     *
C  *       CALL SEQUENCE: CALL SPLE(N,X,A,B,C,D,T,P)             *
C  *       EXTERNAL FUNCTIONS/SUBROUTINES: SUBROUTINE SPLN       *
C  *  PARAMETERS:                                                *
C  *       INPUT:                                                *
C  *          N=NUMBER OF DATA POINTS                            *
C  *          X=N BY 1 ARRAY OF INDEPENDENT DATA POINTS          *
C  *          A,B,C,D=N BY 1 ARRAY OF CUBIC SPLINE COEFFICIENTS  *
C  *          U=GIVEN INDEPENDENT DATA VALUE                     *
C  *       OUTPUT:                                               *
C  *          P=CUBIC SPLINE VALUE P(T)                          *
C  ***************************************************************
C
        DIMENSION X(N),A(N),B(N),C(N),D(N)
C       *** DETERMINE SUBINTERVAL WHICH CONTAINS POINT T ***
        I=2
        DO WHILE(T.GT.X(I))
           I=I+1
        END DO
        I=I-1
C       *** COMPUTE CUBIC SPLINE VALUE AT P(T) ***
        T1=T-X(I)
        P=A(I)+T1*(B(I)+T1*(C(I)+D(I)*T1))
        RETURN
        END
```

─── **EXAMPLE 2.10** ───

Consider the $n = 3$ values

i	1	2	3
x_i	100	121	144
f_i	10	11	12

which are points from the square-root function. Since $n = 3$, $c_1 = c_3 = 0$, and
there is only one unknown coefficient c_2. The value of c_2 can be obtained from
the first equation of (2.31), which has the form

$$2(h_1 + h_2)c_2 = 3 \frac{f_3 - f_2}{h_2} - 3 \frac{f_2 - f_1}{h_1}.$$

Since $a_i = f_i$ $(i = 1, 2, 3)$,

$$a_1 = 10, \quad a_2 = 11, \quad a_3 = 12$$

and

$$h_1 = 121 - 100 = 21, \quad h_2 = 144 - 121 = 23,$$

this equation is

$$88c_2 = -0.012422,$$

which implies that

$$c_2 = -0.000141.$$

Then relations (2.29) and (2.27) imply that

$$b_1 = \frac{f_2 - f_1}{h_1} - \frac{h_1}{3}(2c_1 + c_2) = 0.048607$$

$$b_2 = \frac{f_3 - f_2}{h_2} - \frac{h_2}{3}(2c_2 + c_3) = 0.045640$$

$$d_1 = \frac{c_2 - c_1}{3h_1} = -0.00000224$$

$$d_2 = \frac{c_3 - c_2}{3h_2} = -0.00000205.$$

When these values for a_i, b_i, c_i, and d_i are inserted into (2.24), one may construct the table of spline estimates given in Table 2.15. By way of the calling program

TABLE 2.14   Constructs a Spline for the Square-Root Function

```
C      PROGRAM CSPLIN
C
C      ****************************************************************
C      COMPUTES CUBIC SPLINE ESTIMATE OF F(X)=SQRT(X) AT X=100+4*I
C      I=0 TO 11
C      CALLS:   SPLN;SPLE;GAUS1
C      OUTPUT: X=THE VALUE OF THE INDEPENDENT VARIABLE
C              P= THE SPLINE FUNCTION EVALUATED AT X
C              TRUE=TRUE VALUE OF FUNCTION AS COMPUTED BY THE LIBRARY
C                   FUNCTION SQRT
C              ERR=THE CORRESPONDING ERROR (TRUE-P) AT X
C      ****************************************************************
       DIMENSION Y(3),F(3),A(3),B(3),C(3),D(3)
       N=3
       DATA (Y(I),I=1,3)/100.0,121.0,144.0/
       DATA (F(I),I=1,3)/10.0,11.0,12.0/
C
C      *** SUBROUTINE SPLN COMPUTES THE SPLINE COEFFICIENTS      ***
C
       CALL SPLN(N,Y,F,A,B,C,D)
       M=12
       H=4.0
       X=96.0
C
C      *** THIS LOOP GENERATES  THE 12 VALUES OF X AND PRINTS    ***
C      *** THE SPECIFIED VALUES AT EACH POINT                    ***
C
       DO 8 I=1,M
          X=X+H
C
C      *** SUBROUTINE SPLE COMPUTES SPLINE APPROXIMATION AT X    ***
C
          CALL SPLE(N,Y,A,B,C,D,X,P)
          TRUE=SQRT(X)
          ERR=TRUE-P
          WRITE(10,9),X,P,TRUE,ERR
    9     FORMAT(18X,F5.1,3X,F8.5,3X,F8.5,3X,E9.2)
    8  CONTINUE
       STOP
       END
```

TABLE 2.15 Computations of Example 2.10

x	$s(x)$	$\sqrt{x}$	$\sqrt{x} - s(x)$
100.0	10.00000	10.00000	0.00E+00
104.0	10.19429	10.19804	0.38E-02
108.0	10.38771	10.39230	0.46E-02
112.0	10.57941	10.58300	0.36E-02
116.0	10.76854	10.77033	0.18E-02
120.0	10.95422	10.95445	0.23E-03
124.0	11.13571	11.13553	-0.18E-03
128.0	11.31328	11.31371	0.42E-03
132.0	11.48771	11.48913	0.14E-02
136.0	11.65978	11.66190	0.21E-02
140.0	11.83029	11.83216	0.19E-02
144.0	12.00000	12.00000	0.00E+00

in Table 2.14, we have used the spline subroutines SPLN and SPLE (Tables 2.12 and 2.13) to automate the construct of this table.   ■

ENGINEERING EXAMPLE

Here we forcefully illustrate the utility of splines for interpolating nonmathematical curves (Situation 3 at the beginning of this section). Our draftsman rendered the "generic" executive jet airplane profile that constitutes Figure 2.12 and picked off 20 position points along the top side of this profile. By means of the routines SPLN and SPLE (Tables 2.12 and 2.13), we plotted the spline associated with these data. This spline is shown in Figure 2.13, together with the positions of the original data used in the spline construction. When the routine LAGR (Table 2.5) was applied to the data points, the interpolation polynomial values were completely erratic.

In a similar vein, 32 positions were measured from the back of the polar bear sketched in Figure 2.14. They, together with the associated natural spline, are shown in Figure 2.15.

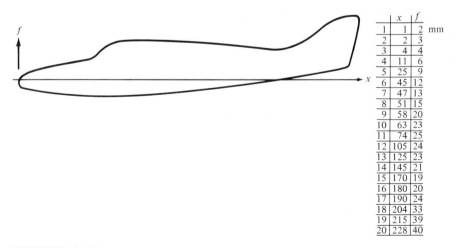

	x	f	
1	1	2	mm
2	2	3	
3	4	4	
4	11	6	
5	25	9	
6	45	12	
7	47	13	
8	51	15	
9	58	20	
10	63	23	
11	74	25	
12	105	24	
13	125	23	
14	145	21	
15	170	19	
16	180	20	
17	190	24	
18	204	33	
19	215	39	
20	228	40	

FIGURE 2.12 A Jet Plane

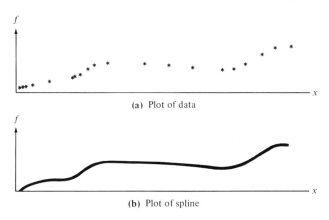

(a) Plot of data

(b) Plot of spline

FIGURE 2.13 Data and Spline of a Jet Plane

	x	f	
1	5	6	m
2	10	11	
3	15	14	
4	20	16	
5	25	18	
6	30	19	
7	35	18	
8	40	17.5	
9	45	17	
10	50	16	
11	55	16.5	
12	60	17.5	
13	65	18.5	
14	70	19	
15	75	18	
16	80	16.5	
17	85	15.5	
18	90	15.5	
19	95	16.5	
20	100	17	
21	105	18	
22	110	19	
23	115	20	
24	120	20	
25	125	19.5	
26	130	18.5	
27	135	17.5	
28	140	15	
29	145	12	
30	150	9	
31	155	3.5	
32	157	0	

FIGURE 2.14 A Polar Bear

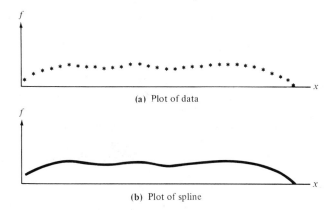

(a) Plot of data

(b) Plot of spline

FIGURE 2.15 Data and Spline of a Polar Bear

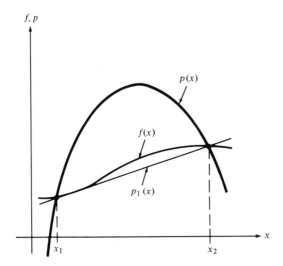

FIGURE 2.16 Large Interpolation Polynomial Deviation

The thrust of this example is to show that splines are effective in economically "presenting" natural shapes to the computer. To our knowledge, for simplicity and effectiveness splines have no close competitor in this capacity. ∎

*2.5.4 Strengths and Limitations of Spline Interpolation

The term *spline* originally referred to a flexible strip that draftsmen use to draw smooth curves between given points. The procedure is to put the strip on its edge and adjust it so that it passes through the points. The draftsman then uses the resulting spline profile as the smooth interconnecting curve. Abstractly, one can say that the more erratic the spline profile, the higher its internal energy. In keeping with the minimal energy principle of physics, therefore, the spline adjusts its shape to become as smooth as possible, subject to its interpolating the desired points.

The mathematical cubic spline is known to retain this minimum strain energy property (e.g., de Boor, 1978, p. 66). A closely related property that seems to us even more interesting is that the cubic spline has minimal integrated squared second derivative among all twice differentiable curves passing through the data. We will prove this assertion, but first let us suggest its significance.

We ask the reader to refer back to Figure 2.6, showing the failure of interpolation polynomials on the Runge function. The failure is characterized by large excursions or "oscillations" of the polynomial between interpolation points. Such large excursions can occur only if the polynomial $p(x)$ has a large second derivative. It is amusing to see that we can substantiate this point by using the interpolation truncation error formula (2.14). For take $p_1(x)$ to be the linear interpolator $p_1(x) = f_1 l_1(x) + f_2 l_2(x)$, which connects the data point (x_1, f_1) with (x_2, f_2), as shown in Figure 2.16. Then by (2.14), with the interpolation polynomial $p(x)$ here playing the role of $f(x)$, we have

$$p(x) = p_1(x) + \tfrac{1}{2} p''(\zeta)(x - x_1)(x - x_2). \tag{2.32}$$

From (2.32) it is clear that $p(x)$ cannot deviate much from the line $p_1(x)$ unless its second derivative is large. In this sense one may say that failure of interpolation polynomials (or any other function, for that matter) manifests itself by large second derivatives. With this in mind, the following statement is interesting. It implies that a natural interpolation spline has an integrated second derivative which is as small as that of any other twice-differentiable interpolation function for the same data set.

Fact. Let $s(x)$ be a natural cubic spline that interpolates the function $f(x)$ at the points $a = x_1 < \cdots < x_n = b$. Let $f''(x)$ be continuous on the interval $[a, b]$. Then

$$\int_a^b [s''(x)]^2 \, dx \le \int_a^b [f''(x)]^2 \, dx. \tag{2.33}$$

PROOF

Let $g(x) = f(x) - s(x)$; then

$$\int_a^b [f''(x)]^2 \, dx = \int_a^b [g''(x) + s''(x)]^2 \, dx = \int_a^b [g''(x)]^2 \, dx \tag{2.34}$$
$$+ \int_a^b [s''(x)]^2 \, dx + 2 \int_a^b [g''(x)s''(x)] \, dx.$$

The last term is zero, since by integrating by parts and by dividing interval $[a, b]$ to the subintervals $[x_1, x_2], \ldots, [x_{n-1}, x_n]$, we get

$$\int_a^b s''(x)g''(x) \, dx = [s''(x)g'(x)] \Big|_a^b - \int_a^b s^{(3)}(x)g'(x) \, dx$$

$$= s''(b)g'(b) - s''(a)g'(a) - \sum_{k=2}^n \int_{x_{k-1}}^{x_k} s^{(3)}(x)g'(x) \, dx.$$

Observe that $s''(b) = s''(a) = 0$ and on each of the subintervals $[x_{k-1}, x_k]$, $s^{(3)}(x)$ is a constant, say C_k. Consequently,

$$\int_a^b s''(x)g''(x) \, dx = -\sum_{k=2}^n C_k \int_{x_{k-1}}^{x_k} g'(x) \, dx$$

$$= -\sum_{k=2}^n C_k[g(x_k) - g(x_{k-1})].$$

But this sum is zero because $s(x)$ interpolates, and thus the summands are each zero:

$$g(x_k) = f(x_k) - s(x_k) = 0, \qquad g(x_{k-1}) = f(x_{k-1}) - s(x_{k-1}) = 0.$$

Then (2.34) implies that

$$\int_a^b [f''(x)]^2 \, dx = \int_a^b [g''(x)]^2 \, dx + \int_a^b [s''(x)]^2 \, dx \geqq \int_a^b [s''(x)]^2 \, dx,$$

which completes the proof.

Experimentally, we have concluded that the spline approach does not fully resolve some approximation issues. To be a candidate for a computer library function, for example, we would anticipate that an algorithm ought to provide an answer with an accuracy of at least 7 to 10 significant decimals after very few arithmetic computations. Sometimes interpolation polynomials fill this need very nicely. We found that by using Chebyshev points, we could interpolate sin (x) on [0, 1] to about 27 significant decimals with a 21st-order interpolation polynomial. The natural spline, using the same 21 data points, gave only about three places of accuracy, and even when we raised the number of data points n to 90, the interpolations were accurate to only about five significant places.

Our conclusion is that spline approximation is perfectly adequate if relative error on the order of 1% is acceptable. On the other hand, if library-function accuracy is required, alternative methods or piecewise polynomials more sophisticated than cubic splines will be needed, unless the intervals are small.

There are undeniably some loose ends to our attempted mastery of the function approximation problem. For example, it is known that there is no polynomial of degree less than 1 million, interpolating or otherwise, which approximates $\sqrt{|x|}$ to four significant places at all points on the interval $[-1, 1]$. In fact, experiments have convinced the authors that interpolation of this function on [0, 1] by either splines or polynomials cannot feasibly be done accurately. Fortunately, we will see that by methods in Chapter 5 direct computation of $\sqrt{|x|}$ can be done quickly, accurately, and simply. One realm in which piecewise polynomials and splines appear to reign supreme is in approximation of "natural" and "nonmathematical" shapes.

2.6
SUPPLEMENTARY NOTES AND DISCUSSIONS

We have considered three basic techniques—Taylor's polynomials, interpolation polynomials, and splines—for approximating functions and "smooth" data. The background developments for Taylor's and interpolation polynomials will serve us well in describing numerical methods for other problem areas. Of these three procedures, Taylor's polynomials have been the least widely used in actual numerical computation (but most widely used in mathematical analysis of numerical algorithms). Our first thought was that the lack of enthusiasm stemmed from the requirement of higher derivatives in (2.6). Higher derivatives of functions other than polynomials are typically at least a nuisance to obtain, and frequently a rich source of error because of calculus mistakes.

Using numerical differentiation ideas such as those offered in Chapter 3, one can in many cases obtain adequate approximations to Taylor's polynomials. Moreover, algebraic computer languages (e.g., MACSYMA and MU-MATH), which automate differentiation, are becoming available. Another drawback then comes to the fore. In contrast to interpolation polynomials, Taylor's polynomial approximation tends to be much more accurate than needed in the vicinity of the expansion point x_0, and defective for large values of $|x - x_0|$. Interpolation error, while also tending to be larger near the endpoints of the interpolation interval, does spread the error out more evenly (particularly if Chebyshev interpolation points are used). In Example 2.4 (see Figure 2.5), this more even distribution of approximation error is clearly in evidence.

For "smooth" functions, interpolation polynomials are adequate for many needs. When the curve or data set has more erratic features, piecewise polynomial approximation and splines have some appealing advantages. Whereas our discussion centered on natural cubic splines, we notify the reader that in recent years, the tendency has been to keep all options open and choose the piecewise polynomial approach most suited to the application at hand. For instance, in interpolating solutions of differential equations, some investigators have found the higher-derivative requirement of the "spline" definition to be overly constricting. Even within the domain of cubic interpolation splines, "natural" splines such as we and most other numerical methods authors propose are known to have certain technical deficiencies. Nevertheless, they do guarantee convergence to any continuous limiting function, as the number of node points x_i become dense, a property that we have noted is not shared by interpolating polynomials.

The techniques given in this chapter are adequate for obtaining computer approximations of common functions, and for transcribing "natural" curves, such as demand/sales curves or transistor operating characteristic curves, into a computer-manageable form. Also, they are useful for "rough and ready" computer graphics displays.

We are hasty to point out, however, that program functions, such as in computer libraries or commercial packages, employ much more sophisticated principles than can be related at the level of this textbook. The subject of computer approximation of functions is a lively and sophisticated research area. Algorithms employing advanced theory have greater accuracy, for a given computational effort, than the prototypical techniques related here. On the other hand, the advanced methods

TABLE 2.16   Match of Interpolation Problem Characteristics and Methods

Function Characteristics	_Suggested Method*_
Few Data Points (<20)	
Derivatives not large	I, S
Large derivatives	S
User can choose data locations x_i, function fairly smooth	IC
Many Data Points	
Data due to "natural" (i.e., nonmathematical) origin	S
Function smooth, regular	IL
Large errors or random origins for data	Go to Chapter 6

*I, interpolation polynomials; IC, interpolation polynomial using Chebyshev points; S, splines; IL, interpolation polynomial using only a moderate number of local (neighboring) points, [i.e., if you wish to approximate at x, construct $p_{n-1}(x)$ on basis of a data pairs (x_i, f_i) such that the x_i's are as close to x as possible.

are, for the most part, an outgrowth of principles presented in this chapter and Chapter 6.

The methods of the present chapter presume that interpolation is truly a sensible activity for the given data. In some cases, interpolation is not the best procedure. Most notably if significant error in the function values f_i is thought to be present, or if the data arise from observations of a random phenomenon, the curve-fitting methods of Chapter 6 should be considered. In curve fitting we do not insist that the approximating function interpolate (pass through the data). Rather we seek a polynomial of specified degree which, according to some measure, is as close as possible to the data points.

In Table 2.16 we have summarized the appropriate problem domains for various techniques.

PROBLEMS

Section 2.2

1. Use Horner's rule to evaluate

$$p(x) = 2x^4 + 3x^3 - 4x^2 + 5x - 6$$

at $x = 2$. Show your calculations.

2. (a) Find a nested formula for computing the derivative $dp(x)/dx$ of a polynomial $p(x)$ at any specified argument x. The rule should look much, but not exactly, like Horner's rule (2.2).
   (b) Check your rule by applying it to evaluation of $dp(x)/dx$ with $p(x)$ as in Problem 1 and $x = 3$.

Section 2.3

3. (a) Show that at $x_0 = 0$, the nth-degree Taylor's polynomial for $f(x) = \exp(x)$ is

$$p_n(x) = 1 + x + \frac{x^2}{2!} + \cdots + \frac{x^n}{n!}.$$

   (b) Find the Taylor's polynomial expansion of degree 3 of exp (x) about $x_0 = 1$.
   (c) For degrees $n = 4$ and 10, evaluate the polynomial in part (b) at values $x = 0.2, 0.4, 0.6, \ldots, 2.0$, comparing these approximations with the values of EXP(X) given at these arguments by your computer. Check the EXP(X) routine by seeing to what accuracy ALOG(EXP(X)) recovers X. Use subroutine TAYLOR (Table 2.2) for evaluating the Taylor's polynomials.

4. If $f(x)$ and $g(x)$ have power series representations $f(x) = a(0) + a(1)x + \cdots + a(k)x^k + \cdots$ and $g(x) = b(0) + b(1)x + \cdots + b(k)x^k + \cdots$, respectively, show that the coefficients for the power series representation of $h(x) = f(x)g(x) = c(0) + c(1)x + \cdots + c(k)x^k + \cdots$, are given by $c(0) = a(0)b(0)$, $c(1) = a(0)b(1) + a(1)b(0)$, and, generally,

$$c(k) = \sum_{j=0}^{k} a(j)b(k - j), \qquad k = 0, 1, \ldots.$$

5. Use the result in Problem 4 to find the Taylor's polynomial of degree 4 of $h(x) = \exp(x) \sin(x)$ about $x_0 = 0$.

6. Find the Taylor's series expansion of $\sqrt{x}$ about x_0, $x_0 > 0$.

7. Use the result in Problem 6 to write a program to approximate $\sqrt{x}$ by a Taylor's polynomial of degree N about $x_0 = 0.5$. Evaluate your polynomial at $x = 0.01$ for degrees $N = 5, 10, 20,$ and 40. Compare the Taylor's polynomial estimate with the value given by your computer's SQRT(X) function at X $= 0.01$. [**NOTE:** It is surprising how poor polynomial approximation of the square-root function can be. More was said on this matter in Section 2.4.4. Check your computer library function SQRT(X) by seeing how well (SQRT(X))**2 recovers X.]

8. Show that the Taylor's polynomial is the only polynomial of degree $n - 1$ satisfying (2.3). (**HINT:** Use the factorization theorem mentioned in Section 2.4.2.)

9. How accurate is the approximation of sin (x) given by the Taylor's polynomial

$$\sin (x) \approx x - \frac{x^3}{3!} + \frac{x^5}{5!}$$

in the interval $[0, \pi/2]$? Give a theoretical bound, by clever use of the remainder formula (2.7). Verify by computation of error at closely spaced points.

10. Verify by mathematical principles that

$$\frac{1}{1 - x} = 1 + x + x^2 + x^3 + \cdots \qquad (-1 < x < 1).$$

11. Use the identity of Problem 10 to show how the division of w by a number z, $0 < z < 2$, can be accomplished on a computer for which the only "built-in" functions are addition, subtraction, multiplication, and conditional branching.

Section 2.4
12. Compare errors of approximating the functions below on $[0, 1]$ by:
(i) Taylor's polynomial about $x_0 = 0.5$.
(ii) Lagrange interpolation at equally spaced points, with $x_1 = 0$ and $x_4 = 1.0$.
(iii) Lagrange interpolation using Chebyshev points (2.17).
    Take the *degree* in all cases to be 3. Check the errors at $0, 0.1, \ldots, 0.9, 1.0$.
(a) sin $(2x)$.
(b) e^x.
(c) $\sqrt{x}$.
(d) $1/(1 + 25x^2)$.
(e) x^4.

13. For the following four points, express the third-degree interpolation polynomial in its:
(a) Lagrange form.
(b) In the form $a_3 x^3 + a_2 x^2 + a_1 x + a_0$.
(c) Newton Form

i	1	2	3	4
x	-2	-1	0	1
f	6	4	3	3

14.   Indirectly check the accuracy of your computer's library functions by seeing how well the following identities hold at values $x = j(\pi/20)$, $j = 1, 2, \ldots, 9$.
   (a) $\sin^2(x) + \cos^2(x) = 1$.
   (b) $\sin(2x) = 2\sin(x)\cos(x)$.
   (c) $\cos(x) = \sin(x + \pi/2)$.
   (d) $\exp(x) \cdot \exp(-x) = 1$.
   (e) $\log_e(e^x) = x$.
   (f) $\operatorname{sqrt}(x) \cdot \operatorname{sqrt}(x) = x$.
   [**HINT FOR PART (c):** Approximate π by 4. *ATAN(1.0). Check to see that $\sin(\pi) \approx 0.0$.]

15.   Consider the interpolating points $x_i = i\,0.5$ ($i = 0, 1, 2, 3, 4$), and the functional values $f_i = \exp(x_i)$. Compute the interpolation polynomial based on these data, and check the accuracy at the test points $t = 0.2, 0.4, \ldots, 2.0$. If you did Problem 3(c), compare the results.

16.   Prove that

$$\sum_{i=1}^{n} l_i(x) \equiv 1, \qquad \text{any } x,$$

where $l_1(x), \ldots, l_n(x)$ are the Lagrange factor polynomials defined in (2.11).

17.   Theoretically bound the error of approximating e^x by the quadratic interpolation polynomial in the interval $[-1, 1]$, using the interpolating points $-1, 0, 1$. Check the error computationally at $x_i = -1 + i/10$, $1 \le i \le 20$, to make sure your bound is correct.

18.   Let $a = x_1, x_2, \ldots, x_{n-1}, x_n = b$ be equally spaced interpolating points on $[a, b]$. Prove that for $x \in [a, b]$,

$$\prod_{i=1}^{n} |x - x_i| = |w_n(x)| \le \tfrac{1}{4}h^n(n-1)!,$$

where $h = (b - a)/(n - 1)$. Prove that if $b - a < 1$, then $w_n(x) \to 0$ as $n \to \infty$, for all $x \in [a, b]$. (**HINT:** Use l'Hospital's rule on ratio of successive terms.)

19.   Repeat the computations of Table 2.8 for $f(x) = \sin(5x)$. Increase the number n of evenly spaced domain points until the error, as determined by

$$\varepsilon = \max_{1 \le i \le 10} |p_{n-1}(0.2i) - f(0.2i)|,$$

increases due to ill-conditioning. Is ε, at the "best" n, commensurate with your machine epsilon?

20.   Apply Chebyshev-point interpolation to $\sqrt{x}$, $0 \le x \le 1$. Measure the error by

$$M_n = \max_{0 \le j \le 100} \left| \sqrt{\frac{j}{100}} - p_n\!\left(\frac{j}{100}\right) \right|.$$

By increasing n, the degree of the polynomial, how small can you make M_n?

Section 2.5

21. Decide whether the following functions are splines.

    (a) $f(x) = \begin{cases} x, & -1 \le x \le 0 \\ 2x, & 0 \le x \le 1 \\ x+1, & 1 \le x \le 2. \end{cases}$

    (b) $f(x) = \begin{cases} x, & -1 \le x \le 0 \\ 2x - 1, & 0 \le x \le 1 \\ x + 1, & 1 \le x \le 2. \end{cases}$

    (c) $f(x) = \begin{cases} 0, & -1 \le x \le 0 \\ x^2, & 0 \le x \le 1 \\ 2x - 1, & 1 \le x \le 2. \end{cases}$

22. By hand calculation, find the linear interpolation spline for the data in the following table.

x	-2	-1	0	1	2	3
f	0	1	0	1	0	1

*23. By hand calculation, find a quadratic ($m = 2$) interpolation spline for the data set of Problem 22. (**HINT:** One condition is missing. We suggest taking c_1, in the notation of Section 2.5.2, to be 0.)

24. Determine the values of a and b such that the function

$$f(x) = \begin{cases} x^3 + x, & -1 \le x \le 0 \\ ax^2 + bx, & 0 \le x \le 1 \end{cases}$$

    is a cubic spline.

25. Set up (but do not solve) the linear equations for finding the coefficients c_i of the natural cubic interpolation spline for the data of Problem 22.

26. Obtain a magazine picture of a reasonably smooth object (e.g., automobile, mountain, reclining human figure, border between states or countries) and measure positions, in the spirit of the engineering example. Now attempt to fill in the profile by cubic splines. Use subroutines SPLN and SPLE.

CHAPTER 3

Numerical Differentiation and Integration

3.1
PRELIMINARIES

Need for numerical integration techniques arises quite extensively and regularly in engineering, physics, and other quantitative sciences. For instance, in analyzing a proposed reservoir site, one must be able to calculate the storage capacity (volume) of a reservoir of some specified height on the basis of a topographic survey of the site. In a more general vein, we remark that probabilistic models are playing an increasingly important role in all areas of engineering. Even the most elementary probabilistic systems analysis leads to integrals that admit no closed-form representation in terms of the common library functions. In this context we mention that $\exp(-x^2)$, $\exp(-x)x^{\alpha-1}$, and $x^{\alpha}(1-x)^{\beta}$, (α, β positive noninteger), are integrands encountered with three of the most popular random variables in probability theory. None of these functions have indefinite integrals expressible in terms of the usual FORTRAN library functions.

Numerical differentiation formulas characteristically seek to estimate the derivative $f'(x)$ of a function at a specified point x_0 through functional values $f(x_j)$ at points x_j near x_0. Numerical differentiation formulas are central to several prominent optimization methods, two-point-boundary-value problems for ordinary differential equations, and in the solution of all types of partial differential equations.

Polynomial approximation methods described in Chapter 2 serve as the foundation for numerical differentiation and integration techniques, as the reader will soon observe. In turn, numerical differentiation and integration developments play fundamental roles in numerical solution of differential equations.

We conclude this introduction by giving some standard notation that will serve us throughout this book. A function $g(h)$ is said to be $O(h)$ (read "big oh of h") at $h = 0$ if there are positive numbers C and D such that whenever $|h| \leq D$, then $|g(h)| \leq C|h|$. Under this circumstance, we sometimes write $g(h) = O(h)$. For example, $\sin(h)$ is $O(h)$ because $|\sin(h)| \leq |h|$ for every number h. More generally, for q positive, $g(h)$ is $O(h^q)$ at $h = 0$ if there exist numbers C and D such that whenever $|h| \leq D$,

$$|g(h)| \leq C|h^q|.$$

For instance, $\cos(h) - 1$ is $O(h^2)$ at $h = 0$, since examination of its Taylor's series expansion reveals that $|\cos(h) - 1| \leq \frac{1}{2}|h|^2$. Similarly, $\sin(h) - h$ is $O(h^3)$ at $h = 0$.

It is customary to express statements such as these in the form

$$\cos(h) = 1 + O(h^2)$$

and

$$\sin(h) = h + O(h^3).$$

Often $O(h^q)$ will refer to the truncation error of some approximating formula. In this context one says that the *order* of the error is q. Usually, the higher the order of truncation error, the more accurate the rule.

The notation $O(h^q)$ is intimately related to Taylor's series expansions. Note that if $p_{n-1}(x)$ is the $(n-1)$st-degree Taylor's polynomial of an n-times continuously differentiable function $f(x)$, then in view of the error formula (2.7) $g(h) = f(x_0 + h) - p_{n-1}(x_0 + h)$ is $O(h^n)$. One may write $f(x) = p_{n-1}(x) + O((x - x_0)^n)$.

3.2
NUMERICAL DIFFERENTIATION

In this section formulas for approximating derivatives of real functions will be introduced. Before deriving the particular approximations, as an example consider a differentiable function $f(x)$ and a particular point x_0 in its domain. Then the definition of $f'(x_0)$ implies that for sufficiently small h,

$$f'(x_0) \approx \frac{f(x_0 + h) - f(x_0)}{h}.$$

By multiplying both sides by h and rearranging the terms, we get the approximating relation

$$f(x_0 + h) \approx f(x_0) + f'(x_0)h.$$

The reader may recognize from (2.6) that this is really the linear Taylor's polynomial approximation of $f(x_0 + h)$. From this simple observation one may conclude that the approximation of the derivative $f'(x_0)$ above and the linear Taylor polynomial about x_0 are clearly related. Taylor's polynomials are the basis for our construction of numerical differentiation formulas.

Assume that the first derivative of a function $f(x)$ is to be computed at the point x_0. The Taylor's formula (2.6) and remainder term (2.7) with $n = 2$ imply that

$$f(x_0 + h) = f(x_0) + f'(x_0)h + \frac{f''(\zeta)}{2}h^2, \qquad (3.1)$$

where ζ is between x_0 and $x_0 + h$. Then by solving (3.1) for $f'(x_0)$, we get

$$f'(x_0) = \frac{f(x_0 + h) - f(x_0)}{h} - \frac{1}{2}f''(\zeta)h. \qquad (3.2)$$

The first term of the right-hand side gives a computational approximation of $f'(x_0)$, and the truncation error of this approximation is $-\frac{1}{2}f''(\zeta)h$. If $f''(x)$ is bounded near x_0, then we may simply write that the error is $O(h)$.

The order of the error can be raised by using also the datum $f(x_0 - h)$. Assume that $f(x)$ is thrice differentiable and consider the quadratic Taylor's polynomials with remainder terms

$$f(x_0 + h) = f(x_0) + f'(x_0)h + \frac{f''(x_0)}{2}h^2 + \frac{f^{(3)}(\zeta_1)}{6}h^3,$$

$$f(x_0 - h) = f(x_0) - f'(x_0)h + \frac{f''(x_0)}{2}h^2 - \frac{f^{(3)}(\zeta_2)}{6}h^3.$$

By subtracting these equations, we find that

$$f(x_0 + h) - f(x_0 - h) = 2f'(x_0)h + \frac{h^3}{6}\left[f^{(3)}(\zeta_1) + f^{(3)}(\zeta_2)\right].$$

After rearrangement,

$$f'(x_0) = \frac{f(x_0 + h) - f(x_0 - h)}{2h} - \frac{h^2}{12}\left[f^{(3)}(\zeta_1) + f^{(3)}(\zeta_2)\right]. \qquad (3.3)$$

The first term of the right-hand side gives the numerical approximation of the derivative and the second term gives the truncation error of this approximation. If $f^{(3)}(x)$ is bounded, the error is $O(h^2)$, and for small values of h, this error is less than the error of the previous approximation (3.2). We illustrate these rules in Figure 3.1.

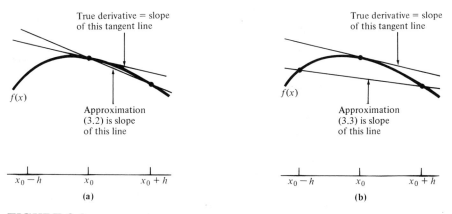

FIGURE 3.1 Numerical Approximation of a First Derivative

Let us seek an approximation for second derivatives. By assuming that function $f(x)$ is four times differentiable, we get

$$f(x_0 + h) = f(x_0) + f'(x_0)h + \frac{f''(x_0)}{2} h^2 + \frac{f^{(3)}(x_0)}{6} h^3 + \frac{f^{(4)}(\zeta_3)}{24} h^4$$

and

$$f(x_0 - h) = f(x_0) - f'(x_0)h + \frac{f''(x_0)}{2} h^2 - \frac{f^{(3)}(x_0)}{6} h^3 + \frac{f^{(4)}(\zeta_4)}{24} h^4.$$

By adding these equations and rearranging terms, we see that

$$\boxed{f''(x_0) = \frac{f(x_0 + h) - 2f(x_0) + f(x_0 - h)}{h^2} - \frac{h^2}{24} \left[f^{(4)}(\zeta_3) + f^{(4)}(\zeta_4) \right],}$$

$$(3.4)$$

where $\zeta_3 \in [x_0, x_0 + h]$ and $\zeta_4 \in [x_0 - h, x_0]$. Consequently, the first term of the right-hand side gives an approximation of $f''(x_0)$ and the truncation error of this approximation is $O(h^2)$ for $f^{(4)}(x)$ bounded.

To illustrate the use of (3.3) and (3.4), consider the following example.

———— **EXAMPLE 3.1** ————————————————————————————

Some values of $f(x) = e^x$ are given in Table 3.1, for $h = 0.1$, along with a list of derivative approximations and the associated errors of (3.2), (3.3), and (3.4) at $x = 0$.

TABLE 3.1   A Derivative Approximation

j	x_j	e^x
-1	-0.1	0.904837
0	0.0	1.000000
1	0.1	1.105172

Error*

(a) Exact value: $f'(0) = 1.000000$
   Approximation by (3.2):

$$\frac{f(x_1) - f(x_0)}{h} = \frac{1.105172 - 1.00000}{0.1}$$

$$= 1.05172 \qquad\qquad -0.05172$$

Approximation by (3.3):

$$\frac{f(x_1) + f(x_{-1})}{2h} = \frac{1.105172 - 0.904837}{0.2}$$

$$= 1.001675 \qquad\qquad -0.00168$$

TABLE 3.1   (Continued)

(b) Exact value: $f''(0) = 1.000000$
     Approximation by (3.4):

$$\frac{f(x_1) + f(x_{-1}) - 2f(x_0)}{h^2} = \frac{1.105172 + 0.904837 - 2.0}{0.01}$$

$$= \frac{0.010009}{0.01} = 1.0009 \qquad\qquad -0.0009$$

*Error = Exact Value − Approximation

One can generalize the foregoing line of reasoning, writing the system of Taylor's series expansions

$$f(x_0 + jh) = \sum_{l=0}^{k} \frac{f^{(l)}(x_0)}{l!} (jh)^l + \frac{f^{(k+1)}(\xi_j)}{(k+1)!} (jh)^{k+1}$$

(3.5)

$$(k_1 \le j \le k_2).$$

One may then neglect the remainder terms, and the resulting system of equations will be linear. If we define $k = k_2 - k_1$, the unknowns $f^{(l)}(x_0)$, $1 \le l \le k$, are uniquely determined by this system and can be routinely computed from the values $f(x_0 + jh)$, $k_1 \le j \le k_2$. The truncation error in estimating $f^{(l)}(x_0)$ is no greater than $O(h^{k+1-l})$. For example, in (3.2), $k_1 = 0$, $k_2 = 1$, and $k = 1$. In (3.3) and (3.4), $k_1 = -1$, $k_2 = 1$, and $k = 2$.

Some popular numerical differentiation formulas for evenly spaced interpolation points are given in Table 3.2. At first glance, the three-point approximation for $f'(x)$ might seem a misnomer, since only two values are used. Actually, it gives highest order among all formulas using the three data points $f(x - h)$, $f(x)$, and $f(x + h)$.

In contrast to integration, discussed in the next section, numerical differentiation is notoriously responsive to the effects of roundoff and data error. Let us refer to inexactness of function values as roundoff error, since it may well stem from floating-point representation approximation. To take a particular case, assume that we use the approximation (3.2). That is,

$$f'(x_0) \approx \frac{f(x_0 + h) - f(x_0)}{h} = \frac{f_1 - f_0}{h}.$$

Let $\bar{f}_0$ and $\bar{f}_1$ be known approximations to the exact unknown values $f_0 = f(x_0)$ and $f_1 = f(x_1)$, respectively, and assume that we have a bound δ so that

$$|f_0 - \bar{f}_0| \le \delta \qquad \text{and} \qquad |f_1 - \bar{f}_1| \le \delta.$$

Then $f'(x_0)$ is approximated by $(\bar{f}_1 - \bar{f}_0)/h$. The error can be bounded as follows:

$$\left| f'(x_0) - \frac{\bar{f}_1 - \bar{f}_0}{h} \right| \le \left| f'(x_0) - \frac{f_1 - f_0}{h} \right| + \left| \frac{f_1 - f_0}{h} - \frac{\bar{f}_1 - \bar{f}_0}{h} \right|$$

$$\le C_1 h + \frac{1}{h} (|f_1 - \bar{f}_1| + |f_0 - \bar{f}_0|) \le C_1 h + \frac{C_2 \delta}{h},$$

TABLE 3.2   Some Popular Derivative Approximations

First derivative approximations

  Three-point

$$f'(x) = \frac{f(x + h) - f(x - h)}{2h} + O(h^2)$$

  Four-point

$$f'(x) = \frac{-f(x + 2h) + 6f(x + h) - 3f(x) - 2f(x - h)}{6h} + O(h^3)$$

  Five-point

$$f'(x) = \frac{-2f(x + 2h) + 16f(x + h) - 16f(x - h) + 2f(x - 2h)}{24h} + O(h^4)$$

Second derivative approximations

  Three-point

$$f''(x) = \frac{f(x + h) - 2f(x) + f(x - h)}{h^2} + O(h^2)$$

  Four-point

    Coincides with three-point formula

  Five-point

$$f''(x) = \frac{-f(x + 2h) + 16f(x + h) - 30f(x) + 16f(x - h) - f(x - 2h)}{12h^2}$$
$$+ O(h^4)$$

where (3.2) gives us the bound

$$C_1 = \tfrac{1}{2}\max_{x_0 \le t \le x_0 + h} |f''(t)|$$

and $C_2 = 2$, since

$$|f_1 - \bar{f}_1| + |f_0 - \bar{f}_0| \le 2\delta.$$

Thus the total absolute error satisfies

$$\left| f'(x_0) - \frac{\bar{f}_1 - \bar{f}_0}{h} \right| \le C_1 h + \frac{2\delta}{h}. \tag{3.6}$$

Observe that for larger values of h the truncation term $C_1 h$ dominates. If the magnitude of h is small, the roundoff error term $2\delta/h$ exerts overriding influence.

This observation implies that judgment must be exercised in choosing the step size h. The effect of a crude approximation of the derivative arising from too large a step size must be balanced against the effect of dividing functional errors by a positive power of the step size. This reasoning applies to all the differentiation formulas mentioned in this section.

━━━ **EXAMPLE 3.2** ━━

By means of the program listed in Table 3.3, we have used the three-point formula of Table 3.2 to estimate the derivative of exp (x) at $x = 0$. Notice in the printout that as the step size h decreases, the approximation first becomes better, and then deteriorates; from (3.6), such behavior should be anticipated. The sources of error here are those due to computer word roundoff and inaccuracy in the library function for exp (x). These errors are machine dependent, of course. Our computation was performed by a DEC 10. Notice that even at the best step size h, the relative error is about 1000 times larger than machine epsilon.

An important phenomenon is exemplified in the error column of Table 3.4. From (3.3), the error, in theory, is proportional to h^2, since f'' is smooth. This tells us that if we use two step sizes, h_1 and h_2, where $h_2 = h_1/10$, the associated errors err (h_1) and err (h_2) ought to be related by

$$\text{err } (h_2) \approx \text{err } (h_1)\left(\frac{h_2}{h_1}\right)^2 = \frac{\text{err } (h_1)}{100}.$$

Examination of the error column in Table 3.4 reveals that this relation is satisfied very nicely in the range in which truncation error is the limiting factor. The

TABLE 3.3   Program for Derivative Computation

```
C      PROGRAM DERIVATIVE
C
C      ************************************************************
C      CALCULATES DERIVATIVE OF F(X)=EXP(X) AT X=0 BY THREE POINT
C      RULE USING 10 DIFFERENT STEP SIZES
C      OUTPUT: H=STEP SIZE
C                   DERIV=DERIVATIVE EXSTIMATED USING THREE POINT RULE
C                   ERROR=THE DIFFERENCE BETWEEN THE TRUE VALUE OF THE
C                         DERIVATIVE AND THE APPROXIMATION DERIVATIVE
C      ************************************************************
C
C      *** STATEMENT FUNCTION FOR CALCULATING COMPONENTS OF   ***
C      *** APPROXIMATION                                      ***
C
       F(X)=EXP(X)
C
C      *** THIS LOOP ITERATES THE STEP SIZE H AND COMPUTES    ***
C      *** THE CORRESPONDING DERIVATIVE APPROXIMATION         ***
C
       DO 1 N=1,10
         H=0.1**N
         DERIV=(F(H)-F(-H))/(2.0*H)
         ERROR=1.0-DERIV
         WRITE(10,*)H,DERIV,ERROR
     1 CONTINUE
       STOP
       END
```

TABLE 3.4   Printout for Derivative Computation

h	Estimate $f'(0)$	Error	
0.1000	1.001668	-1.6676188E-03	
1.0000E-02	1.000017	-1.6808510E-05	Truncation error
1.0000E-03	1.000017	-1.6927719E-05	dominates
1.0000E-04	1.000166	-1.6582012E-04	
1.0000E-05	1.001358	-1.3579130E-03	
1.0000E-06	0.9834764	1.6523600E-02	Roundoff error
1.0000E-07	1.192093	-0.1920927	dominates
1.0000E-08	0.0000000E+00	1.000000	
1.0000E-09	0.0000000E+00	1.000000	
1.0000E-10	0.0000000E+00	1.000000	

TABLE 3.5   Derivative Printout for Inaccurate Function Values

h	Estimate of $f'(0)$	Error
0.1000	1.001936	-0.1936242E-02
1.0000E-2	0.9985861	0.1413949E-02
1.0000E-3	1.003999	-0.3999293E-02
1.0000E-4	1.325831	-0.3258308
1.0000E-5	-1.946837	2.946837
1.0000E-6	-12.93421	13.93421

important fact here is that our truncation error formulas do give accurate indications of what to expect in practical situations.

Now that we have a more mature understanding of finite difference approximations, the reader may be interested in turning back to Figure 1.2, where he or she will see that the log plot associated with the error of approximation (3.2) has a slope of approximately 45° when truncation error dominates. This is confirmation that the order q is 1 for that formula.

Next we replaced the declared function of the program in Table 3.3 by

$$F(X) = EXP(X) + 0.0001*(RAN(0) - 0.5),$$

which effectively adds a little random error to the function values. The expression "RAN(0)" calls the DEC 10 random number routine. The function values are still correct to three or four significant figures. But from the listing in Table 3.5 we see that even this miniscule noise devastates our derivative approximations. ∎

An alternative development of numerical differentiation formulas can be made by observing that if $p(x)$ is a polynomial approximation of $f(x)$ (such as obtained by methods of Chapter 2), then $p^{(j)}(x)$, the jth derivative of $p(x)$, affords an approximation of $f^{(j)}(x)$. Such a viewpoint also yields those formulas given in Table 3.2. This viewpoint is explored in Problem 4. Clearly, spline derivatives can also be used in the context of numerical differentiation.

3.3
INTERPOLATORY QUADRATURE

In interpolatory quadrature, one uses an interpolation polynomial $p(x)$ to approximate the intergrand $f(x)$ over the domain $[a, b]$ of integration. Then the desired integral

$$\int_a^b f(x) \, dx$$

is approximated by the easily computed value of the integral

$$\int_a^b p(x) \, dx$$

of the interpolation polynomial. In Figure 3.2 we illustrate the principle of interpolatory quadrature. The desired integral is the area under $f(x)$, and its interpolatory quadrature approximation is the crosshatched area under the interpolation polynomial.

Assume that we wish to approximate the integral of $f(x)$ over a finite interval $[a, b]$. Let $x_1, \ldots, x_n$ be distinct points and define $p(x)$ to be the interpolation polynomial for $f(x_i)$, $1 \leq i \leq n$. Then (2.14) implies that

$$\int_a^b f(x) \, dx = \int_a^b p(x) \, dx + \int_a^b \frac{f^{(n)}(\zeta(x))}{n!} w_n(x) \, dx. \tag{3.7}$$

The first term of the right-hand side can be viewed as a computation approximation of the integral and the second term gives the truncation error of this approximation.

By using the Lagrangian representation (2.11) for the interpolation polynomial, we have

$$\int_a^b p(x) \, dx = \int_a^b \sum_{k=1}^n f(x_k) l_k(x) \, dx = \sum_{k=1}^n f(x_k) \int_a^b l_k(x) \, dx.$$

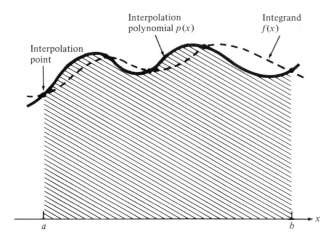

FIGURE 3.2 Interpolatory Quadrature

Thus *interpolatory-type* integration rules have the form

$$\int_a^b f(x)\,dx \approx \sum_{k=1}^n f(x_k)A_k, \tag{3.8}$$

with

$$A_k = \int_a^b l_k(x)\,dx, \qquad 1 \le k \le n. \tag{3.9}$$

The coefficients A_k depend only on the choice of the interpolating points x_k, $1 \le k \le n$, and the domain $[a, b]$ of integration. Regardless of the rationale for choosing the A_k's, any numerical integration rule of the form (3.8) is known as a *quadrature formula*. Essentially all integration formulas are of this type. The numbers x_k are called *quadrature points*. The values A_k are called *quadrature weights*.

EXAMPLE 3.3

Let $[a, b] = [-1, 1]$, $n = 3$, $x_1 = -1$, $x_2 = 0$, and $x_3 = 1$. Then

$$l_1(x) = \frac{(x - 0)(x - 1)}{(-1 - 0)(-1 - 1)} = \frac{1}{2}(x^2 - x)$$

$$l_2(x) = \frac{(x + 1)(x - 1)}{(0 + 1)(0 - 1)} = -x^2 + 1$$

$$l_3(x) = \frac{(x + 1)(x - 0)}{(1 + 1)(1 - 0)} = \frac{1}{2}(x^2 + x),$$

which implies that

$$A_1 = \int_{-1}^1 \tfrac{1}{2}(x^2 - x)\,dx = \tfrac{1}{3}$$

$$A_2 = \int_{-1}^1 (-x^2 + 1)\,dx = \tfrac{4}{3}$$

$$A_3 = \int_{-1}^1 \tfrac{1}{2}(x^2 + x)\,dx = \tfrac{1}{3}.$$

Thus we have derived the integration formula

$$\int_{-1}^1 f(x)\,dx \approx \tfrac{1}{3}f(-1) + \tfrac{4}{3}f(0) + \tfrac{1}{3}f(1).$$

As an example, this formula is applied to approximate the integral $\int_{-1}^1 e^x\,dx$. The formula gives the estimate

$$\frac{e^{-1} + 4e^0 + e^{+1}}{3} = 2.352.$$

To three decimal places the exact value equals 2.350.

If $f(x)$ is a polynomial of degree $n - 1$ or less, the uniqueness of the interpolation polynomial implies that for all x, $f(x) = p(x)$. Consequently,

$$\int_a^b f(x) \, dx = \int_a^b p(x) \, dx = \sum_{k=1}^n f(x_k) A_k. \tag{3.10}$$

By choosing $f_j(x) = x^j$ ($j = 0, 1, \ldots, n - 1$), this relation gives a system of linear equations for the unknown interpolatory quadrature weights $A_1, \ldots, A_n$:

$$A_1 x_1^0 + A_2 x_2^0 + \cdots + A_n x_n^0 = \int_a^b x^0 \, dx = b - a$$

$$A_1 x_1^1 + A_2 x_2^1 + \cdots + A_n x_n^1 = \int_a^b x^1 \, dx = \frac{b^2 - a^2}{2}$$

$$\cdot \qquad \cdot \qquad \cdot \qquad \cdot \qquad \cdot$$

$$A_1 x_1^{n-1} + A_2 x_2^{n-1} + \cdots + A_n x_n^{n-1} = \int_a^b x^{n-1} \, dx = \frac{b^n - a^n}{n}.$$

$$(3.11)$$

If the points $x_1, \ldots, x_n$ are distinct, the linear system of equations (3.11) has a unique solution (Bellman, 1970, p. 193). Subroutine COEFF (Table 3.6) evaluates quadrature weights by this method.

TABLE 3.6   Subroutine COEFF for Interpolatory Integration Weights

```
      SUBROUTINE COEFF(N,X,A,B,AA)
C
C     ****************************************************************
C     *   FUNCTION: THIS SUBROUTINE COMPUTES THE QUADRATURE         *
C     *             WEIGHTS FOR INTERPOLATORY QUADRATURE OVER       *
C     *             THE INTERVAL [A,B]                              *
C     *   USAGE:                                                    *
C     *        CALL SEQUENCE: CALL COEFF(N,X,A,B,AA)                *
C     *        EXTERNAL FUNCTIONS/SUBROUTINES:                      *
C     *                      SUBROUTINE GAUS1(N,M,ND,T,EPS)        *
C     *   PARAMETERS:                                               *
C     *      INPUT:                                                 *
C     *          N=NUMBER OF QUADRATURE POINTS (NOT LESS THAN 2,   *
C     *            NOT MORE THAN 30)                                *
C     *          X=N BY 1 ARRAY OF QUADRATURE POINTS                *
C     *          A=INTERVAL LEFT ENDPOINT                           *
C     *          B=INTERVAL RIGHT ENDPOINT                          *
C     *      OUTPUT:                                                *
C     *          AA=N BY 1 ARRAY OF QUADRATURE WEIGTS               *
C     ****************************************************************
C
      DIMENSION X(N),AA(N),T(30,31)
C     *** INITIALIZATION ***
      U=B
      V=A
      DO 1 J=1,N
         T(1,J)=1.0
    1 CONTINUE
```

TABLE 3.6 (Continued)

```
C       *** COMPUTE QUADRATURE WEIGHTS ***
        T(1,N+1)=U-V
        DO 2 I=2,N
          DO 3 J=1,N
            T(I,J)=T(I-1,J)*X(J)
 3        CONTINUE
          U=U*B
          V=V*A
          T(I,N+1)=(U-V)/I
 2      CONTINUE
C       *** COMPUTE COEFFICIENTS BY GAUSSIAN ELIMINATION ***
        M=1
        ND=30
        EPS=0.000001
        CALL GAUS1(N,M,ND,T,EPS)
        DO 4 I=1,N
          AA(I)=T(I,N+1)
 4      CONTINUE
        RETURN
        END
```

EXAMPLE 3.4

From the linear equation approach, we reconstruct the coefficients obtained in Example 3.3, where $a = -1$, $b = 1$ and $x_1 = -1$, $x_2 = 0$, $x_3 = 1$. Equations (3.11) take the values

$$A_1 + A_2 + A_3 = 2$$

$$-A_1 \qquad + A_3 = 0$$

$$A_1 \qquad + A_3 = \tfrac{2}{3}.$$

The solution of these equations is

$$A_1 = \tfrac{1}{3}, \qquad A_2 = \tfrac{4}{3}, \qquad \text{and} \qquad A_3 = \tfrac{1}{3}.$$

This yields the numerical integration formula

$$\int_{-1}^{1} f(x)\,dx \approx \tfrac{1}{3}f(-1) + \tfrac{4}{3}f(0) + \tfrac{1}{3}f(1).$$

This coincides with the formula obtained in Example 3.3, as it must. ∎

If the interpolating points $a = x_1 < x_2 < \cdots < x_n = b$ are equally spaced, (3.8) is called a *Newton–Cotes formula*. Newton–Cotes formulas are a popular class of interpolatory rules, but Gauss interpolatory quadrature (Section 3.6) has many advantages, as we will see.

Note that once we have established an integration rule of the form (3.8) for any finite interval $[a, b]$, it is a simple matter to convert and apply it to any desired interval $[c, d]$ of integration. For suppose that we wish to evaluate $\int_c^d f(t)\, dt$. From elementary integration theory, by the substitution of variable

$$x = \frac{t - Q}{V},\qquad (3.12)$$

with

$$V = \frac{d - c}{b - a}, \qquad Q = \frac{cb - ad}{b - a},$$

we have that

$$\int_c^d f(t)\, dt = V \int_a^b f(Vx + Q)\, dx.$$

To verify this relation, observe that for $t = c$,

$$x = \frac{c - Q}{V} = \frac{c - \dfrac{cb - ad}{b - a}}{\dfrac{d - c}{b - a}} = a.$$

For $t = d$,

$$x = \frac{d - Q}{V} = \frac{d - \dfrac{cb - ad}{b - a}}{\dfrac{d - c}{b - a}} = b$$

and

$$dx = \frac{1}{V}\, dt.$$

Consequently, our approximation of $\int_c^d f(t)\, dt$ corresponding to a given rule with points and weights (x_k, A_k), $k = 1, \ldots, n$, for the interval $[a, b]$ is

$$\boxed{\int_c^d f(t)\, dt \approx \frac{d - c}{b - a} \sum_{k=1}^{n} A_k f\left(\frac{d - c}{b - a} x_k + \frac{cb - ad}{b - a}\right).} \qquad (3.13)$$

This simple technique for conversion of integration domains makes quadrature point and weight tables such as in (Abramowitz and Stegun, 1965, Chap. 25) universally applicable.

EXAMPLE 3.5

The task is to integrate

$$\int_{2.0}^{2.5} e^t \, dt,$$

using the quadrature formula of Example 3.3, which is,

$$\int_{-1}^{1} f(x) \, dx \approx \tfrac{1}{3}f(-1) + \tfrac{4}{3}f(0) + \tfrac{1}{3}f(1).$$

Here, in the notation of (3.13), $a = -1$, $b = 1$, $c = 2$, and $d = 2.5$, so

$$\frac{d-c}{b-a} = \frac{1}{4} \quad \text{and} \quad \frac{cb-ad}{b-a} = \frac{9}{4}.$$

Thus, by using relation (3.13), we calculate

$$\int_{2}^{2.5} e^t \, dt \approx \tfrac{1}{4}\{\tfrac{1}{3} \exp\left[\tfrac{1}{4} \cdot (-1) + \tfrac{9}{4}\right] + \tfrac{4}{3} \exp\left(\tfrac{1}{4} \cdot 0 + \tfrac{9}{4}\right)$$

$$+ \tfrac{1}{3} \exp\left(\tfrac{1}{4} \cdot 1 + \tfrac{9}{4}\right)\}$$

$$\approx 4.793541.$$

The actual value is $e^{2.5} - e^2 \approx 4.793437.$ ∎

3.4
COMPOUND QUADRATURE FORMULAS

3.4.1 Midpoint, Trapezoidal, Simpson's, and Related Rules

In Section 2.4.4 and elsewhere in Chapter 2, we observed that polynomial interpolation is, in some instances, surprisingly difficult. Approximations for some continuous functions can get worse, rather than better, as more data points are used. Other functions can require impossibly high-degree polynomials for reasonable accuracy, no matter which interpolation points are chosen, and the use of even moderately high degree polynomials can engender instabilities from roundoff error. Unfortunately, as one might anticipate, these effects carry over to interpolatory-type quadrature rules offered in the preceding section: Newton–Cotes truncation error need not go to zero as the number n of points increases. Even for rules having as few as 10 points, computation of weights A_j can be numerically unstable.

In the case of interpolation, to some extent we were able to bypass these difficulties by resorting to piecewise polynomials such as splines. Several of the most popular quadrature rules can be motivated by the viewpoint that they integrate piecewise interpolation polynomials. One advantage of these "compound" quadrature formulas, as they are called, is that they do converge to the integral as more points are used (Szidarovszky and Yakowitz, 1978, Sec. 3.2).

Let N be an integer greater than 1, $[a, b]$ be the interval of integration, and assume that we have a numerical integration formula available. Then the associated compound formula requires that the interval $[a, b]$ be divided into N subintervals, usually of equal lengths, and has us apply our integration formula over each of the subintervals. Specifically, N being the number of subintervals, let

$$a = z_0 < z_1 < \cdots < z_{N-1} < z_N = b$$

denote the partition points. The *compound quadrature formula* requires that for each k, $0 \le k < N$, we apply the given integration rule to estimate

$$\int_{z_k}^{z_{k+1}} f(x) \, dx$$

on each of the subintervals $[z_k, z_{k+1}]$ and then sum these estimates. In Figure 3.3 we suggest how points of the given quadrature formula are distributed into each subinterval to obtain the quadrature points of the associated compound formula.

The advantage of this procedure is that, as we will see, we can get fairly accurate estimates of the integral without having to solve a formidable linear equation for the coefficients. A disadvantage is that often we can achieve greater accuracy by a single high-order interpolatory rule, for a given set of function values, than by a compound formula using this same set of values. This will be shown in Examples 3.6 and 3.10.

We now reveal the most common of the compound rules. They are obtained by compounding the simplest possible interpolatory-type formulas. Specifically, the reader will observe that in the notation of (3.8), the midpoint and trapezoidal rules are compound interpolatory rules for $n = 1$ and 2. The Simpson and Bode rules, to be given later, can be viewed as compound rules with $n = 3$ and 5, respectively. This hierarchy could, of course, be extended.

Assume that the interval $[a, b]$ is divided into N equal subintervals by points $a = z_0 < z_1 < \cdots < z_{N-1} < z_N = b$. Consider now one of the subintervals, say $[z_k, z_{k+1}]$.

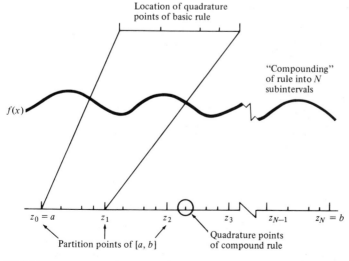

FIGURE 3.3 Compounding of a basic Quadrature Formula

If x_{k+1} denotes the midpoint of this subinterval,

$$x_{k+1} = \frac{z_k + z_{k+1}}{2},$$

and if $f(x)$ is approximated by the constant interpolation polynomial $p_0(x) = f(x_{k+1})$ based on this only interpolating point, we obtain the one-point interpolatory-type formula

$$\int_{z_k}^{z_{k+1}} f(x)\, dx \approx \int_{z_k}^{z_{k+1}} f(x_{k+1})\, dx = hf(x_{k+1}) \qquad (3.14)$$

for $k = 0, 1, \ldots, N - 1$, and $h = z_{k+1} - z_k$.

Alternatively, by using the endpoints $x_k = z_k$, $x_{k+1} = z_{k+1}$ as interpolating points and by constructing a linear interpolation polynomial $p_1(x)$, we get the functional approximation

$$f(x) \approx p_1(x) = \frac{x - x_{k+1}}{x_k - x_{k+1}} f(x_k) + \frac{x - x_k}{x_{k+1} - x_k} f(x_{k+1}),$$

which leads to the two-point interpolatory-type formula

$$\int_{z_k}^{z_{k+1}} f(x)\, dx \approx f(x_k) \int_{z_k}^{z_{k+1}} \frac{x - x_{k+1}}{-h}\, dx + f(x_{k+1}) \int_{z_k}^{z_{k+1}} \frac{x - x_k}{h}\, dx$$

$$= \frac{h}{2}(f(x_k) + f(x_{k+1})). \qquad (3.15)$$

Adding the integrals (3.14) and (3.15), respectively, over all subintervals, we obtain the compound formula approximations of $\int_a^b f(x)\, dx$ given by

$$\boxed{\begin{array}{l} m_N = h[f(x_1) + f(x_2) + \cdots + f(x_N)] \\[2mm] \qquad \text{for } h = \dfrac{b - a}{N},\ x_j = a + \left(j - \dfrac{1}{2}\right)h,\ j = 1, 2, \ldots, N, \end{array}} \qquad (3.16)$$

and

$$\boxed{\begin{array}{l} t_N \equiv h[\tfrac{1}{2}f(x_0) + f(x_1) + \cdots + f(x_{N-1}) + \tfrac{1}{2}f(x_N)] \\[2mm] \qquad \text{for } h = \dfrac{b - a}{N},\ x_j = a + jh,\ j = 0, 1, \ldots, N. \end{array}} \qquad (3.17)$$

The formulas m_N and t_N are called the *midpoint formula* and the *trapezoidal rule*, respectively. The trapezoidal rule can be viewed as an integral of the linear spline

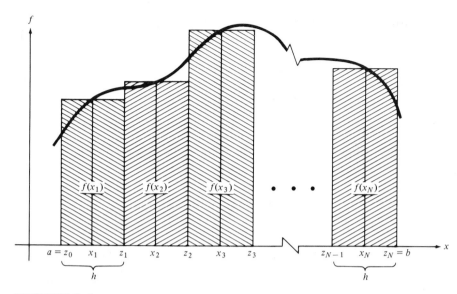

FIGURE 3.4   The Midpoint Formula

(or broken-line process) which interpolates the data $(x_k, f(x_k))$, $k = 0, \ldots, N$. Many textbooks, for example, Davis and Rabinowitz (1975), use the notation t_N and refer to t_N as the N-point trapezoidal rule, although strictly speaking, it uses $N + 1$ quadrature points. The midpoint formula m_N uses exactly N points.

The graphical representations of the midpoint formula and the trapezoidal rule are given in Figures 3.4 and 3.5. The approximating value m_N of the integral is given by the sum of the areas of rectangles shown in Figure 3.4, and similarly the trapezoidal rule t_N is the total area of the trapezoids in Figure 3.5.

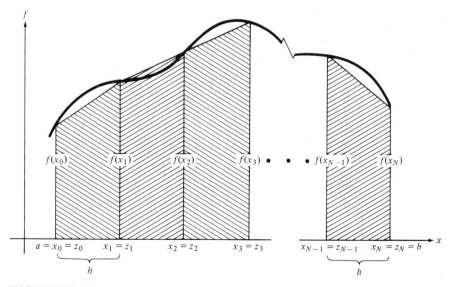

FIGURE 3.5   The Trapezoidal Rule

━━━━ **EXAMPLE 3.6** ━━━

We apply the trapezoidal rule to integrate

$$\int_{-1}^{1} e^x \, dx,$$

the integral in Example 3.3. The approximation t_2, according to (3.17), with x_0, x_1, and x_2 being -1, 0, and 1, respectively, is

$$t_2 = \tfrac{1}{2}e^{-1} + e^0 + \tfrac{1}{2}e^1 = 2.54.$$

To three places, the correct value is 2.35. Note that the pure interpolatory rule gave a more accurate answer, using the very same data.

━━━ ∎

Subroutine TRAP for computing the trapezoidal formula t_N is given in Table 3.7.

TABLE 3.7   Subroutine TRAP for the Trapezoidal Rule

```
        SUBROUTINE TRAP(A,B,N,E)
C
C     ****************************************************************
C     *   FUNCTION: APPROXIMATES THE INTEGRAL OF F(X) OVER THE      *
C     *             INTERVAL [A,B] BY THE TRAPEZOIDAL RULE          *
C     *   USAGE:                                                    *
C     *         CALL SEQUENCE: CALL TRAP(A,B,N,E)                   *
C     *         EXTERNAL FUNCTIONS/SUBROUTINES: FUNCTION F(X)       *
C     *   PARAMETERS:                                               *
C     *       INPUT:                                                *
C     *             A=INTERVAL LEFT ENDPOINT                        *
C     *             B=INTERVAL RIGHT ENDPOINT                       *
C     *             N=NUMBER OF SUBINTERVALS                        *
C     *       OUTPUT:                                               *
C     *             E=ESTIMATE OF THE INTEGRAL                      *
C     ****************************************************************
C
C     *** INITIALIZATION ***
        H=(B-A)/N
        E=(F(A)+F(B))/2.0
        IF (N.GT.1) THEN
C     *** THE TRAPEZOIDAL FORMULA ***
          X=A
          DO 1 I=1,N-1
            X=X+H
            E=E+F(X)
    1     CONTINUE
        END IF
        E=E*H
        RETURN
        END
```

─── **EXAMPLE 3.7** ───

By means of the FORTRAN program shown in Table 3.8, the value of the 10-point trapezoidal rule applied to approximating

$$\int_0^{10} 10 \sin (1 - 0.1x) \, dx$$

is found. The output of this computation is presented in Table 3.9.

TABLE 3.8   Program for Example 3.7

```
C       PROGRAM TRAPEZOID
C
C       ****************************************************************
C       CALCULATES THE INTEGRAL OF F(X)=10*SIN(1-0.1*X) OVER [0,10]
C       BY THE TRAPEZOIDAL RULE
C       CALLS: TRAP
C       OUTPUT: TRUE=TRUE VALUE OF THE INTEGRAL AS CALCULATED USING
C                    LIBRARY FUNCTION COS
C               EST=ESTIMATED VALUE CALCULATED BY THE TRAPEZOID RULE
C               ERROR=TRUE-EST
C       ****************************************************************
C
        A=0.
        B=10.
        N=10
        TRUE=100.*(1-COS(1.0))
C
C       *** SUBROUTINE TRAP ESTIMATES THE INTEGRAL BY TRAPEZOIDAL    ***
C       *** RULE                                                     ***
C
        CALL TRAP(A,B,N,EST)
        ERROR=TRUE-EST
        WRITE(10,*)TRUE,EST,ERROR
        STOP
        END
C
C       *** SUBROUTINE FUNCTION F(X) COMPUTES THE INTEGRAND          ***
C
        FUNCTION F(X)
        F=10.*SIN(1.-0.1*X)
        RETURN
        END
```

TABLE 3.9   Output of Example 3.7

True	10-Point Trapezoidal	Error
45.96977	45.93146	3.8314819E-02

■

Consider again one subinterval $[z_k, z_{k+1}]$, and approximate the function $f(x)$ by a quadratic interpolation polynomial based on the endpoints and the midpoint of this subinterval. Then, by recalling Example 3.3 and relation (3.13), we get the approximation

$$\int_{z_k}^{z_{k+1}} f(x)\,dx \approx \frac{h}{3}\left[f(z_k) + 4f\left(\frac{z_k + z_{k+1}}{2}\right) + f(z_{k+1})\right],$$

where now $h = \frac{1}{2}(z_{k+1} - z_k)$. The quadrature points $x_0, x_1, \ldots, x_M$ are related to the compounding interval endpoints $z_0, z_1, \ldots, z_N$ by $M = 2N$ and

$$x_0 = z_0, \quad x_1 = \frac{z_0 + z_1}{2}, \quad x_2 = z_1,$$

$$x_3 = \frac{z_1 + z_2}{2}, \quad \ldots, x_{M-2} = z_{N-1},$$

$$x_{M-1} = \frac{z_{N-1} + z_N}{2}, \quad x_M = z_N.$$

In terms of these quadrature points and after adding the quadratures over each subinterval, we have

$$\boxed{\begin{aligned} s_M &= \frac{h}{3}\left[f(x_0) + 4(f(x_1) + f(x_3) + \cdots + f(x_{M-1})) + 2(f(x_2)\right. \\ &\quad \left. + f(x_4) + \cdots + f(x_{M-2})) + f(x_M)\right] \\ &\text{where } h = \frac{b-a}{M}, \quad \text{and for } j = 0, 1, \ldots, M, \quad x_j = a + jh. \end{aligned}}$$

(3.18)

For any even number M, s_M is called the M-point *Simpson's rule*, although, in fact, it has $M + 1$ quadrature points. Simpson's rule amounts to compounding the three-point Newton–Cotes formula over $M/2$ subintervals.

Subroutine SIMP for Simpson's rule is given in Table 3.10, where the calling parameter M is defined as above.

The graphical representation of Simpson's rule is given in Figure 3.6. For $k = 0, 2, \ldots, 2N - 2$, consider the interpolating points x_k, x_{k+1}, x_{k+2}, the corresponding functional values $f(x_k), f(x_{k+1}), f(x_{k+2})$, and the quadratic interpolating polynomial $p_2(x)$ based on these data. Then the integral of $f(x)$ over the interval $[x_k, x_{k+2}]$ is approximated by the area under this quadratic interpolation polynomial, and the integral $\int_a^b f(x)\,dx$ is approximated by summing the areas under these quadratic polynomials, for $k = 0, 2, \ldots, 2N - 2$. The integral approximation is the crosshatched area shown in the figure.

Simpson's rule can be viewed as the integral of a continuous piecewise quadratic interpolation function. The quadratic pieces are the second-degree polynomials that interpolate the data $(x_k, f(x_k))$, $(x_{k+1}, f(x_{k+1}))$, and $(x_{k+2}, f(x_{k+2}))$, $k = 0, 2, \ldots, M - 2$, where $M = 2N$. The resultant piecewise polynomial is not, technically speaking, a quadratic spline because the first derivatives need not exist at the points x_k.

TABLE 3.10 Subroutine SIMP for Simpson's Rule

```
          SUBROUTINE SIMP(A,B,M,E)
C
C    ****************************************************************
C    *   FUNCTION: APPROXIMATES THE INTEGRAL OF F(X) OVER THE      *
C    *             INTERVAL [A,B] BY SIMPSON'S RULE                *
C    *   USAGE:                                                    *
C    *         CALL SEQUENCE: CALL SIMP(A,B,M,E)                   *
C    *         EXTERNAL FUNCTIONS/SUBROUTINES: FUNCTION F(X)       *
C    *   PARAMETERS:                                               *
C    *         INPUT:                                              *
C    *             A=INTERVAL LEFT ENDPOINT                        *
C    *             B=INTERVAL RIGHT ENDPOINT                       *
C    *             M=NUMBER OF SUBINTERVALS(POSITIVE EVEN INTEGER)*
C    *         OUTPUT:                                             *
C    *             E=ESTIMATE OF THE INTEGRAL                      *
C    ****************************************************************
C
C         *** INITIALIZATION ***
          E=F(A)+F(B)
          H=(B-A)/M
          X=A-H
C         *** COMPUTE THE SUM OF ODD INDEX TERMS ***
          DO 1 I=1,M-1,2
             X=X+2.0*H
             E=E+4.0*F(X)
        1 CONTINUE
C         *** COMPUTE THE SUM OF EVEN INDEX TERMS ***
          X=A
          DO 2 I=2,M-2,2
             X=X+2.0*H
             E=E+2.0*F(X)
        2 CONTINUE
          E=E*H/3.0
          RETURN
          END
```

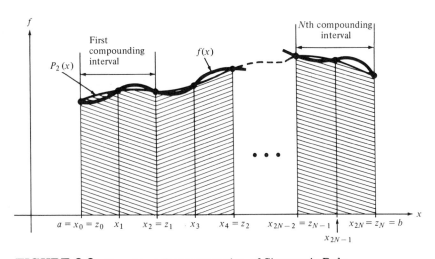

FIGURE 3.6 Graphical Representation of Simpson's Rule

EXAMPLE 3.8

The integral

$$\int_0^{10} 10 \sin (1 - 0.1x) \, dx$$

in Example 3.7 was computed by Simpson's rule, with $M = 2N = 10$. Subroutine SIMP given in Table 3.10 was applied to get the approximation

$$s_{10} = 45.96979,$$

which is much closer to the true value of 45.96977 than the value obtained by the trapezoidal rule t_{10}. Note that both t_{10} and s_{10} used the same quadrature points and function values.

$\blacksquare$

*3.4.2   Error Analysis of Compound Formulas

The truncation error of the trapezoidal rule will be derived in detail. Other compound formulas can be examined similarly. They all spring from the idea of integrating (2.14).

From the error formula for linear interpolating polynomials [equation (2.14) with $n = 2$] we have

$$f(x) - p_1(x) = \frac{f''(\zeta(x))}{2!} (x - x_k)(x - x_{k+1}),$$

where $p_1(x)$ is the linear interpolation polynomial based on the abscissas x_k, x_{k+1}. Then formally

$$\int_{x_k}^{x_{k+1}} f(x) \, dx - \int_{x_k}^{x_{k+1}} p_1(x) \, dx = \int_{x_k}^{x_{k+1}} \frac{f''(\zeta(x))}{2} (x - x_k)(x - x_{k+1}) \, dx.$$

The second term of the left-hand side is given in (3.15), and by substituting the variable $t = (x - x_k)/h$ into the integral, we get

$$\int_{x_k}^{x_{k+1}} f(x) \, dx - \frac{h}{2} (f(x_k) + f(x_{k+1})) = \int_0^1 \frac{h^3 f''(\zeta(t))}{2} t(t - 1) \, dt. \quad (3.19)$$

Since $t(t - 1)$ has no roots in $(0, 1)$, the mean-value theorem for integral calculus [equation (B.19) of Appendix B] implies that the right-hand side is equal to

$$\frac{h^3 f''(\eta_k)}{2} \int_0^1 t(t - 1) \, dt = -\frac{h^3 f''(\eta_k)}{12}, \quad (3.20)$$

where $\eta_k \in (x_k, x_{k+1})$. After adding (3.19) for $k = 0, 1, \ldots, N - 1$ and using (3.20), the intermediate-value theorem for derivatives applied to $f''(x)$ implies that

$$\int_a^b f(x) \, dx = t_N - \sum_{k=0}^{N-1} \frac{h^3 f''(\eta_k)}{12} = t_N - \frac{h^3 N}{12} \frac{1}{N} \sum_{k=0}^{N-1} f''(\eta_k)$$

$$= t_N - \frac{h^3 N f''(\eta)}{12},$$

where $\eta \in (a, b)$. Since $h = (b - a)/N$, we conclude that the error of the trapezoidal formula can be rewritten as

$$\int_a^b f(x) \, dx - t_N = -\frac{(b - a)h^2 f''(\eta)}{12} \tag{3.21}$$

and the absolute value of the truncation error can be bounded as follows:

$$\left| \int_a^b f(x) \, dx - t_N \right| \le \frac{M_2 (b - a)h^2}{12}, \tag{3.22}$$

where M_2 is selected so that

$$|f''(x)| \le M_2$$

for all $x \in (a, b)$.

We emphasize that the bound (3.22) accounts only for truncation error. For N large (above 1000, say), and computer word length short, roundoff error can cause further degradation.

───── **EXAMPLE 3.9** ─────────────────────────────────

We bound the error of the integral computed in Example 3.7. Since $N = 10$, $a = 0$, $b = 10$, $h = 1.0$ and

$$f(x) = 10 \sin (1 - 0.1x),$$

we have

$$f'(x) = -\cos (1 - 0.1x)$$

and

$$f''(x) = -0.1 \sin (1 - 0.1x).$$

Thus $M_2 = 0.1$, and the truncation error can be bounded by

$$\frac{M_2 (b - a)h^2}{12} = \frac{1}{12} = 0.08383. \ldots$$

To five significant decimals, the exact value of the integral is 45,970, and in Example 3.7 we have seen that the actual error is 0.039.

Formula (3.22) can be used to find the number N of subintervals needed to assure six-decimal-place accuracy. In this case the error will not exceed $\frac{1}{2} \times 10^{-6}$ if the value of N satisfies

$$\frac{M_2(b - a)^3}{12N^2} \leq \frac{1}{2} \times 10^{-6},$$

that is,

$$\frac{10^2}{12N^2} \leq \frac{1}{2} \times 10^{-6}.$$

Solving for N, we calculate

$$N \geq \left(\frac{10^2 \times 10^6 \times 2}{12}\right)^{1/2} = \frac{10^4}{\sqrt{6}} \approx 4082.48,$$

which implies that $N = 4083$ is satisfactory. In fact, by applying subroutine TRAP (Table 3.7), with $N = 4083$ to the integration problem above, the authors found the actual error to be about -2.3×10^{-7}. The calculations were done on a VAX computer in double precision. In single precision, roundoff error is the limiting factor, and the error is about -9×10^{-7}. ∎

Table 3.11 summarizes the specific Newton–Cotes integration rules and associated errors of methods in Section 3.3, and we have included a few other popular Newton–Cotes formulas. Further rules may be found in Abramowitz and Stegun (1965, pp. 885–887). Table 3.12 similarly recapitulates popular compound formulas and associated errors.

TABLE 3.11   Some Popular Interpolatory-Type Formulas and Their Errors

Notation	Step size $h > 0$, $x_{k+1} = x_k + h$, $f_k = f(x_k)$, ζ a point in the domain of integration
Simple trapezoidal rule	$\displaystyle\int_{x_0}^{x_1} f(x)\,dx \approx \frac{h}{2}(f_0 + f_1)$
	$\displaystyle\text{error} = -\frac{h^3}{12}f''(\zeta)$
Simple Simpson's rule	$\displaystyle\int_{x_0}^{x_2} f(x)\,dx \approx \frac{h}{3}(f_0 + 4f_1 + f_2)$
	$\displaystyle\text{error} = -\frac{h^5}{90}f^{(4)}(\zeta)$

TABLE 3.11   (Continued)

Simple Simpson's $\frac{3}{8}$ rule	$\int_{x0}^{x3} f(x)\, dx \approx \frac{3h}{8} (f_0 + 3f_1 + 3f_2 + f_3)$
	$\text{error} = -\frac{3h^5}{80} f^{(4)}(\zeta)$
Simple Bode's rule	$\int_{x0}^{x4} f(x)\, dx \approx \frac{2h}{45} (7f_0 + 32f_2 + 32f_3 + 7f_4)$
	$\text{error} = -\frac{8h^7}{945} f^{(6)}(\zeta)$

TABLE 3.12   Some Popular Compound Formulas and Their Errors

Notation	Step size $h = x_{i+1} - x_i = \dfrac{b - a}{N}$, $x_{k+1} = x_k + h$, $f_k = f(x_k)$, ζ a point in domain of integration
(Compound) trapezoidal rule	$\int_a^b f(x)\, dx \approx h \left(\dfrac{f_0}{2} + f_1 + \cdots + f_{N-1} + \dfrac{f_N}{2} \right) = t_N$
	$\text{error} = -\dfrac{(b - a)h^2}{12} f''(\zeta)$
Modified trapezoidal rule	$\int_a^b f(x)\, dx \approx h \left(\dfrac{f_0}{2} + f_1 + \cdots + f_{N-1} + \dfrac{f_N}{2} \right)$
	$\qquad\qquad + \dfrac{h}{24} (-f_{-1} + f_1 + f_{N-1} - f_{N+1})$
	$\text{error} = -\dfrac{11(b - a)h^4}{720} f^{(4)}(\zeta)$
(Compound) Simpson's rule	$\int_a^b f(x)\, dx \approx \dfrac{h}{3} [f_0 + 4(f_1 + f_3 + \cdots + f_{M-2})$
	$\qquad\qquad + 2(f_2 + f_4 + \cdots + f_{M-1}) + f_M] = s_M$
	$\text{error} = -\dfrac{(b - a)h^4}{180} f^{(4)}(\zeta) \qquad \left(h = \dfrac{b - a}{M} \right)$

EXAMPLE 3.10

Here we compare the performances of Simpson's rule with that of the Newton–Cotes formula using the very same data values (x_k, f_k), $k = 1, \ldots, n$. By means of the program and function subroutine listed in Table 3.13, we applied subroutines SIMP and COEFF to evaluate the "usual" integral

$$\int_0^{10} 10 \sin (1 - 0.1x)\, dx.$$

For reasons to be noted in a moment, these computations were done in double precision.

TABLE 3.13   Program to Compare Simpson's and Newton–Cotes Integration

```
C       PROGRAM COMPARISON
C
C       *************************************************************
C       COMPARES NEWTON-COTES AND SIMPSON'S RULE ON THE
C       INTEGRAL OF F(X)=10*SIN(1-0.1*X) OVER [0.,10.]
C       CALLS:COEFF,GAUS1,SIMP(MODIFIED FOR DOUBLE PRECISION)
C       OUTPUT: NN(I)=NUMBER OF POINTS
C               TRUE=TRUE INTEGRAL OF F(X)
C               S=NEWTON-COTES INTEGRAL ESTIMATE OF F(X)
C               ERRS=NEWTON-COTES ERROR
C               E=SIMPSON'S INTEGRAL ESTIMATE OF F(X)
C               ERRE=SIMPSON'S ERROR
C       *************************************************************
C
        IMPLICIT DOUBLE PRECISION(A-H,O-Z)
        DIMENSION X(250),AA(250),NN(20)
        DATA (NN(I),I=1,9)/3,5,7,9,11,13,15,19,21/
        V=1.0D0
        A=0.0D0
        B=10.0D0
        TRUE=100.0D0*(1.0D0-DCOS(V))
C
C       *** ITERATES ON THE NUMBER OF POINTS               ***
C
        DO 6 I=1,9
           H=(B-A)/(NN(I)-1)
           X(1)=A
           DO 3 J=2,NN(I)
              X(J)=X(J-1)+H
    3      CONTINUE
C
C       *** SUBROUTINE COEFF COMPUTES INTEGRATION COEFFICIENTS ***
C       *** OF NEWTON-COTES METHOD                          ***
C
           CALL COEFF(NN(I),X,A,B,AA)
           S=0.0D0
           DO 4 J=1,NN(I)
C
C       *** COMPUTES INTEGRAL BY NEWTON-COTES RULE          ***
C
              S=S+AA(J)*F(X(J))
    4      CONTINUE
           ERRS=TRUE-S
C
C       *** SUBROUTINE SIMP COMPUTES INTEGRAL BY SIMPSON'S RULE***
C
           CALL SIMP(A,B,NN(I)-1,E)
           ERRE=TRUE-E
           WRITE(10,7)NN(I),TRUE,S,ERRS,E,ERRE
    7      FORMAT(10X,I2,4(3X,F17.10))
    6   CONTINUE
        STOP
        END
C
C       *** SUBROUTINE FUNCTION F COMPUTES THE FUNCTIONAL  ***
C       *** VALUES AT THE POINT X                          ***
C
        REAL*8 FUNCTION F(X)
        IMPLICIT DOUBLE PRECISION(A-H,O-Z)
        V=1.0D0-0.1D0*X
        F=10.0D0*DSIN(V)
        RETURN
        END
```

TABLE 3.14 Error Comparison for Example 3.10

Number of Points	Error, Newton–Cotes	Error, Simpson
3	-0.0164495739	-0.0164495739
5	0.0000245534	-0.0010050795
7	-0.0000000304	-0.0001977119
9	-0.0000000003	-0.0000624667
11	0.0000000000	-0.0000255692
13	-0.0000000021	-0.0000123263
15	0.0000017216	-0.0000066520
19	0.0590227117	-0.0000024337
21	27.1303181091	-0.0000015966
31		-0.0000003153
91		-0.0000000039
151		-0.0000000005
201		-0.0000000002

From the output, summarized in Table 3.14, we see that in the case of three points, the Newton–Cotes and Simpson's rules give exactly the same value. This is because the corresponding Simpson's rule *is* the three-point, Newton–Cotes formula. As the number of points increases, the Newton–Cotes formula becomes much more accurate than Simpson's rule, but after the 11-point formula, its performance starts deteriorating until by 21 points, the answer is quite erroneous. The cause of deterioration is not the Newton–Cotes method itself, but that as the number of points increases the solution of the linear equation (3.11) tends to become subject to extreme errors as a result of roundoff, even in double precision. A reason why we can confidently ascribe blame to the linear equation, rather than instabilities in polynomial approximation, is that (as the reader can easily verify) polynomial approximation of sin (x) is quite stable and accurate to at least six significant decimals, for 20 data points. Small error in interpolation polynomial approximation implies small error in interpolatory quadrature. Ill-conditioned linear equations are discussed further in Chapter 4. Observe that the error of the 11-point Newton–Cotes rule is less than that of the 201-point Simpson estimate. ∎

The practical significance of truncation error analysis may be assessed from Table 3.14. From Table 3.12 we know that the error of Simpson's rule is proportional to h^4. So if $h_2 = h_1/2$, for instance, the errors err (h_1) and err (h_2) ought to be related by

$$\text{err } (h_2) = \text{err } (h_1)\left(\frac{h_2}{h_1}\right)^4 = \frac{\text{err } (h_1)}{16}.$$

For the number of points 5 and 9, the ratio of step sizes equals $\frac{1}{2}$, and the ratio of errors is in fact, very close to $\frac{1}{16}$. Similarly, the reader can check that for 11 and 21 points, the ratio of errors is again very close to $\frac{1}{16}$.

The essence of the situation is that Newton–Cotes integration, if done carefully, can be appreciably more accurate than the Simpson method, but because of the instability of the associated system of linear equations for a large number of points, care must be exercised. In this regard it is useful to note from examination of (3.11) that the solution of the system of linear equations does not depend on the integrand. So before accepting an answer concerning an unknown integral, we can check the Newton–Cotes coefficients computed by COEFF by applying them to integrals having known values.

Let us examine the effect of rounding errors of the functional values. Assume that instead of the true values $f(x_k)$, only their approximations $\bar{f}_k$ are known, and that for all values of k, we have an absolute error bound δ. That is,

$$|f(x_k) - \bar{f}_k| \le \delta. \tag{3.23}$$

Then the roundoff error of the trapezoidal rule can be bounded as follows:

$$\left| h[\tfrac{1}{2}f(x_0) + f(x_1) + \cdots + f(x_{N-1}) + \tfrac{1}{2}f(x_N)] - h(\tfrac{1}{2}\bar{f}_0 + \bar{f}_1 + \cdots \right.$$
$$\left. + \bar{f}_{N-1} + \tfrac{1}{2}\bar{f}_N) \right|$$
$$\le h[\tfrac{1}{2}|f(x_0) - \bar{f}_0| + |f(x_1) - \bar{f}_1| + \cdots + |f(x_{N-1}) - \bar{f}_{N-1}|$$
$$+ \tfrac{1}{2}|f(x_N) - \bar{f}_N|] \le hN\,\delta = (b - a)\,\delta.$$

The bound $(b - a)\,\delta$ is valid for all quadrature formulas having no negative coefficients. For any such rule, (this includes trapezoidal, midpoint, and Simpson's rules)

$$\left| I_N(\bar{f}) - \int_a^b f(x)\,dx \right| \le O(h^p) + (b - a)\,\delta$$

where p is positive, $I_N(\bar{f})$ is the value of the numerical integration formula computed using imperfect functional values $\bar{f}_k$, and $O(h^p)$ is a bound for the error of the integration formula when the correct functional values were used. From its definition,

$$O(h^p) \to 0 \quad \text{as} \quad h \to 0.$$

So, ignoring roundoff effects, no harm is done (except perhaps some wasted computational expense) if a very small step size h is used. By contrast, we saw in Section 3.2 that in numerical differentiation formulas, the step size shows up in the denominator of the term representing the effect of functional error. As the step size decreases, therefore, the error of the computed derivative can become arbitrarily large. (If h is so small that an enormously large N is needed, smearing can occur, as illustrated in Problem 14.)

3.5
EXTRAPOLATION AND ROMBERG INTEGRATION

Richardson extrapolation, described in the optional section that follows, is a powerful but subtle concept serving as the foundation for Romberg quadrature and other numerical methods. The section after that, on the implementation of the popular and efficient Romberg quadrature method, may be read independently.

*3.5.1   Richardson's Extrapolation and Corrected Formulas

The notion of Richardson's extrapolation will be described in the context of "correcting" the compound trapezoidal formula t_N given by (3.17). As in the preceding section, we wish to approximate $\int_a^b f(x)\,dx$. Let $h = (b - a)/N$ and assume that $f(x)$ is a function with a Taylor's series representation valid over the interval $[a, b]$. Then (Szidarovszky and Yakowitz, 1978, Sec. 3.2.4) there exist constants $c_1, c_2, \ldots$, which depend on $f(x)$, the interval $[a, b]$, but not on N or h, such that

$$\int_a^b f(x)\,dx - t_N = c_2 h^2 + c_4 h^4 + c_6 h^6 + \cdots. \qquad (3.24)$$

Note that if N is doubled, the step size becomes $h/2$, and therefore

$$\int_a^b f(x)\,dx - t_{2N} = c_2 \frac{h^2}{4} + c_4 \frac{h^4}{16} + c_6 \frac{h^6}{64} + \cdots. \qquad (3.25)$$

By subtracting the multiple 4 of this equation from (3.24), we get the relation

$$-3 \int_a^b f(x)\,dx + 4t_{2N} - t_N = c_4(1 - \tfrac{1}{4})h^4 + c_6(1 - \tfrac{1}{16})h^6 + \cdots, \qquad (3.26)$$

which implies that

$$\int_a^b f(x)\,dx - \tfrac{1}{3}(4t_{2N} - t_N) = d_4 h^4 + d_6 h^6 + \cdots, \qquad (3.27)$$

where $d_1, d_2, \ldots$, are also independent of h or N.
    The quadrature formula

$$\tilde{t}_{2N} = \tfrac{1}{3}(4t_{2N} - t_N) \qquad (3.28)$$

is called the *corrected* trapezoidal formula. It has the characteristic that its truncation error is, in view of (3.27), $O(h^4)$ instead of merely $O(h^2)$, the order of the "uncorrected" trapezoidal formula. It turns out that the corrected trapezoidal formula coincides with the Simpson rule, so at this stage we have only a new idea, not a new formula. The idea of adding together two formulas dependent on h so as to remove the lowest term in the power series expansion of the truncation error is known as *Richardson's extrapolation*. Its repeated application to remove successively higher terms in the truncation error expansion of the trapezoidal formula leads us to the Romberg integration method of the following section. Let us proceed one more step along this pathway. Noting that $\tilde{t}_{2N}$ defined by (3.28) is really the Simpson formula s_{2N}, as in (3.27), we obtain

$$\int_a^b f(x)\,dx - s_{2N} = d_4 h^4 + d_6 h^6 + \cdots,$$

where the coefficients d_4, d_6, . . . are independent of N and h. And as in the first extrapolation, a judicious sum of s_N and s_{2N} can rid us of the h^4 term. Specifically, the Richardson correction of the Simpson formula gives us the representation

$$\int_a^b f(x)\ dx = \tfrac{1}{15}(16s_{2N} - s_N) + O(h^6). \tag{3.29}$$

The corrected Simpson rule is

$$\tilde{s}_{2N} = \tfrac{1}{15}(16s_{2N} - s_N),$$

which is known as *Bode's rule*. From (3.29), its truncation error is $O(h^6)$, whereas the uncorrected rule (3.27) was $O(h^4)$. As seen in the next example, increasing the order of the truncation error can have dramatic practical significance.

EXAMPLE 3.11

We apply the corrected Simpson's formula immediately above, with $N = 2$, to our familiar integral

$$10 \int_0^{10} \sin (1 - 0.1x)\ dx.$$

In our computations using subroutine SIMP, in double precision, we found the following:

Formula	Estimate	Error
s_2	45.9862	0.016450
s_4	45.9707	0.001005
$\tilde{s}_4 = \tfrac{1}{15}(16s_4 - s_2)$	45.969769	0.000022

The improvement in accuracy achieved by Richardson extrapolation is remarkable. This idea is fundamental to the integration procedure in the next section. ∎

3.5.2 Romberg Integration

The error of the trapezoidal rule (3.17) is known to satisfy

$$\int_a^b f(x)\ dx - t_N = c_2 h^2 + c_4 h^4 + \cdots. \tag{3.30}$$

Here $h = (b - a)/N$ and the constants $c_2, c_4, \ldots$ do not depend on N. We have used this observation and Richardson extrapolation to construct the Simpson method as the corrected trapezoidal rule, wherein the h^2 term in the error is eliminated. Repeated use of this idea, eliminating successively the h^2, h^4, h^6, . . . terms in the truncation error, leads to *Romberg integration*. The method can be described as follows.

First define the numbers

$$T_{0,k} = t_{2^k} \qquad (k = 0, 1, 2, \ldots, M), \tag{3.31}$$

where M is a given positive integer and t_{2^k} is the 2^k-point trapezoidal rule. The successive values of $T_{0,k}$ can be found efficiently from the recursion

$$T_{0,k} = \frac{T_{0,k-1}}{2} + h_k[f(a + h_k) + f(a + 3h_k) + f(a + 5h_k) + \cdots$$
$$+ f(a + (2^k - 1)h_k)], \tag{3.32}$$

where $h_k = (b - a)/2^k$. Then, for $l = 1, 2, \ldots, M$, recursively obtain

$$T_{l,k} = \frac{1}{4^l - 1} (4^l T_{l-1,k} - T_{l-1,k-1}) \qquad (k = l, l + 1, \ldots, M). \tag{3.33}$$

The value $T_{M,M}$ will be designated as the approximating value of the integral. In practice, it is convenient to compute the array $\{T_{l,k}\}$ column by column in the order indicated by Figure 3.7. The tails of the arrows pointing to a term show the quantities needed for the calculation of that term. Problem 15 suggests a way to use successive $T_{0,k}$ values as a stopping criterion (i.e., to find the value M beyond which roundoff error becomes the limiting factor).

Table 3.15 gives a listing of subroutine ROMB, which performs Romberg integration. The array **T** gives the values of $t_{l,k}$, $l \le k \le M$. The value of $T_{M,M}$ is returned to the main program by the calling parameter E.

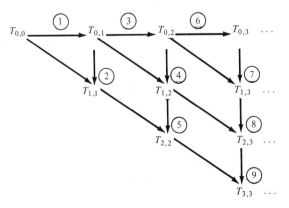

FIGURE 3.7 Order of Computations for Romberg Integration

TABLE 3.15   Subroutine ROMB for Romberg Integration

```
      SUBROUTINE ROMB(A,B,M,E)
C
C     ***************************************************************
C     *  FUNCTION: APPROXIMATES THE INTEGRAL OF F(X) OVER THE       *
C     *            INTERVAL [A,B] BY THE ROMBERG METHOD             *
C     *  USAGE:                                                     *
C     *        CALL SEQUENCE: CALL ROMB(A,B,M,E)                    *
C     *        EXTERNAL FUNCTIONS/SUBROUTINES: FUNCTION F(X)        *
C     *  PARAMETERS:                                                *
C     *     INPUT:                                                  *
C     *        A=INTERVAL LEFT ENDPOINT                             *
C     *        B=INTERVAL RIGHT ENDPOINT                            *
C     *        M=NUMBER OF ITERATIVE STEPS                          *
C     *          (NOT GREATER THAN 100)                             *
C     *     OUTPUT:                                                 *
C     *        E=ESTIMATE OF THE INTEGRAL                           *
C     ***************************************************************
C
      DIMENSION T(101,101)
C     *** INITIALIZATION ***
      K=1
      H=B-A
      T(1,1)=(F(A)+F(B))/2.0*H
      DO 1 I=2,M+1
C     *** COMPUTE THE TRAPEZOIDAL TERM ***
         K=2*K
         H=H/2.0
         X=A-H
         E=0.0
         T(1,I)=T(1,I-1)/2.0
         DO 2 J=1,K-1,2
            X=X+2.0*H
            E=E+F(X)
2        CONTINUE
C     *** RICHARDSON EXTRAPOLATION ***
         T(1,I)=T(1,I)+E*H
         LA=1
         DO 3 J=2,I
            LA=LA*4
            T(J,I)=(LA*T(J-1,I)-T(J-1,I-1))/(LA-1)
3        CONTINUE
1     CONTINUE
      E=T(M+1,M+1)
      RETURN
      END
```

──────── **EXAMPLE 3.12** ────────

The Romberg subroutine was called upon to evaluate the integral

$$\int_0^{10} 10 \sin (1 - 0.1x) \, dx,$$

which has served us in earlier examples. The number of extrapolations was set to $M = 5$. To the number of places shown, 45.96977 is the correct answer. From the printout shown in Table 3.16, one sees that the values of $T_{l,k}$ tend to increase in accuracy quite dramatically with increase in l, in accordance with the discussion that immediately follows this example.

TABLE 3.16   Romberg Integration Terms $T_{l,k}$

				k		
	0	1	2	3	4	5
0	42.07355	45.00806	45.73010	45.90990	45.95481	45.96603
1	0.00000	45.98623	45.97078	45.96984	45.96978	45.96978
l 2	0.00000	0.00000	45.96975	45.96978	45.96978	45.96978
3	0.00000	0.00000	0.00000	45.96978	45.96978	45.96978
4	0.00000	0.00000	0.00000	0.00000	45.96978	45.96978
5	0.00000	0.00000	0.00000	0.00000	0.00000	45.96978

Let us examine the error of Romberg integration. Results in the preceding subsection imply that $T_{1,k}$ is a Simpson's formula and $T_{2,k}$ is Bode's rule. Repeated application of the Richardson extrapolation technique and finite induction implies that for some array of constants $\{\gamma_{l,k}\}$,

$$\int_a^b f(x)\,dx - T_{l,k} = \sum_{j=l+1}^{\infty} \gamma_{l,j}\left(\frac{b-a}{2^k}\right)^{2j}, \qquad (3.34)$$

a formula derived in Szidarovszky and Yakowitz (1978), Sec. 3.2.5.

The effectiveness of Romberg integration, in terms of its order of error, simplicity, and stability, is quite remarkable. The accuracy cannot readily be seen from the limited number of digits displayed in Table 3.16, but in fact, the error of $T_{2,3}$ is 2.45×10^{-5}; $T_{3,3}$, -9.6×10^{-9}; and $T_{4,4}$, 99.5×10^{-13}. $T_{2,2}$ is a Newton–Cotes rule, and the error of $T_{3,3}$ is only about a factor of 30 greater than the (nine-point) Newton–Cotes rule using exactly the same data as reported in Table 3.14. Linear equation instabilities, which plague the Newton–Cotes procedure, are not a factor with Romberg formulas. Moreover, it is known (Davis and Rabinowitz, 1975, Chap. 6) that in the absence of roundoff, as M grows, $T_{M,M}$ converges to the exact integral for any (Riemann) integrable function. On the other hand, Newton–Cotes quadrature may diverge as the number of points increases, for some continuous integrands. However, interpolatory integration is not to be discarded. The supreme integration rule, in terms of accuracy and stability, Gaussian quadrature, is an interpolatory rule. This method occupies the following section.

ENGINEERING EXAMPLE

In Figure 3.8, we have designed a sail for a sailboat. The exact equation of our sail's upper edge is

$$f(x) = 20 \sin\left(\frac{x^2}{100}\right) m, \qquad 0 \le x \le 5.$$

The problem is to find the amount of material needed. That is, we need to calculate the area of the sail,

$$I = \int_0^5 20 \sin\left(\frac{x^2}{100}\right) dx.$$

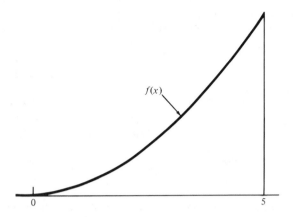

FIGURE 3.8 Design of a Sail

To our knowledge, this integral does not have a closed-form representation in terms of the common functions. We resort, therefore, to numerical methods. Our plan is to use Romberg quadrature to achieve high accuracy and relative stability, without recourse to tables or sophistication. We let the Romberg parameter M in subroutine ROMB (Table 3.15) vary between 2 and 5, and if the associated estimates lie close to each other, we take that as a sign that the truncation error is correspondingly small. For so few points, one would not anticipate round-off to be a problem.

For $M = 3$, 4, and 5, the computed integral "held steady" at 8.2962048 m^2. We accept this as our approximation of the area of material needed. ∎

3.6
GAUSS QUADRATURE

In Section 3.3 we saw that for any choice of interpolating points $x_1, \ldots, x_n$, the interpolatory-type integration formula (3.10) gives the true value of the integral when $f(x)$ is a polynomial of degree not exceeding $n - 1$. We will see that by choosing $x_1, \ldots, x_n$ wisely, the degree of polynomials that can be integrated exactly by an n-point rule is more than doubled, going from $n - 1$ to $2n - 1$.

In view of the discussion at the close of Section 3.3, without loss of generality we can assume that the interval for integration is $[-1, 1]$, since in view of (3.13), any finite interval of integration can be transformed into $[-1, 1]$.

In (3.11), the interpolating points were assumed to be given and the values $A_1, \ldots, A_n$ were unknowns. If we are also allowed to select the interpolating points, the number of unknowns in (3.11) is equal to $2n$, since the interpolating points are also variables. Consequently, we can add on more equations to (3.11) in order to have the same number of variables (i.e., $A_1, \ldots, A_n, x_1, \ldots, x_n$) and equations. These additional equations are determined by the condition that the numerical integration formulas should given the exact value of the integral for polynomials of as high a degree as possible. This idea leads us to the set of $2n$ equations:

$$A_1x_1^0 + A_2x_2^0 + \cdots + A_nx_n^0 = \int_{-1}^{1} 1 \, dx = (+1) - (-1) = 2$$

$$A_1x_1^1 + A_2x_2^1 + \cdots + A_nx_n^1 = \int_{-1}^{1} x \, dx = \frac{(+1)^2 - (-1)^2}{2} = 0$$

$$\cdots \qquad \cdots$$

$$A_1x_1^k + A_2x_2^k + \cdots + A_nx_n^k = \frac{(+1)^{k+1} - (-1)^{k+1}}{k+1}$$

$$= \begin{cases} 0 & \text{if } k \text{ is odd} \\ \dfrac{2}{k+1} & \text{if } k \text{ is even} \end{cases}$$

$$\cdots \qquad \cdots$$

$$A_1x_1^{2n-1} + A_2x_2^{2n-1} + \cdots + A_nx_n^{2n-1} = \frac{(+1)^{2n} - (-1)^{2n}}{2n} = 0.$$

$$(3.35)$$

Notice how this generalizes (3.11), with $a = -1$ and $b = 1$. Relation (3.35) is a system of $2n$ equations, and (3.11) a system of n equations. In (3.35) the x_j's as well as the A_j's are regarded as variables, whereas in (3.11), only the A_j's are to be found.

For any $n = 1, 2, \ldots$, this system (3.35) of nonlinear equations has a unique solution satisfying $-1 < x_k < 1$, and $A_k > 0 (1 \le k \le n)$ (Davis and Rabinowitz, 1975, Sec. 2.7). The interpolatory-type numerical quadrature formula

$$\int_{-1}^{1} f(x) \, dx \approx A_1f(x_1) + A_2f(x_2) + \cdots + A_nf(x_n), \qquad (3.36)$$

having the coefficients A_k and interpolating points x_k ($1 \le k \le n$) which satisfy (3.35), is called *Gauss quadrature*.

-------- **EXAMPLE 3.13** --------

For $n = 1$, equations (3.35) have the form

$$A_1x_1^0 = A_1 = 2$$

$$A_1x_1^1 = A_1x_1 = 0,$$

which implies that $A_1 = 2$ and $x_1 = 0$. The one-point Gaussian formula, then, is

$$\int_{-1}^{1} f(x) \, dx = 2f(0).$$

For $n = 2$, equations (3.35) have the form

$$A_1 x_1^0 + A_2 x_2^0 = 2$$

$$A_1 x_1 + A_2 x_2 = 0$$

$$A_1 x_1^2 + A_2 x_2^2 = \tfrac{2}{3} \tag{3.37}$$

$$A_1 x_1^3 + A_2 x_2^3 = 0.$$

From the first equation, $A_2 = 2 - A_1$, and by substituting A_2 into the second equation, we see that

$$A_1 x_1 + (2 - A_1)x_2 = 0,$$

so

$$x_1 = \frac{(A_1 - 2)x_2}{A_1}. \tag{3.38}$$

Then the third equation in (3.37) can be rewritten as

$$A_1 \frac{(A_1 - 2)^2 x_2^2}{A_1^2} + (2 - A_1)x_2^2 = \frac{2}{3}.$$

Solving for x_2^2, we find that

$$x_2^2 = \frac{A_1}{6 - 3A_1}.$$

The fourth equation implies that

$$A_1 \frac{(A_1 - 2)^3 x_2^3}{A_1^3} + (2 - A_1)x_2^3 = 0,$$

which is equivalent to

$$x_2^3(A_1 - 2)[(A_1 - 2)^2 - A_1^2] = 0. \tag{3.39}$$

If x_2 were 0, then from (3.38), we would have $x_1 = 0$. But $x_1 = x_2 = 0$ violates the third equation in (3.37). If $A_1 = 2$, then $A_2 = 0$, and the second and third equations in (3.37) contradict each other. For

$$A_1 x_1 + A_2 x_2 = 2x_1 = 0$$

$$A_1 x_1^2 + A_2 x_2^2 = 2x_1^2 = (2x_1)x_1 = 0 \cdot x_1 = 0 \neq \tfrac{2}{3}.$$

Thus (3.39) implies that

$$(A_1 - 2)^2 - A_1^2 = 0,$$

which has only one solution: $A_1 = 1$. Now we substitute this into earlier equations to find that

$$A_2 = 2 - A_1 = 1$$

$$x_2^2 = \frac{1}{6 - 3} = \frac{1}{3}, \qquad x_2 = \pm\frac{\sqrt{3}}{3},$$

$$x_1 = \frac{1 - 2}{1} x_2 = \mp\frac{\sqrt{3}}{3}.$$

If we assume that $x_1 < x_2$, then

$$x_1 = -\frac{\sqrt{3}}{3} \qquad \text{and} \qquad x_2 = \frac{\sqrt{3}}{3}.$$

Thus we have derived the Gauss quadrature formula for $n = 2$, namely

$$\int_{-1}^{1} f(x)\, dx \approx f\left(-\frac{\sqrt{3}}{3}\right) + f\left(\frac{\sqrt{3}}{3}\right).$$

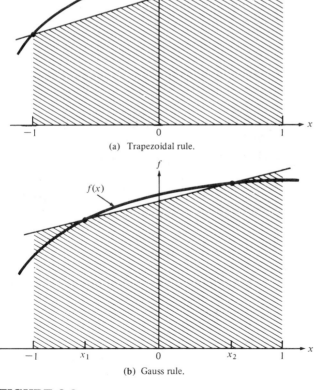

(a) Trapezoidal rule.

(b) Gauss rule.

FIGURE 3.9 Comparison of 2-point Gauss and Trapezoidal Rules

TABLE 3.17 Gauss Quadrature Weights and Coefficients

Number, n, of Points	Gauss Quadrature Weights A_j*			
	A_1	A_2	A_3	A_4
1	2.0000000000000000			
2	1.0000000000000000			
3	0.5555555555555555	0.8888888888888888		
4	0.3478548451374538	0.6521451548625461		
5	0.2369268850561890	0.4786286704993664	0.5688888888888888	
6	0.1713244923791703	0.3607615730481386	0.4679139345726910	
7	0.1294849661688696	0.2797053914892766	0.3818300505051189	0.4179591836734693

Number, n, of Points	Gauss Quadrature Points x_j†			
	x_1	x_2	x_3	x_4
1	0.0000000000000000			
2	-0.5773502691896257			
3	-0.7745966692414833	0.0000000000000000		
4	-0.8611363115940525	-0.3399810435848562		
5	-0.9061798459386639	-0.5384693101056830	0.0000000000000000	
6	-0.9324695142031520	-0.6612093864662645	-0.2386191860831969	
7	-0.9491079123427585	-0.7415311855993944	-0.4058451513773971	0.0000000000000000

*$A_{n+1-j} = A_j$.
†$x_{n+1-j} = -x_j$

In Figure 3.9 we have compared the areas associated with the two-point trapezoidal and Gauss rules. One sees that with its centered data points, the Gauss rule does have some capacity to account for changing slopes.

The points and weights of several Gauss-quadrature formulas are given in Table 3.17. A more extensive table, with values stated to 20 significant decimal places, is given as Table 25.4 of Abramowitz and Stegun (1965).

The computer subroutine GAQU for Gauss quadrature is given in Table 3.18. The Gaussian quadrature weights A_k and quadrature points x_k, such as in Table 3.17, must be supplied.

TABLE 3.18   Subroutine GAQU for Gauss Quadrature

```
        SUBROUTINE GAQU(A,B,N,AA,X,E)
C
C     ***************************************************************
C     *  FUNCTION: THIS SUBROUTINE COMPUTES THE AREA (INTEGRAL)      *
C     *            UNDER THE CURVE F(X) OVER THE INTERVAL [A.B]      *
C     *            USING GAUSSIAN QUADRATURE                         *
C     *  USAGE:                                                      *
C     *      CALL SEQUENCE: CALL GAQU(A,B,N,AA,X,E)                  *
C     *      EXTERNAL FUNCTIONS/SUBROUTINES: FUNCTION F(X)           *
C     *  PARAMETERS:                                                 *
C     *      INPUT:                                                  *
C     *          A=INTERVAL LEFT ENDPOINT                            *
C     *          B=INTERVAL RIGHT ENDPOINT                           *
C     *          N=NUMBER OF INTERPOLATING POINTS                    *
C     *          (NOT GREATER THAN 100)                              *
C     *          AA=1 BY N ARRAY SPECIFIED IN TABLES.  THE ELEMENT   *
C     *          AA(J) IS THE J-TH WEIGHT OF THE GAUSSIAN            *
C     *          QUADRATURE WITH N POINTS                            *
C     *          X=1 BY N ARRAY SPECIFIED IN TABLES.  THE ELEMENT    *
C     *          X(J) IS THE J-TH ABSCISSA OF THE GAUSSIAN           *
C     *          QUADRATURE WITH N POINTS                            *
C     *      NOTE   :IT IS SUFFICIENT TO INPUT ONLY THE FIRST        *
C     *          (N+1)/2 OF AA(J) AND X(J)                           *
C     *      OUTPUT:                                                 *
C     *          E=VALUE OF THE INTEGRAL                             *
C     ***************************************************************
C
        DIMENSION X(50),AA(50)
C     *** INITIALIZATION ***
        E=0.0
        W=2.0/(B-A)
        V=(A+B)/(A-B)
C     *** COMPUTE GAUSSIAN QUADRATURE FACTORS ***
        DO 1 I=1,(N+1)/2
          X1=(X(I)-V)/W
          XX1=(-X(I)-V)/W
          IF(X1.EQ.XX1)THEN
            E=E+F(X1)*AA(I)
          ELSE
            E=E+(F(X1)+F(XX1))*AA(I)
          ENDIF
      1 CONTINUE
        E=E/W
        RETURN
        END
```

━━━━━ **EXAMPLE 3.14** ━━

By way of comparison with earlier quadrature examples, we computed the approximating values of the integral

$$\int_0^{10} 10 \sin (1 - 0.1x) \, dx$$

using Gauss quadrature rules with two, three and five points. Gauss quadrature is at its best when, as in this case, the function is smooth (i.e., f has many derivatives and the magnitudes of these derivatives do not grow quickly with increasing order). From Table 3.19, one sees how phenomenal the performance of Gauss quadrature can be. Since the answer is about 45.969759, with but two points, the approximation is correct to four significant decimal places, and with but three points, the estimate is correct to seven places.

TABLE 3.19   Comparison of Accuracy for Different Quadrature Rules

Number of Points	Quadrature Errors			
	Simpson	Romberg	Newton–Cotes	Gauss
2	*	*	3.9	1.1×10^{-2}
3	1.6×10^{-2}	*	1.6×10^{-2}	-2.4×10^{-5}
5	-1.0×10^{-3}	-2.4×10^{-5}	-2.4×10^{-5}	-1.9×10^{-11}

*There is no n-point rule of this type.

━━━ ■

It can be proven (see Szidarovszky and Yakowitz, 1978, pp. 93–94) that for continuous functions the Gauss quadrature formulas converge to the exact value of the integral as $n \to \infty$. If $f(x)$ is $2n$-times differentiable, the error term of Gauss quadrature is given by

$$\int_{-1}^{1} f(x) \, dx - \sum_{i=1}^{n} A_i f(x_i) = \frac{(n!)^4 2^{2n+1} f^{(2n)}(\eta)}{((2n)!)^3 (2n + 1)}, \qquad (3.40)$$

where $\eta \in (-1, 1)$.

From the preceding discussion and example, we have evidence concerning the impressive power of Gauss quadrature. Without question, when the integrand possesses derivatives of all orders, Gauss quadrature stands supreme in its accuracy for a given number of interpolation points. Moreover, in comparison to Newton–Cotes rules, Gauss rules have reduced sensitivity to smearing (Section 1.4.2) because the coefficients A_k are positive.

A drawback to the Gaussian quadrature approach, however, is that the points and coefficients must be entered or computed. The computation option is not appealing because even for moderate n, sophistication is needed.

*3.7
IMPROPER INTEGRALS

In the preceding sections we have assumed that the interval $[a, b]$ of integration is finite and the integrand remains bounded. In many applications we need to compute integrals for which at least one of these assumptions fails. In such cases the integral is said to be *improper,* and its calculation needs special consideration. We discuss two cases here: (1) The interval is finite but $f(x)$ is unbounded, and (2) the interval is infinite.

CASE 1.   Consider the integral

$$\int_0^1 \frac{\cos (x)}{\sqrt{x}} \, dx,$$

for which the integrand is infinite at $x = 0$. If δ is a small positive number, the integral

$$\int_\delta^1 \frac{\cos (x)}{\sqrt{x}} \, dx$$

gives a good approximation of the original integral. Observe that in the interval $[\delta, 1]$ the integrand is continuous. Consequently, the value of this approximating integral can in principal be computed by the use of the methods discussed earlier in this chapter. We can also bound the error introduced by this truncation:

$$0 < \int_0^1 \frac{\cos (x)}{\sqrt{x}} \, dx - \int_\delta^1 \frac{\cos (x)}{\sqrt{x}} \, dx = \int_0^\delta \frac{\cos (x)}{\sqrt{x}} \, dx < \int_0^\delta \frac{1}{\sqrt{x}} \, dx = [2\sqrt{x}]\big|_0^\delta$$

$$= 2\sqrt{\delta} - 0 = 2\sqrt{\delta}.$$

It happens, however, that at this stage the integration problem is far from solved. For example, if one blithely sets δ to 10^{-3} and applies Simpson's rule s_{20} (which under "normal" circumstances, ought to be pretty accurate), the integral estimate turns out to be 18, and $s_{50} = 8.3$. As we will see below, the correct integral value, to six places, is 1.80905.

The difficulty stems from the fact that whereas the integrand is, in fact, continuous on the interval $[10^{-3}, 1]$, it undergoes a large variation in values near the lower end of the interval. The more erratic the integrand, the less reliable the numerical approximation. For erratic behavior is associated with large fluctuations in derivatives, and error bounds such as in Table 3.12 are similarly large.

There are often clever ways to overcome the effects of integrand singularities, but, of course, inventiveness will be required, and each problem has to be handled on its own terms. For the problem at hand, a substitution of variables is effective. Let $t = \sqrt{x}$. Then $x = t^2$; consequently, $dx = 2t \, dt$, and

$$\int_0^1 \frac{\cos (x)}{\sqrt{x}} \, dx = \int_0^1 \frac{\cos (t^2)}{t} \, 2t \, dt = \int_0^1 2 \cos (t^2) \, dt.$$

Now the integrand $2 \cos (t^2)$ is well behaved in every respect. When Simpson's rule s_{20} is applied to the equivalent integrand over the interval $[0, 1]$, the approximation of the integral is 1.809048, which is correct to six significant places.

For many unbounded integrands, the use of Taylor's series methods can be warmly endorsed as a means of getting rid of singularities such as $1/\sqrt{x}$. In our example we have

$$
\int_0^1 \frac{\cos (x)}{\sqrt{x}} \, dx = \int_0^1 \frac{1}{\sqrt{x}} \left(1 - \frac{x^2}{2!} + \frac{x^4}{4!} - \frac{x^6}{6!} + \cdots \right) dx
$$

$$
= \int_0^1 \left(\frac{1}{\sqrt{x}} - \frac{x^{3/2}}{2!} + \frac{x^{7/2}}{4!} - \frac{x^{11/2}}{6!} + \cdots \right) dx
$$

$$
= \left[2\sqrt{x} - \frac{2}{5} \cdot \frac{x^{5/2}}{2!} + \frac{2}{9} \cdot \frac{x^{9/2}}{4!} - \frac{2}{13} \cdot \frac{x^{13/2}}{6!} + \cdots \right]_0^1
$$

$$
= 2 - \frac{2}{5 \cdot 2!} + \frac{2}{9 \cdot 4!} - \frac{2}{13 \cdot 6!} + \cdots .
$$

The four terms explicitly shown above add to 1.809046, and this Taylor's series is alternating and decreasing in magnitude, so that the error is bounded by the magnitude of the first truncated term, which is $2/(17 \cdot 8!)$, or less than 10^{-4}.

CASE 2.   For integration over infinite domains, a substitution of variable may suffice to transform the domain to a finite region. For example, consider the integral

$$
\int_1^\infty \frac{\cos (1/x^2)}{x^2} \, dx.
$$

By introducing the transformation $t = 1/x$, we obtain the proper integral

$$
\int_0^1 \cos (t^2) \, dt.
$$

We now turn our attention to an approach more in the tradition of numerical analysis than calculus. The interpolatory quadrature approach in Section 3.3 can be extended to infinite domains $[a, b]$ of integration by the artifice of introducing a "weighting" function $w(x)$ which diminishes for large $|x|$ fast enough to assure that for any $k \geq 0$,

$$
\int_a^b |w(x)x^k| \, dx < \infty.
$$

Then following the linear equation representation (3.11), for quadrature points x_j, $1 \leq j \leq n$, one computes quadrature coefficients A_j so that if $f(x)$ is a polynomial of degree less than n, then $\int_a^b w(x)f(x) \, dx$ is integrated exactly. This condition is enforced by choosing the A_j's to satisfy the linear system of equations:

$$A_1 x_1^0 + A_2 x_2^0 + \cdots + A_n x_n^0 = \int_a^b w(x) x^0 \, dx$$

$$A_1 x_1^1 + A_2 x_2^1 + \cdots + A_n x_n^1 = \int_a^b w(x) x^1 \, dx$$

$$\cdot \qquad \cdot \qquad \cdot \qquad \cdot$$

$$A_1 x_1^{n-1} + A_2 x_2^{n-1} + \cdots + A_n x_n^{n-1} = \int_a^b w(x) x^{n-1} \, dx.$$

(3.41)

As in standard interpolatory quadrature (3.41) is subject to numerical instabilities for n larger than 10.

To use (3.41), rewrite the integrand as $f(x) = w(x)g(x)$, and then the quadrature approximant is

$$\int_a^b f(x) \, dx \approx \sum_{k=1}^n A_k g(x_k).$$

EXAMPLE 3.15

By means of the calling program listed in Table 3.20, we constructed a five-point quadrature rule for integrating $\int_0^\infty w(x)f(x) \, dx$, with $w(x) = e^{-x}$. The points x_i in

TABLE 3.20 Interpolatory Quadrature for an Infinite Domain of Integration

```
C      PROGRAM INFDOM
C
C      ****************************************************************
C      THIS PROGRAM DEMONSTRATES THE USE OF INTERPOLATORY
C      QUADRATURE FOR AN INTEGRAL OVER AN INFINITE DOMAIN
C      THE INTEGRAL OF F(X)=EXP(-X)*SIN(X) OVER [0.,INF.) IS
C      COMPUTED BY THE CONSTRUCTION OF A FIVE POINT RULE
C      CALLS:GAUSS
C      OUTPUT: EST=ESTIMATED VALUE OF THE INTEGRAL
C      ****************************************************************
C
       DIMENSION A(5,6),X(5)
       EST=0.
       H=.5
       FACT=1.
C
C      *** COLUMN 6 OF THE ARRAY A CONTAINS THE RIGHT HAND    ***
C      *** SIDE VALUES IN OUR SYSTEM OF EQUATIONS.  THESE     ***
C      *** VALUES AND THE NECESSARY FACTORIALS AS WELL AS OUR ***
C      *** X VALUES ARE COMPUTED HERE                         ***
C
       DO 1 J=1,5
          A(J,6)=FACT
          FACT=FACT*J
          X(J)=(J-1)*H+.1
     1 CONTINUE
C
C      *** VALUES FOR THE LEFT HAND SIDE OF OUR SYSTEM OF     ***
C      *** EQUATIONS ARE COMPUTED                             ***
C
       DO 2 J=1,5
          DO 2 I=1,5
             A(I,J)=X(J)**(I-1)
     2 CONTINUE
```

TABLE 3.20   (Continued)

```
C
C     *** SUBROUTINE GAUSS WILL SOLVE THE SYSTEM OF EQUATIONS***
C     *** THE SOLUTION IS RETURNED IN THE LAST COLUMN OF A    ***
C
      CALL GAUSS(5,1,A,EPS)
C
C     *** SUCCESSIVE SUMMANDS AT EACH QUADRATURE POINT ARE    ***
C     *** COMPUTED AND ADDED TO GET OUR APPROXIMATION         ***
C
      DO 3 J=1,5
        EST=EST+A(J,6)*SIN(X(J))
    3 CONTINUE
      WRITE(10,*) EST
      STOP
      END
```

(3.41) were taken to be $x_i = 0.1 + 0.5i$, and the coefficient vector of the linear equation made use of the fact that $n! = \int_0^\infty e^{-x} x^n \, dx$. The program applies this rule to the integral

$$\int_0^\infty e^{-x} \sin(x) \, dx,$$

whose integral is known by analytic means [Laplace transform of sin (x), evaluated at $s = 1$] to be 0.5. The computed approximation was 0.57. ∎

The Gauss principle (Section 3.6) for optimal location of quadrature points may be extended in an obvious manner to weighted quadrature for improper integrals (Davis and Rabinowitz, 1975, Chap. 3). When this is done for the preceding example problem, the three-point rule gives the estimate 0.496 and for four points, 0.5048.

*3.8
ON-LINE ERROR ESTIMATION AND ADAPTIVE QUADRATURE

*3.8.1   Estimation and Control of Error

In Section 3.4.2 we gave a technique for bounding and controlling integration error. That technique required that a derivative bound be available. Often, such bounds are difficult to find, and in many cases, they lead to pessimistic error estimates. Here we sidestep the derivative estimation problem and get error estimates and control strategies that are often quite accurate, as in an example to follow.

Suppose that one has in mind some specific integration problem of the form

$$I(a, b) = \int_a^b f(x) \, dx. \tag{3.42}$$

Let $I = I(a, b)$ denote the exact value of this integral, and $I(h)$ denote its numerical approximation obtained by some compound integration formula, with h the common lengths of the subintervals. That is, $h = (b - a)/N$, N being the number of "compoundings." Assume that the order of error of the rule is k. From the considerations that led to the error formulas in Table 3.12, one may conclude that, provided $f(x)$ is $k + 1$ times continuously differentiable, for some constant C, the error of the integral approximation is given by

$$I(h) - I = Ch^k + O(h^{k+1}) \tag{3.43}$$

In other words as step size h becomes small, the dominant portion of the error is Ch^k. In this section we explain how the representation (3.43) can be employed to get a good approximation of the error. This can be useful information in itself, and it can also be a vital input to a procedure for error control, so as to obtain a final integral estimate that falls within a prescribed tolerance bound.

The idea we will explore is very closely related to Richardson's extrapolation (Section 3.5.1) and Romberg integration (Section 3.5.2). The latter techniques were motivated by the obvious idea that having a means of estimating what the error is, subtract it from the integral approximation itself, to get a more accurate answer.

Let h_1 and h_2 be two distinct step sizes. Then from (3.43), and ignoring the $O(h^{k+1})$ term, we can estimate C by

$$I(h_1) - I(h_2) \approx C(h_1^k - h_2^k). \tag{3.44}$$

Solve for C, and insert h_1 in place of h in (3.43) to get that the integration error for step size h_1, for example, is

$$\boxed{I(h_1) - I \approx \frac{I(h_1) - I(h_2)}{1 - (h_2/h_1)^k}.} \tag{3.45}$$

─────── **EXAMPLE 3.16** ───────

Here we estimate the error of Simpson's rule with $[a, b]$ divided into four intervals [$M = 4$, in the notation of (3.18)]. In terms of the preceding discussion, we set $h_1 = (b - a)/2$ and $h_2 = (b - a)/4$. The integral studied is our usual one:

$$10 \int_0^{10} \sin (1. - 0.1x) \, dx$$

and the error estimate provided by (3.45), with $k = 4$, in accordance with Table 3.12, was computed to be 0.001005. The exact error of Simpson's rule using h_1 is 0.00103. When you consider that the integral value itself is on the order of 45, you must appreciate that the guess provided by our technique is not bad. ■

An automatic *error control routine* for integration is a program for which the user specifies the integrand $f(x)$, the domain $[a, b]$ of integration, and a positive number ε. The program is expected to provide an estimate $I(h)$ of the exact integral I satisfying

$$|I(h) - I| < \varepsilon.$$

The formulas (3.43) and (3.45) can serve as a basis for construction of an error control routine. For in view of (3.44), one may approximate C by

$$C \approx \frac{I(h_1) - I(h_2)}{h_1^k - h_2^k}.$$

Then from (3.43), and neglecting the $O(h^{k+1})$ term, one can conclude that if h_0 is defined as

$$h_0 = \left| \frac{\varepsilon(h_1^k - h_2^k)}{I(h_1) - I(h_2)} \right|^{1/k}, \tag{3.46}$$

then for any $h < h_0$, the condition

$$|I(h) - I| < \varepsilon$$

should be satisfied.

───── **EXAMPLE 3.17** ─────

Suppose that by means of Simpson's rule we wish to approximate our "usual" integral

$$I(0, 10) = 10 \int_0^{10} \sin(1. - 0.1x) \, dx$$

to five significant decimal places. Then we set $\varepsilon = 5 \times 10^{-6}$. With h_1 and h_2 as in Example 3.16, by (3.46) one calculates h_0 to be 0.66. Since $b - a$ is 10, the smallest even positive integer M for which $(b - a)/M < h_0$ is $M = 16$. Take M in subroutine SIMP (Table 3.10) to be 16 and the actual error, when the calculations are performed on the VAX in double precision, turns out to be 3.9×10^{-6}. ■

The reader should bear in mind that there are some critical assumptions that must be in force before the error estimate (3.45) can be trusted, and in many applications, these conditions are difficult to check. First, the integrand $f(x)$ must be k times continuously differentiable, and second, the step size h must be sufficiently small that the $O(h^{k+1})$ term really is negligible. Davis and Rabinowitz (1975, Chap. 6) give amusing examples of innocent-looking integration problems in which automatic error control routines mistakenly announce very faulty answers as being within a prescribed tolerance region.

*3.8.2   Adaptive Quadrature

We saw in the preceding section that on-line error estimation is possible in numerical integration. This capability can be used in an alternative way to the one sketched above. The alternative has some appeal if the integrand is well behaved in some portions of the domain and fluctuates drastically in others. For the function illustrated in Figure 3.10, for example, one would think it reasonable to select quadrature points more densely in the right-hand portion of the interval than in the left. Methods we have discussed up to this point do not allow for such function-dependent flexibility.

However, in the preceding section we established techniques for obtaining, for any x and positive h:

1. A quadrature estimate $I_x(h)$ of

$$I(x, x + h) = \int_x^{x+h} f(x)\, dx.$$

2. An estimate $E_x(h)$ of the truncation error

$$\left| I(x, x + h) - I_x(h) \right|.$$

With these constructs it is not difficult to "adaptively" allocate quadrature points so that they will typically be dense where the integrand $f(x)$ is erratic and sparse where it is smooth. Moreover, the adaptive technique can meanwhile attempt to keep the total quadrature error within a prescribed bound, which we will designate as TOL (for "tolerance").

The following general idea is representative of a family of rules referred to collectively as *adaptive quadrature*. Assume an integrand $f(x)$; an interval $[a, b]$

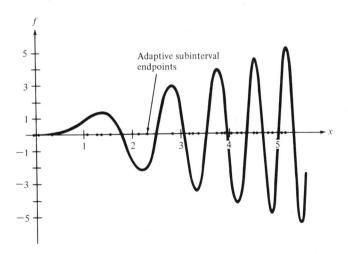

FIGURE 3.10   Integrand and ASIMP Quadrature Points

TABLE 3.21 Subroutine ASIMP for Adaptive Quadrature

```
         SUBROUTINE ASIMP(A,B,M,TOL,SUM)
C
C    *************************************************************
C    *  FUNCTION: THIS SUBROUTINE COMPUTES THE ESTIMATE OF THE   *
C    *            INTEGRAL OF THE FUNCTION F(X) USING THE        *
C    *            ADAPTIVE SIMPSON'S METHOD                      *
C    *  USAGE:                                                   *
C    *       CALL SEQUENCE: CALL ASIMP(A,B,M,TOL,SUM)            *
C    *       EXTERNAL FUNCTIONS/SUBROUTINES:                     *
C    *                      SUBROUTINE SIMP(A,B,M,E)             *
C    *  PARAMETERS:                                              *
C    *       INPUT:                                              *
C    *            A=INTERVAL LEFT ENDPOINT                       *
C    *            B=INTERVAL RIGHT ENDPOINT                      *
C    *            M=MAXIMUM NUMBER OF ITERATIONS                 *
C    *          TOL=TOLERANCE                                    *
C    *       OUTPUT:                                             *
C    *            SUM=ESTIMATE OF THE INTEGRAL                   *
C    *************************************************************
C
C        *** INITIALIZATION ***
         H=.10E-02
         N1=4
         X=A
         N2=8
         SUM=0.0
C        *** COMPUTE SOLUTION ITERATIVELY ***
         DO WHILE(X.LT.B)
            N=N+1
            CALL SIMP(X,X+H,N1,E1)
            CALL SIMP(X,X+H,N2,E2)
            CORR=(16.0*E2-E1)/15.0
            ERR=CORR-E1
C        *** TEST IF THE NUMBER OF ITERATIONS EXCEEDED ***
            IF(N.GT.M) THEN
               WRITE(6,1)
    1          FORMAT(1X,'PROGRAM STOPPED TOO MANY ITERATIONS')
               STOP
            END IF
C        *** TEST STEP SIZE ***
            IF(ABS(ERR).LT.TOL*H) THEN
               X=X+H
               H=3.0*H
               SUM=SUM+E2
               WRITE(3,2) X
    2          FORMAT(33X,1PE14.7)
            ELSE
               H=H/10.0
            END IF
         END DO
         X=X-H/3.0
         SUM=SUM-E2
         H=B-X
         CALL SIMP(X,X+H,N2,E2)
         SUM=SUM+E2
         RETURN
         END
```

of integration; a quadrature rule giving an estimate $I_x(h)$ and error approximation $E_x(h)$, as in constructs 1 and 2 above, and that a positive number TOL is specified. Initialize by setting $x = a$, SUM = 0, and taking some small but positive number for h. Then proceed recursively as follows:

A. Compute $I_x(h)$ and $E_x(h)$.
B. If $E_x(h) \geq h * \text{TOL}/(b - a)$, reduce h and go to step A.
C. Otherwise, set SUM = SUM + $I(x, x + h)$, and $x = x + h$, and if $x < b$, increase the step size h and return to step A.

By this procedure, if $E_x(h)$ really does bound the error, then by summing over accepted quadrature points x,

$$|I(a, b) - \text{SUM}| \leq \sum_x |I(x, x + h) - I_x(h)| \leq \sum_x E_x(h)$$
$$< \sum_x h \, \text{TOL}/(b - a) = \text{TOL}. \tag{3.47}$$

We may anticipate that satisfaction of the condition $E_x(h) < h \, \text{TOL}/(b - a)$ will require h to be relatively small where $f(x)$ is fluctuating dramatically.

In Table 3.21, we present subroutine ASIMP, which applies the Simpson formula (3.18), via subroutine SIMP, to realize the quadrature rule $I_x(h)$. The estimate (3.45), with $h_2 = h_1/2$, is employed as $E_x(h)$. In the case of the Simpson rule, from Table 3.12, we have that $k = 4$. If the condition in step B fails, h is reduced by a factor of 10, and otherwise, h is set to three times its former value.

─────── **EXAMPLE 3.18** ───────

The subroutine ASIMP was used to approximate $\int_a^b x \sin (x^2) \, dx$. TOL was set to 10^{-3} and $a = 0$, $b = 3\sqrt{\pi}$. By substitution of variables, with $t = x^2$, one may confirm that the true value is exactly 1. In the printout (Table 3.22), we have given the values of x at which acceptance (passage to step C in the adaptive quadrature algorithm) occurred. One confirms from this printout that initially the quadrature points are dense and then, toward the end of the integration interval, they again become dense. To begin with, h was set to a relatively small number (10^{-3}) to try to ensure that the h^k term in the truncation error expansion dominates. The algorithm comes to recognize that h is small because the estimated error is far below the tolerance threshold. Thus h is allowed to increase. Later, as the integrand becomes more oscillatory, the subinterval length must and does become smaller. In subinterval [1, 2], where the integrand is smooth and the step size has adaptively been increased, there are only four steps. As the integrand becomes increasingly ragged, the steps get smaller. There are three, eight and nine integration subintervals in [2, 3], [3, 4], and [4, 5], respectively. Figure 3.10 displays the integrand as well as the interval steps accepted by ASIMP.

TABLE 3.22   Partition Points of Adaptive Quadrature Subintervals

Accepted x Values
1.0000000E-03
4.0000002E-03
1.3000001E-02
4.0000003E-02
1.2100001E-01
3.6400005E-01
1.0930002E+00
1.3117002E+00
1.3773103E+00
1.5741403E+00
2.1646304E+00
2.3417773E+00
2.8732183E+00
3.0326507E+00
3.0804803E+00
3.2239695E+00
3.2670162E+00
3.3961563E+00
3.7835770E+00
3.8998032E+00
3.9346709E+00
4.0392747E+00
4.0706558E+00
4.1647992E+00
4.1930423E+00
4.2777710E+00
4.5319576E+00
4.6082134E+00
4.8369813E+00
4.9056115E+00
5.1115026E+00
5.1732702E+00
5.3585725E+00
Computed value of the integral = 1.0000324

The adaptive quadrature subroutine ASIMP is intended for illustrative purposes only. It lacks the efficiency and truncation-error sensing safeguards of professional library-quality programs.

The design of adaptive quadrature algorithms is as much an art as it is a science. Because of their heuristics and intricacies, their analysis in terms of precise convergence statements is difficult. For every such adaptive algorithm, there exists some perverse integrand that misleads it and forces the algorithm to be satisfied with a highly erroneous estimate. Nevertheless, for many applications, these adaptive techniques are quite satisfactory, and their use is becoming increasingly popular. The IMSL* library has an adaptive quadrature routine, DCADRE, which is built around the Romberg integration technique.

3.9
SUPPLEMENTARY NOTES AND DISCUSSIONS

We saw that standard numerical differentiation formulas and their error bounds can be derived directly from either the Taylor's or interpolation polynomial ap-

---

*IMSL is the acronym of the International Mathematical and Statistical Libraries, Inc., NBC Building, 7500 Bellaire Boulevard, Houston, TX 77031.

proaches. It was noted that numerical differentiation tended to be unstable in the sense that as the step size h between successive domain points decreases, error resulting from roundoff and inexact measurement can become overwhelmingly large. Such a phenomenon does not occur for any compound rule, or Romberg or Gauss quadrature or any other interpolatory system for which the weights are all nonnegative.

The popular quadrature formulas were derived by integrating interpolation polynomials or piecewise polynomials. The most popular compound formulas, the trapezoidal, midpoint, and Simpson rules, were described and their error formulas were given. At this point we caution the reader that these particular popular formulas are holdovers from the age of hand calculation. They are simple to understand and easy to implement. But if the integrand is "smooth," one can obtain a more accurate estimate of the integral by using Gaussian or Romberg quadrature. We noted a danger arising from high-order Newton–Cotes rules: The linear equations for the quadrature weights tend to be sensitive to roundoff error. There are further drawbacks we did not mention.

Heuristically speaking, for integrands that possess several derivatives (we need not calculate what they are), much more information is contained in the data points (x_1, f_1), (x_2, f_2), . . ., (x_n, f_n) than is extracted by the popular compound formulas. Romberg's method does effectively achieve a high order of accuracy. The mechanism for this success was explained through the principle of Richardson's extrapolation.

If we are constrained to use an n-point quadrature rule, but the quadrature points x_1, . . ., x_n can be placed at our discretion, then from a certain reasonable point of view, Gauss quadrature constitutes the best possible integration method. Our derivation of quadrature domain points and weights is, pedagogically, as simple as any we know of. But from a numerical standpoint, more sophisticated techniques (some of which are described in Szidarovszky and Yakowitz, 1978, Sec. 3.2.3) are available.

In Table 3.23, which is meant to be suggestive rather than canonical, we have tried to indicate the appropriate domains for various techniques studied in this chapter.

TABLE 3.23 Match of Integration Problem Characteristics and Methods

Integrand Characteristic	Method*
Few Data Points (< 10)	
Points evenly spaced, accuracy not as critical, as simplicity, or no computer available	M, S, T
Points equally spaced, accuracy critical	RO, NC
Points may be chosen at discretion of user	G
Points irregularly spaced but fixed	I
Many Data Points (≥ 10)	
Simplicity above all	S, T, M
Accuracy important	High order G, RO

*G, Gauss quadrature; I, interpolatory rule; M, midpoint rule; NC, Newton–Cotes; RO, Romberg integration; S, Simpson's rule; T, trapezoidal formula.

PROBLEMS

Section 3.1

1. Find the order q [in $O(h^q)$] of the following functions at the points x_0 indicated:
   (a) $\cos(x) - e^x$ at $x_0 = 0$.
   (b) $x^2 [\cos(x) - e^x]$ at $x_0 = 0$.
   (c) $\cos(x)$ at $x_0 = \pi/2$.
   (**HINT:** Taylor's expansions about x_0 provide a systematic approach.)

Section 3.2

2. Apply the three- and four-point formulas of Table 3.2 to estimate the derivatives of e^x, $\sin(x)$, $x^3 - x^2 + x - 1$, and $\log_e(1 + x)$ at $x_0 = 0$. Use step sizes $h = 10^{-j}$, $j = 1, 2, \ldots, 10$. Make a table comparing the actual errors, as a function of h, of the two formulas. Indicate the ranges in which truncation error of the derivative formulas is greater than roundoff.

3. Derive the stated order of error (i.e., $q = 3$) of the four-point rule for $f'(x)$ of Table 3.2.

4. Derive the three-point formula

$$f'(x_0) = \frac{f(x_0 + h) - f(x_0 - h)}{2h},$$

by the Lagrange interpolation polynomial approach. [**HINT:** Take $x_0 = 0$, for convenience, and use interpolation points $-h$, 0, and h, to obtain the polynomial

$$p(x) = f(-h)l_1(x) + f(0)l_2(x) + f(h)l_3(x).$$

Now differentiate this polynomial and evaluate at $x_0 = 0$.] If you are an advanced or precocious student, try to derive the order of error.

*5. Derive the formula

$$f'(x_0) = \frac{4f(x_0 + h) - 3f(x_0) - f(x_0 + 2h)}{2h} + O(h^2).$$

[**HINT:** Use the Taylor's polynomial representation of the terms $f(x_0 + h)$, $f(x_0 + 2h)$.]

6. Using the values of the function $f(x) = e^x - \sin(x)$ at points $x_i = 0.01(i - 1)$ ($i = 1, \ldots, 101$), compute $f'(x)$ and $f''(x)$ at the points $x_2, \ldots, x_{100}$ by (3.3) and (3.4). Compare the results and the true values.

Section 3.3

7. Find the interpolatory rule for integrating over $[1, 2]$ using the quadrature points 1.0 and 1.5. Apply it to integrating $f(x) = x^3$. Show your work.

8. Construct the Newton–Cotes quadrature formula for the interval $[0, 1]$ and nodes $x_1 = 0$, $x_2 = \frac{1}{4}$, $x_3 = \frac{1}{2}$, $x_4 = \frac{3}{4}$, $x_5 = 1$.
   [**HINT:** Use subroutine COEFF, Table 3.6.]

9. Apply the integral formula obtained in Problem 8 to the integral given in Example 3.5. Also evaluate

$$\int_0^1 e^x \, dx, \qquad \int_0^1 \sin{(3x)} \, dx, \qquad \text{and} \qquad \int_{-1}^2 \frac{1}{x^2 + 1} \, dx.$$

   Find the errors of your approximations.

Section 3.4

10. Consider the quadrature formula

$$\int_0^1 f(x) \, dx \approx \tfrac{1}{4}f(0.2) + \tfrac{1}{2}f(0.5) + \tfrac{1}{4}f(0.8).$$

   (a) What is the degree of the lowest-degree polynomial *not* integrated exactly by this rule?
   (b) Is this rule an interpolatory-type quadrature formula?
   (c) In the sense of (3.13), find the weights of this rule so that it is a quadrature formula for $\int_1^3 f(t) \, dt$.

*11. Assume that the value of the integral

$$\int_0^1 e^{x^2} \, dx$$

   has to be determined to six significant decimals. Determine the necessary number of subintervals required by Simpson's rule.

12. Approximate the following integrals by s_2, t_2, s_4, t_4, s_{10}, and t_{10}.
    (a) $\int_0^\pi \sin{(x)} \, dx$.
    (b) $\int_0^\pi \sin{(2x)} \, dx$.
    (c) $\int_0^\pi \sin{(5x)} \, dx$.
    (d) $\int_0^\pi \sin{(10x)} \, dx$.
    Make a table of actual approximation errors.

13. Approximate the integral

$$\int_0^\pi \sin{(5x)} \, dx$$

   as accurately as possible (since you know what it should be) by Simpson's and the trapezoidal rules. In each case, adjust the number of quadrature points to minimize the actual error. Which rule yields the more accurate approximation? Explain why.

*14. Davis and Rabinowitz (1975) use the term "overcomputation" to refer to the situation in which one has so many quadrature points that because of roundoff error, the estimate is less accurate than it would be had fewer points been used.

The following program, run on a PDP11, and printout provide an illustration of this phenomenon.

```
A = 0.0
B = 2.0
TRUE = EXP(2.0) − 1.0
DO 1 K = 4,20
    N = 2**K
    CALL TRAP(A,B,N,EST)
    ERROR = TRUE − EST
    TYPE *,ERROR,EST,N/2
1  CONTINUE
STOP
END

FUNCTION F(X)
F = EXP(X)
RETURN
END
```

Actual Quadrature Errors	Estimate	Number, N, of Points
−8,3174706E−03	6.397374	16
−2.0799637E−03	6.391136	32
−5.1975250E−04	6.389576	64
−1.2922287E−04	6.389185	128
−3.0994415E−05	6.389087	256
−8.1062317E−06	6.389064	512
−3.8146973E−06	6.389060	1024
−2.3841858E−06	6.389059	2048
−9.5367432E−07	6.389057	4096
0.0000000	6.389056	8192
6.6757202E−06	6.389050	16384
−1.4305115E−06	6.389058	32768
4.2915344E−06	6.389052	65536
1.7166138E−05	6.389039	131072
3.1948090E−05	6.389024	262144
8.4400177E−05	6.388972	524288

Exhibit this effect on your computer. Run the program and then use a different compound rule. Finally, use a different integrand. Explain whether you would anticipate the effect to be more or less pronounced if the basic compound rule is of higher order than the trapezoidal rule used here.

Section 3.5

*15. For thrice-differentiable integrands $f(x)$, in the absence of roundoff effects, if $I = \int_0^b f(x)\,dx$ is the integral value, then

$$I - t_N = C\left(\frac{1}{N}\right)^2 + O\left(\left(\frac{1}{N}\right)^4\right),$$

where C is a constant that does not depend on the number N of trapezoidal quadrature points. The relation above implies that for Romberg array elements $T_{l,k}$,

$$r_k = \frac{I - T_{0,k+1}}{I - T_{0,k}} \approx \frac{1}{4}.$$

This, in turn, can be used to adaptively stop calculations in construction of the Romberg array. For example, one can hope that ratio estimate r_k is valid for $k \geq 5$, say, and so if the array $\{T_{l,k}\}$ is computed one column at a time, as in subroutine ROMB, one can stop whenever

$$|T_{k+1,k+1} - T_{0,k+1}| > \tfrac{1}{2}|T_{k+1,k+1} - T_{0,k}|$$

with the idea in mind that roundoff error in evaluating $f(x)$ is becoming a limiting factor. Try this idea out on each of the following expressions.

(a) $\int_0^1 \sin(x)\, dx$.

(b) $\int_0^1 \sin(10x)\, dx$.

(c) $\int_0^{1/2} \sqrt{x}\, dx$.

Use single and double precision, and see how well the stopping rule finds the point at which further refinement is futile.

16. Construct a compound rule from the basic rule (Problem 10)

$$\int_0^1 f(x)\, dx = \tfrac{1}{4}f(0.2) + \tfrac{1}{2}f(0.5) + \tfrac{1}{4}f(0.8).$$

*17. Construct a Richardson extrapolation correction for the compound rule obtained in Problem 16.

Section 3.6

18. We have remarked that Gauss quadrature is, in fact, an interpolatory-type rule. For $n = 3$ and 4, calculate the weights A_j for Gaussian quadrature from knowledge of the quadrature points x_j (Table 3.17) and use of subroutine COEFF. Verify that these computed weights are those of Table 3.17.

19. Compare the errors associated with the six-point Gauss quadrature, trapezoidal, and Simpson's rules on the integrals in Problem 12. Also compare on $\int_0^1 g(x)\, dx$, where $g(x)$ is the sawtooth function, $g(x) = (-1)^{[10x]}(1 - 2(10x - [10x]))$. As usual, $[\cdot]$ is the "integer part of" operator.

20. In view of our findings in Section 2.4.4 that Chebyshev points [defined in (2.17)] provide more accurate interpolation polynomial approximation than do equally spaced points, an appealing idea is to choose Chebyshev points as quadrature points. It turns out that this is indeed a sensible path, as we avoid the pain of finding the optimal (Gauss) quadrature points, since Cheybshev points are simple to compute, yet it is indeed true that Cheybyshev-point quadrature tends to be more accurate than Newton–Cotes rules of like order. Furthermore, unlike Newton–Cotes, Chebyshev-point interpolatory quadrature is guaranteed to have vanishingly small truncation error as the number of points increases. By use of subroutines COEFF and GAQU, compare the accuracy of Newton–Cotes, Chebyshev-interpolatory, and Gauss quadrature of orders $n = 3, 4, 5$, and 6 on the following integrals.

(a) $\int_0^\pi \sin(x) \, dx$.

(b) $\int_{-1}^1 e^x \, dx$.

(c) $\int_0^1 \dfrac{1}{x^2 + 25} \, dx$.

(d) $\int_0^\pi \sin(10x) \, dx$.

21. Prove that the weights A_j in Gauss quadrature are always strictly positive. (**HINT:** If $A_1 \leq 0$, consider the polynomial $p(x) = [\Pi_{j=2}^n (x - x_j)]^2$.)

*Section 3.7

22. Compute the values of the following improper integrals to three significant digits:

(a) $\int_0^\infty e^{-x} \sin x^2 \, dx$.

(b) $\displaystyle \int_1^\infty \frac{2x \, dx}{1 + x^4}$

CHAPTER 4

The Solution of Simultaneous Linear Equations

4.1 PRELIMINARIES

Simultaneous linear equations are equations of the form

$$
\begin{aligned}
a_{11}x_1 + a_{12}x_2 + \cdots + a_{1n}x_n &= b_1 \\
a_{21}x_1 + a_{22}x_2 + \cdots + a_{2n}x_n &= b_2 \\
\vdots \qquad \vdots \qquad\qquad \vdots \quad\;\; \vdots & \\
a_{m1}x_1 + a_{m2}x_2 + \cdots + a_{mn}x_n &= b_m.
\end{aligned}
\tag{4.1}
$$

The coefficients a_{ij} and b_i are given real or complex numbers. Equation (4.1) is a system of m equations with n variables, or "unknowns," $x_1, x_2, \ldots, x_n$. Our job is to determine their values.

We have already seen that many numerical techniques require, as a key step, solving simultaneous linear equations. For example, linear equations were encountered in the construction (2.20) of natural cubic splines, the determination of weights in numerical differentiation through (3.5) and interpolatory-type quadrature (3.11). We will find in later chapters that the solution of systems of nonlinear equations, the method of least squares for function approximation, and popular methods for solving certain types of differential equations also entail simultaneous linear equation problems. Many engineering and economic applications, such as current response to fixed-frequency voltage sources in electric networks, heat transfer in static thermal models, and equilibrium problems of economic sectors involve linear equations. More can be said: Linear equations are fundamental to virtually every field of numerical computation and, as a consequence, to every scientific and engineering discipline, even (and sometimes, especially) those which on the surface seem most remote from "linearity." In brief, it is virtually impossible to overestimate the value of mastery of the theoretical and computational aspects of linear algebra.

Equation (4.1) can be written compactly in matrix form as

$$\mathbf{Ax} = \mathbf{b}, \tag{4.2}$$

where $\mathbf{A} = (a_{ij})$ is an $m \times n$ matrix of coefficients, and $\mathbf{b} = (b_i)$ and $\mathbf{x} = (x_j)$ are column vectors of dimension m and n, respectively. The usual case of interest is when m, in (4.1), equals n. In this standard case, the *order* of the linear system is n. Appendix A supplies the elementary definitions and results of matrix theory needed for understanding the material of this chapter.

Some familiar methods for solving "textbook-sized" linear equations by hand calculations are totally inadequate for higher-order systems. For example, the solution of a 20-variable ($m = n = 20$) system by the popular Cramer's rule requires about 5×10^{19} multiplications, and such an effort would require on the order of 1 million years by a modern computer. Yet people routinely solve systems having several hundred variables by techniques described in this chapter.

In many cases, "naive" codes are sufficient for solving fairly large ($n > 100$) order systems of linear equations. But, solution of some important linear equations is subject to ill-conditioning effects. By way of illustration, we saw in Example 3.10 that linear equations for weights in Newton–Cotes quadrature characteristically are unstable, and this instability leads to deterioration of accuracy for orders n as low as 20, even with double-precision number representation. Thus in the course of this chapter, we are obliged to cast an eye toward the effects of roundoff error and partial remedies of these effects.

4.2
GAUSSIAN ELIMINATION

4.2.1 An Example of Elimination

The most popular linear equation methods are the different variants of the Gaussian elimination algorithm. By "elimination" here we refer to a procedure for adding multiples of one equation in (4.1) to other equations so as to set the coefficients of one of the variables, say x_j, in these other equations to zero. This is repeated until the resulting system of equations is characterized by having its coefficient matrix in upper triangular form. That is (Appendix A, property 3), all the elements a_{ij}, $i > j$, are zero. When the linear system is in this form, it is very simple to solve for the unknowns x_j.

Because it is a purely arithmetic process, Gaussian elimination has the interesting feature that in contrast to other major computational methods we have discussed, there is no truncation error. Limitations on accuracy are due solely to roundoff error. The mathematical formulas of Gaussian elimination are given in the next section. But this technique is one of those procedures most readily understood by example, and with this conviction in mind, let us examine such an example.

───── **EXAMPLE 4.1** ─────────────────────────

We now solve the system of equations

$$\begin{aligned}
2x_1 + x_2 + x_3 &= 4 \\
x_1 + 3x_2 + 2x_3 &= 6 \\
x_1 + 2x_2 + 2x_3 &= 5.
\end{aligned} \tag{4.3}$$

By subtracting the multiple $\frac{1}{2}$ of the first equation from the second and third equations, the first "derived system" is obtained:

$$2x_1 + x_2 + x_3 = 4$$
$$\tfrac{5}{2}x_2 + \tfrac{3}{2}x_3 = 4$$
$$\tfrac{3}{2}x_2 + \tfrac{3}{2}x_3 = 3.$$

Let us look in detail at the computation of the second or middle equation above. The first equation, after being multiplied by $\frac{1}{2}$, becomes

$$x_1 + \tfrac{1}{2}x_2 + \tfrac{1}{2}x_3 = 2.$$

Subtract this from the middle equation in (4.3) to see that the result is

$$(1 - 1)x_1 + (3 - \tfrac{1}{2})x_2 + (2 - \tfrac{1}{2})x_3 = (6 - 2).$$

After performing the arithmetic indicated by the parenthetical expressions above, we have the second equation of the first derived system. The same rationale yields the third equation in the first derived system.

Continue the computation by subtracting the multiple $\frac{3}{5}$ of the second equation of the first derived system from the third equation to obtain the second derived system:

$$2x_1 + x_2 + x_3 = 4$$
$$\tfrac{5}{2}x_2 + \tfrac{3}{2}x_3 = 4$$
$$\tfrac{3}{5}x_3 = \tfrac{3}{5}.$$

The derivation of this system of equations having an upper triangular matrix of coefficients is called the *forward elimination process*.

From the third equation of the system above, we immediately calculate

$$x_3 = \frac{3/5}{3/5} = 1.$$

Then the second equation, with 1 substituted for x_3, implies that

$$x_2 = \frac{4 - 3/2 \times 1}{5/2} = 1.$$

Finally, from the first equation, with their numerical values replacing x_2 and x_3, we see that $x_1 = 1$, and thereby obtain the solution which is

$$x_1 = x_2 = x_3 = 1.$$

The computation of the unknowns from the upper triangular system is known as *back substitution*.

4.2.2   Naive Gaussian Elimination

We now formalize the elimination and back-substitution algorithms illustrated in the preceding example.

Let us assume that in equations (4.1) the matrix of coefficients is nonsingular. That is, $m = n$, and a solution exists and is unique. Then (4.1) can be rewritten as

$$a_{11}x_1 + a_{12}x_2 + \cdots + a_{1n}x_n = a_{1,n+1}$$

$$a_{21}x_1 + a_{22}x_2 + \cdots + a_{2n}x_n = a_{2,n+1}$$

$$\cdots$$

$$a_{n1}x_1 + a_{n2}x_2 + \cdots + a_{nn}x_n = a_{n,n+1},$$

(4.4)

where, for later notational convenience, we have defined $a_{i,n+1} = b_i (1 \le i \le n)$. Assume that $a_{11} \ne 0$, and subtract the multiple a_{i1}/a_{11} of the first equation from the ith equation for $i = 2, \ldots, n$. The coefficient of x_1 in the ith equation then becomes 0, and we thereby obtain the first derived system, which has the form

$$a_{11}x_1 + a_{12}x_2 + \cdots + a_{1n}x_n = a_{1,n+1}$$

$$a_{22}^{(1)}x_2 + \cdots + a_{2n}^{(1)}x_n = a_{2,n+1}^{(1)}$$

$$\vdots \qquad\qquad \vdots$$

$$a_{n2}^{(1)}x_2 + \cdots + a_{nn}^{(1)}x_n = a_{n,n+1}^{(1)},$$

(4.5)

where

$$a_{ij}^{(1)} = a_{ij} - \frac{a_{i1}}{a_{11}} a_{1j} \qquad (2 \le i \le n, \ 2 \le j \le n + 1).$$

(4.6)

There is no loss of generality in making the assumption that $a_{11} \ne 0$, since in any nonsingular matrix **A** there is at least one nonzero entry in the first column. If $a_{11} = 0$ but for index i, $a_{i1} \ne 0$, then the first and ith equations may be interchanged.

One can see intuitively that if $x_1, \ldots, x_n$ is the solution to (4.1) and some multiple of the first ($i = 1$) equation is added to the jth equation, then $x_1, \ldots, x_n$ is also the solution of the new system. More justification is offered in Appendix A, property 23.

Note that in the first derived system, the variable x_1 has been "eliminated" in all but the first equation of (4.4). The matrix of coefficients of the first derived system (4.5) is nonsingular, since row operations such as interchanging rows or adding a multiple of one row to another, as in (4.5), cannot make the matrix singular (Appendix A, property 23).

Assume next that $a_{22}^{(1)} \ne 0$, and subtract the multiple $a_{i2}^{(1)}/a_{22}^{(1)}$ of the second equation of (4.5) from the ith equation ($i = 3, \ldots, n$). We thereby obtain the second derived system:

$$a_{11}x_1 + a_{12}x_2 + a_{13}x_3 + \cdots + a_{1n}x_n = a_{1,n+1}$$

$$a_{22}^{(1)}x_2 + a_{23}^{(1)}x_3 + \cdots + a_{2n}^{(1)}x_n = a_{2,n+1}^{(1)}$$

$$a_{33}^{(2)}x_3 + \cdots + a_{3n}^{(2)}x_n = a_{3,n+1}^{(2)} \qquad (4.7)$$

$$\vdots \qquad\qquad \vdots \qquad \vdots$$

$$a_{n3}^{(2)}x_3 + \cdots + a_{nn}^{(2)}x_n = a_{n,n+1}^{(2)}.$$

By repeating this process until the $(n - 1)$st derived system has been constructed, we obtain

$$a_{11}x_1 + a_{12}x_2 + a_{13}x_3 + \cdots + a_{1n}x_n = a_{1,n+1}$$

$$a_{22}^{(1)}x_2 + a_{23}^{(1)}x_3 + \cdots + a_{2n}^{(1)}x_n = a_{2,n+1}^{(1)}$$

$$a_{33}^{(2)}x_3 + \cdots + a_{3n}^{(2)}x_n = a_{3,n+1}^{(2)} \qquad (4.8)$$

$$\vdots \qquad \vdots$$

$$a_{nn}^{(n-1)}x_n = a_{n,n+1}^{(n-1)},$$

where the relation for obtaining the coefficients of the kth derived system from the coefficients of the preceding system has the general form

$$\boxed{a_{ij}^{(k)} = a_{ij}^{(k-1)} - \frac{a_{ik}^{(k-1)}}{a_{kk}^{(k-1)}} a_{kj}^{(k-1)} \qquad (i = k + 1, \ldots, n; \\ j = k + 1, \ldots, n + 1).} \qquad (4.9)$$

In (4.9) k ranges from 1 to $n - 1$. The process is started by assigning

$$\boxed{a_{ij}^{(0)} = a_{ij} \qquad (i = 1, \ldots, n; j = 1, \ldots, n + 1).}$$

By inspection of (4.8), we see that the coefficient matrix of the $(n - 1)$st derived system is in upper triangular form. The remaining step in solving this system is easy. The value of x_n can be obtained from the final equation of (4.8) since nonsingularity of **A** implies that necessarily $a_{nn}^{(n-1)} \neq 0$. Specifically,

$$\boxed{x_n = \frac{a_{n,n+1}^{(n-1)}}{a_{nn}^{(n-1)}},}$$

and the (backward) recursive formula for obtaining the values of the unknowns x_k in terms of the previously calculated values $x_j (j > k)$ is

$$\boxed{x_k = \frac{1}{a_{kk}^{(k-1)}}\left[a_{k,n+1}^{(k-1)} - \sum_{j=k+1}^{n} a_{kj}^{(k-1)}x_j\right] \qquad (k = n - 1, \ldots, 1).} \qquad (4.10)$$

This relation is obtained from the kth equation of system (4.8). Formula (4.10) is called *back substitution*. Process (4.9) is referred to as *forward elimination*. The entire process of forward elimination and back substitution is called *Gaussian elimination*. The adjective "naive" is sometimes applied to this procedure since, as will be seen in our discussion of pivoting, modifications are often advised for improving accuracy in the face of roundoff error. The subroutine for Gaussian elimination, which is deferred to the next section, will employ a pivoting strategy.

───── **EXAMPLE 4.2** ───

The coefficients of the successive derived systems associated with Example 4.1 are given in Table 4.1.

TABLE 4.1   Derived Systems in Forward Elimination

Original System	2	1	1	4
	1	3	2	6
	1	2	2	5
First Derived System	2	1	1	4
		$\frac{5}{2}$	$\frac{3}{2}$	4
		$\frac{3}{2}$	$\frac{3}{2}$	3
Second Derived System	2	1	1	4
		$\frac{5}{2}$	$\frac{3}{2}$	4
			$\frac{3}{5}$	$\frac{3}{5}$

■

4.2.3   Gaussian Elimination with Pivoting

Because of the effect of propagated rounding errors, naive Gaussian elimination is sometimes unsatisfactory. This fact can be illustrated by the following example.

───── **EXAMPLE 4.3** ───

Assume that in the absence of roundoff error, the last two equations of the $(n - 2)$nd derived system are

$$0x_{n-1} + x_n = 1$$

$$2x_{n-1} + x_n = 3,$$

where the zero coefficient is obtained from previous calculations. The solution is $x_n = x_{n-1} = 1$. As we have seen in Chapter 1, resultants of arithmetic operations usually have roundoff errors: Therefore, the computed derived system of equations is actually

$$\varepsilon x_{n-1} + x_n = 1$$

$$2x_{n-1} + x_n = 3.$$

Here ε is a number having small magnitude, and the rounding errors of the other coefficients have been neglected. Since $\varepsilon \neq 0$, in the computation of the $(n-1)$st derived system, the naive elimination process is continued with the nonzero element ε, and the last two equations of the final derived system are

$$\varepsilon x_{n-1} + x_n = 1$$

$$\left(1 - \frac{2}{\varepsilon}\right)x_n = 3 - \frac{2}{\varepsilon}.$$

Back substitution applied to these equations results in the values

$$x_n = \frac{3 - 2/\varepsilon}{1 - 2/\varepsilon}$$

$$x_{n-1} = \frac{1 - x_n}{\varepsilon}.$$

For values of ε near zero, the values $3 - 2/\varepsilon$ and $1 - 2/\varepsilon$ are both very close to $-2/\varepsilon$, and the value of x_n is very close to 1, which is the correct value. But the computed solution $x_{n-1} = (1 - x_n)/\varepsilon$ will be nonsense on two counts: First, the numerator $1 - x_n$ will suffer from subtractive cancellation (Section 1.4.2). Second, and worse, the denominator ε, which arises purely from roundoff error, is an arbitrary number unrelated to the problem. The method of back substitution (4.10) implies that the values of all the other unknowns $x_{n-2}, \ldots x_1$ obtained by the use of the erroneous value of x_{n-1} are also suspect. ∎

The difficulty in the system discussed above is not due simply to ε being small but rather to its being small relative to other coefficients in the same column. This difficulty can be remedied if we choose as the kth pivot element the coordinate $a_{i^*j^*}^{(k-1)}$ having the largest magnitude among all $a_{ij}^{(k-1)}$, $k \le i \le n$, $k \le j \le n$. The element $a_{i^*j^*}^{(k-1)}$ is then put in the diagonal position by interchanging rows i^* and k and columns j^* and k. That is, equation i^* and k, and unknowns x_{j^*} and x_k are interchanged. This operation is called *pivoting for maximal size* or, more simply, *maximal pivoting* (see Figure 4.1). We need to find the maximum of $(n - k + 1)^2$ numbers before computing the kth derived system.

At the expense of some increase in roundoff error propagation, a popular alternative procedure is to perform *partial pivoting,* where the maximal element is chosen from only the kth column. That is, we choose the kth pivot element, to be any coordinate $a_{i^*k}^{(k-1)}$ that maximizes $|a_{ik}^{(k-1)}|$, $i \ge k$. Then the kth and i^*th rows are interchanged to put the pivot element in the diagonal position. Figure 4.1 illustrates the partial and maximal pivoting strategies.

The partial pivoting strategy in this present form has a theoretical weakness, since by rescaling the equations of the derived systems (i.e., by multiplying them by sufficiently large scalars), any nonzero element could be made to satisfy the pivot condition. A heuristic procedure known as *equilibration* is sometimes used to sidestep this problem. In equilibration, at each step of elimination, before choosing the pivot element, we scale each of the remaining ($i \ge k$) rows by scalars

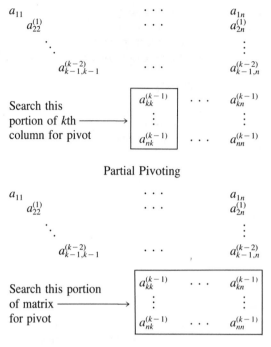

FIGURE 4.1 Partial and Maximal Pivoting

α_i so that, for example,

$$\alpha_i(|a_{i,k}| + |a_{i,k+1}| + \cdots + |a_{in}|) = 1.$$

Although this is sometimes found to be useful, there exist cases when equilibration makes roundoff effects even worse.

EXAMPLE 4.4

Let us examine what happens in the situation of Example 4.3 under the partial pivoting regime. In constructing the $(n - 1)$st derived system, we examine the last two rows of the $(n - 1)$st column,

$$\varepsilon x_{n-1} + x_n = 1$$

$$②x_{n-1} + x_n = 3$$

and find that the encircled term has the largest magnitude. Then this encircled term is selected as the pivot, and the equations are rearranged as

$$2x_{n-1} + x_n = 3$$

$$\varepsilon x_{n-1} + x_n = 1.$$

Now proceeding with elimination, we calculate,

$$2x_{n-1} + x_n = 3$$

$$\left(1 - \frac{\varepsilon}{2}\right)x_n = 1 - \frac{3\varepsilon}{2}.$$

Now for small ε, $x_n \approx 1.0$, and

$$x_{n-1} = \frac{3 - x_n}{2} \approx 1.0,$$

and our answer is accurate. ∎

Subroutine GAUSS, which performs Gaussian elimination with partial pivoting, is listed in Table 4.2. The calling parameter M represents the number of linear systems with the same coefficient matrix to be solved and should not be confused with m in (4.1). If, as is the usual case, only one linear system of equations is to be solved, as in (4.1), then $M = 1$. The calling parameter DELT is the positive threshold below which the pivot element is regarded as zero and the system declared singular. Machine epsilon as described in Section 1.4.1 is probably a sensible value for DELT.

The structure of subroutine GAUSS is as follows. The DO loop with label 1 computes the derived systems 1 through $N - 1$, and the DO loop with label 6 performs back substitution. The DO loop with label 2 finds the row index IN and value U of the element of the largest magnitude among A(K,K), . . ., A(N,K), and DO the loop with label 3 interchanges rows K and IN.

**TABLE 4.2   Subroutine GAUSS for Gauss Elimination
with Partial Pivoting**

```
      SUBROUTINE GAUSS(N,M,A,DELT)
C
C     ***************************************************************
C     *  FUNCTION: THIS SUBROUTINE COMPUTES THE SOLUTIONS FOR M    *
C     *            SYSTEMS WITH N EQUATIONS AND N UNKNOWNS USING   *
C     *            GAUSSIAN ELIMINATION                            *
C     *  USAGE:                                                    *
C     *      CALL SEQUENCE: CALL GAUSS(N,M,A,DELT)                 *
C     *  PARAMETERS:                                               *
C     *      INPUT:                                                *
C     *          N=NUMBER OF EQUATIONS AND UNKNOWNS                *
C     *          M=NUMBER  OF SYSTEMS (RIGHT HAND SIDE VECTORS)    *
C     *          A=N BY M+N ARRAY OF COEFFICIENTS AUGMENTED        *
C     *            WITH EACH RIGHT SIDE VECTOR                     *
C     *          DELT=MACHINE ZERO (TOLERANCE)                     *
C     *      OUTPUT:                                               *
C     *          A(1,N+J),...,A(N,N+J)                             *
C     *            =SOLUTION OF THE J-TH SYSTEM (J=1,...,M)        *
C     ***************************************************************
C
      DIMENSION A(N,M+N)
      IF(N.GT.1) THEN
        DO 1 K=1,N-1
          U=ABS(A(K,K))
          KK=K+1
          IN=K
```

TABLE 4.2   (Continued)

```
C     *** SEARCH FOR INDEX IN OF MAXIMUM PIVOT VALUE ***
            DO 2 I=KK,N
            IF(ABS(A(I,K)).GT.U) THEN
               U=ABS(A(I,K))
               IN=I
            END IF
      2     CONTINUE
            IF(K.NE.IN) THEN
C     *** INTERCHANGE ROWS K AND INDEX IN ***
            DO 3 J=K,M+N
               X=A(K,J)
               A(K,J)=A(IN,J)
               A(IN,J)=X
      3     CONTINUE
            END IF
C     *** CHECK IF PIVOT TOO SMALL ***
            IF(U.LT.DELT) THEN
               WRITE(6,4)
      4        FORMAT(2X,'THE MATRIX IS SINGULAR. GAUSSIAN'
      1             ' ELIMINATION CANNOT BE PERFORMED.')
            RETURN
            END IF
C     *** FORWARD ELIMINATION STEP ***
            DO 5 I=KK,N
            DO 5 J=KK,M+N
               A(I,J)=A(I,J)-A(I,K)*A(K,J)/A(K,K)
      5     CONTINUE
      1  CONTINUE
         IF(ABS(A(N,N)).LT.DELT) THEN
            WRITE(6,4)
            RETURN
         END IF
C     *** BACK SUBSTITUTION ***
         DO 6 K=1,M
            A(N,K+N)=A(N,K+N)/A(N,N)
            DO 6 IE=1,N-1
               I=N-IE
               IX=I+1
               DO 7 J=IX,N
                  A(I,K+N)=A(I,K+N)-A(J,K+N)*A(I,J)
      7        CONTINUE
               A(I,K+N)=A(I,K+N)/A(I,I)
      6     CONTINUE
         RETURN
      ELSE IF(ABS(A(1,1)).LT.DELT) THEN
         WRITE(6,4)
         RETURN
      END IF
      DO 8 J=1,M
         A(1,N+J)=A(1,N+J)/A(1,1)
      8 CONTINUE
      RETURN
      END
```

EXAMPLE 4.5

We call on subroutine GAUSS to solve our example problem

$$2x_1 + x_2 + x_3 = 4$$
$$x_1 + 3x_2 + 2x_3 = 6$$
$$x_1 + 2x_2 + 2x_3 = 5.$$

Here $N = 3$ and $M = 1$. We take the singularity threshold DELT to be approximately machine epsilon. The calling program and output are given in Tables 4.3 and 4.4. The calling program copies the original matrix into array **B**, and **B** is subsequently passed on to GAUSS. This matrix is altered as GAUSS performs eliminations. Then the solution computed by GAUSS (and stored in the fourth column of the returned matrix **B**) is checked. The simple check of comparing A$\tilde{x}$ with **b**, $\tilde{x}$ being the computed solution, is a popular test. The difference A$\tilde{x}$ − **b** is called the *residual vector,* and it will be examined in Section 4.2.4. In our case we can see that the first component of the residual vector is on the order of machine epsilon.

TABLE 4.3   Calling Program for a Linear Equation Solution

```
C       PROGRAM RESIDUAL
C
C       ************************************************************
C       THIS PROGRAM SOLVES A 3 BY 3 SYSTEM OF LINEAR EQUATIONS,
C       AND COMPUTES RESIDUALS ASSOCIATED WITH THE SOLUTION
C       CALLS:  GAUSS
C       OUTPUT: A=3 BY 4 ARRAY CONTAINING COEFFICIENTS AND RIGHT
C               HAND SIDE VECTOR
C               B(I,4)=3 BY 1 SOLUTION VECTOR
C               R=1 BY 3 ARRAY OF RESIDUALS ASSOCIATED WITH
C               COMPUTED SOLUTION OF THE SYSTEM
C       ************************************************************
C
        DIMENSION A(3,4),B(3,4),R(3)
        DATA A /2,1,1,1,3,2,1,2,2,4,6,5/
C
C       *** COPIES THE ORIGIONAL MATRIX IN B                 ***
        DO 1 I=1,3
           DO 1 J=1,4
              B(I,J)=A(I,J)
      1 CONTINUE
        WRITE(10,7)
      7 FORMAT(' COEFFICIENTS AND RIGHT HAND SIDE')
        WRITE(10,8)
      8 FORMAT(//)
        WRITE(10,4)((B(I,J),J=1,4),I=1,3)
      4 FORMAT(4X,4F9.5)
        EPS=1.E-7
C
C       *** SUBROUTINE GAUSS COMPUTES SOLUTION OF THE LINEAR ***
C       *** SYSTEM                                           ***
C
        CALL GAUSS(3,1,B,EPS)
        WRITE(10,8)
        WRITE(10,*)'ANSR ',(B(I,4),I=1,3)
        WRITE(10,8)
C
C       *** COMPUTES RESIDUALS ASSOCIATED WITH SOLUTION      ***
C
        DO 3 KK=1,3
           S=0
           DO 2 J=1,3
              S=S+A(KK,J)*B(J,4)
      2    CONTINUE
           R(KK)=S-A(KK,4)
           WRITE(10,*)'RESID',R(KK)
      3 CONTINUE
        STOP
        END
```

TABLE 4.4   Output for a Linear Equation Solution

ANSR	1.000000	1.000000	0.9999998
RESID	4.7683716E-07		
RESID	0.0000000E+00		
RESID	0.0000000E+00		

■

A number of subroutines in this book call for GAUS1 instead of GAUSS, when a linear system of equations requires solution. GAUS1 is identical to GAUSS, except for the added calling parameter ND. The need for this alternative subroutine arises in cases in which the matrix of the linear system to be solved is not a calling parameter of the main program and consequently its dimension cannot be varied. In such cases, the subroutine calling GAUS1 sets ND to an appropriate application upper bound. The first few lines of GAUS1 are given in Table 4.5. Below the dimension declaration, GAUS1 and GAUSS are identical.

TABLE 4.5   Subroutine GAUS1 *

```
      SUBROUTINE GAUS1(N,M,ND,A,DELT)
C
C  **************************************************************
C  *  FUNCTION: THIS SUBROUTINE COMPUTES THE SOLUTIONS FOR M    *
C  *            SYSTEMS WITH N EQUATIONS AND N UNKNOWNS USING    *
C  *            GAUSSIAN ELIMINATION                            *
C  *  USAGE:                                                    *
C  *       CALL SEQUENCE: CALL GAUS1(N,M,ND,A,DELT)             *
C  *  PARAMETERS:                                               *
C  *     INPUT:                                                 *
C  *           N=NUMBER OF EQUATIONS AND UNKNOWNS               *
C  *           M=NUMBER  OF SYSTEMS (RIGHT HAND SIDE VECTORS)   *
C  *           ND=UPPER BOUND TO THE LINEAR EQUATION ORDER      *
C  *           A=N BY M+N ARRAY OF COEFFICIENTS AUGMENTED       *
C  *              WITH EACH RIGHT SIDE VECTOR                   *
C  *           DELT=MACHINE ZERO (TOLERANCE)                    *
C  *     OUTPUT:                                                *
C  *        A(1,N+J),...,A(N,N+J)                               *
C  *           =SOLUTION OF THE J-TH SYSTEM (J=1,...,M)         *
C  **************************************************************
C
      DIMENSION A(ND,ND+M)
```

*Subroutine GAUS1 is identical to GAUSS, below dimension declaration.

The consensus of experts is that among the alternative pivoting strategies—no pivoting, partial pivoting, and maximal pivoting—partial pivoting is the most attractive for solution of "routine" linear equations. Any sensible code should have some check to avoid zero pivots. Partial pivoting provides such a check and as the order n of the linear system increases, the relative expense of this check becomes inconsequential. For the reader may confirm that fewer than $\frac{1}{2}n^2$ comparisons need to be made during the course of a calculation, whereas in Section 4.2.5 we will see that the number of arithmetic operations grows as n^3. The number of comparisons by maximal pivoting is also proportional to n^3, so the effort of this alternative is commensurate with the computation time of the elimination process itself.

In theory (e.g., Stewart, 1973, p. 152), the roundoff error with partial pivoting can grow much faster with increasing number n of equations than with maximal pivoting (2^n versus $n^{1/2}$). But in practice, the error growth with partial pivoting is typically far smaller than its theoretical limit. The extra computing expense associated with maximal pivoting is significant for large matrices. For comparable additional expense, one can typically achieve much greater improvement in accuracy by performing Gaussian elimination with partial pivoting in double precision.

───── **ENGINEERING EXAMPLE** ─────────────────────────

Ohm's law, a primary law of circuit theory, tells us that the voltage V across a resistor is related to the current i passing through it, and the resistance value R, by

$$V = iR.$$

As illustrated in Figure 4.2, the voltage drop is in the direction of current flow. Two famous circuit theory laws, Kirchhoff's laws, tell us that the net voltage drop around a closed path in a circuit must be zero, and that the net current flow into a junction of connecting wires must be zero. From these laws one can determine the currents and voltages at any location in a configuration of resistors, voltage sources, and current pumps, once one is given the interconnection diagram and the values of the resistances and current and voltage sources. In fact, these laws lead to a linear system of equations for the unknown voltages and currents.

In Figure 4.2 we have presented a simple resistor and battery circuit. The implication of Kirchhoff's and Ohm's laws are that the currents i_1, i_2, and i_3 must satisfy the following relations:

$$(R_1 + R_2 + R_4)i_1 \quad -R_2 i_2 \quad\quad -R_4 i_3 = 0$$
$$-R_2 i_1 + (R_2 + R_3)i_2 \quad\quad -R_3 i_3 = V_1$$
$$-R_4 i_1 \quad -R_3 i_2 + (R_3 + R_4 + R_5)i_3 = V_2.$$

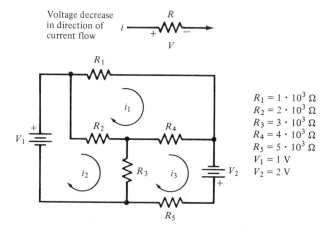

$R_1 = 1 \cdot 10^3 \ \Omega$
$R_2 = 2 \cdot 10^3 \ \Omega$
$R_3 = 3 \cdot 10^3 \ \Omega$
$R_4 = 4 \cdot 10^3 \ \Omega$
$R_5 = 5 \cdot 10^3 \ \Omega$
$V_1 = 1 \ V$
$V_2 = 2 \ V$

FIGURE 4–2 Resistive Network

Upon substituting numerical values, we have

$$7 \times 10^3 i_1 \quad -2 \times 10^3 i_2 \quad -4 \times 10^3 i_3 = 0$$
$$-2 \times 10^3 i_1 \quad +5 \times 10^3 i_2 \quad -3 \times 10^3 i_3 = 1$$
$$-4 \times 10^3 i_1 \quad -3 \times 10^3 i_2 \quad +12 \times 10^3 i_3 = 2.$$

A program too trivial to include calling subroutine GAUSS (Table 4.2) gives the solution that

$$i_1 = 0.486 \text{ mA}, \quad i_2 = 0.696 \text{ mA}, \quad i_3 = 0.503 \text{ mA},$$

where "mA" denotes 10^{-3} ampere.

4.2.4 Error Checks

In the preceding subsection we acknowledged that the cumulative effects of round-off error in naive Gaussian elimination can be troublesome, although pivoting techniques can, to some extent, reduce these effects. Now rules for on-line assessment of error are offered.

A popular procedure for checking the computed solution of a linear system is to examine the residual. The *residual error* (or simply *residual*) associated with a computed solution $\tilde{x}$ of (4.1) is defined in matrix notation by

$$\mathbf{r} = \mathbf{A}\tilde{\mathbf{x}} - \mathbf{b}. \tag{4.11}$$

Clearly, if $\tilde{x}$ is the exact solution, $\mathbf{r}$ is zero. The residual gives some indication of how reliable a computed solution $\tilde{x}$ is. If the coordinates of $\mathbf{r}$ are large, $\tilde{x}$ is suspect. However, a small residual does not guarantee that $\tilde{x}$ is accurate, as will be demonstrated in Section 4.4. On the positive side, it is possible (Section 4.4.2) to bound the error in terms of the residual.

Another accuracy check is the following. Suppose that we choose a vector $\mathbf{y} = (y_1, y_2, \ldots, y_n)^T$ to be a vector of 1's. That is, $y_i = 1$, for $i = 1, 2, \ldots, n$. Then define the vector $\mathbf{c} = (c_1, c_2, \ldots, c_n)^T$ by the condition that

$$c_i = \sum_{j=1}^{n} a_{ij}, \quad 1 \le i \le n, \tag{4.12}$$

By construction, therefore, the solution of

$$\mathbf{Ay} = \mathbf{c} \tag{4.13}$$

is the vector of $\mathbf{y} = (1, 1, \ldots, 1)^T$. Now by setting $M = 2$ and defining $a_{i,n+2} = c_i$, $i = 1, \ldots, n$, one may employ subroutine GAUSS (Table 4.2) to solve both equations $\mathbf{Ax} = \mathbf{b}$ and $\mathbf{Ay} = \mathbf{c}$ at essentially the same expense as either one of them, since forward elimination (4.8) needs to be performed only once. When the solution $\tilde{y}$ is returned, one can see whether the computed solution of (4.13) is close to the known solution vector $\mathbf{y} = (1, 1, \ldots, 1)^T$. A rule of thumb is that if for a given matrix $\mathbf{A}$, the solution of $\mathbf{Ax} = \mathbf{b}$ is sensitive to roundoff, it will be sensitive for every coefficient vector **b**. The procedure of solving equation (4.13) allows us to assess this sensitivity. (The use of 1's here is arbitrary. Any vector **y** could be chosen, with the understanding that **c** = **Ay**.)

We compare some results based on computational experimentation in the next two examples.

────── **EXAMPLE 4.6** ──────────────────────────

By means of the calling program in Table 4.6, subroutine GAUSS is used to solve a linear equation of order 80. We assess the accuracy by the measures just discussed. The coordinates of the coefficient matrix **A** and the vector **b**, in the equation **Ax** = **b** are chosen by the DEC 10 random number generator. The vector $A(j, 82)$, $1 \le j \le 80$, is defined as **c** in (4.12). The program in Table 4.6 executes the calculation the results of which are given in Table 4.7. The 80 components of the outputted residual **Ax̃** − **b** were of the order of 10^{-6} or 10^{-7}, and the computed coordinates of **y** were all equal to 1, to at least four significant decimals. In view of this evidence, we feel confident that the computed solution vector **x̃** was accurate to at least three significant decimals.

TABLE 4.6   Calling Program for a Linear Equation of Order 80

```
C       PROGRAM RAN
C
C       ****************************************************************
C       GENERATES A SYSTEM OF 80 LINEAR EQUATIONS WITH RANDOM
C       COEFFICIENTS, SOLVES THE SYSTEM AND COMPUTES THE RESIDUALS
C       ASSOCIATED WITH THE SOLUTION
C       CALLS:  GAUSS
C       OUTPUT: A=80 BY 1 ARRAY CONTAINING THE SOLUTION TO THE
C               SYSTEM
C               R=80 BY 1 ARRAY OF RESIDUALS
C       ****************************************************************
C
        DIMENSION A(80,82),B(80,82),R(80)
C
C       *** GENERATE THE INPUT DATA AND STORE IT IN B         ***
C
        KSEED=12345
        DO 1 I=1,80
        DO 1,J=1,81
          A(I,J)=RAN(KSEED)
          B(I,J)=A(I,J)
      1 CONTINUE
C
C       *** GENERATES A DATA VECTOR SO THAT Y(J)=1.0 IS      ***
C       *** A SOLUTION                                       ***
C
        DO 2 I=1,80
        DO 2 J=1,80
          A(I,82)=A(I,82)+A(I,J)
      2 CONTINUE
        EPS=1.E-7
C
C       *** SUBROUTINE GAUSS COMPUTES SOLUTION OF THE LINEAR ***
C       *** SYSTEM                                           ***
C
        CALL GAUSS(80,2,A,EPS)
C
C       *** COMPUTES RESIDUALS ASSOCIATED WITH SOLUTION      ***
C
```

TABLE 4.6 (Continued)

```
      DO 3 I=1,80
        S=0.
        DO 4 J=1,80
          S=S+B(I,J)*A(J,81)
    4     CONTINUE
        R(I)=S-B(I,81)
    3 CONTINUE
      DO 5 I=1,40
        J=I+40
        WRITE(10,6)I,A(I,82),R(I),J,A(J,82),R(J)
    6   FORMAT(1X,I3,2X,F10.6,2X,E12.5,6X,I3,2X,F10.6,2X,E12.5)
    5 CONTINUE
      STOP
      END
```

TABLE 4.7 Application of GAUSS to a System of 80 Variables

Coordinate Number k	Computed Test Coordinate $\tilde{y}_k$	Residual r_k	k	$\tilde{y}_k$	r_k
1	0.999992	0.17881E-06	41	1.000012	-0.14901E-05
2	0.999998	-0.21458E-05	42	0.999997	-0.65565E-06
3	0.999981	0.36359E-05	43	1.000014	-0.26822E-05
4	1.000000	-0.54464E-05	44	1.000028	-0.29802E-06
5	1.000007	0.45896E-05	45	1.000001	0.17881E-05
6	1.000007	0.26822E-05	46	0.999999	0.15795E-05
7	1.000005	0.22650E-05	47	1.000009	-0.59605E-06
8	0.999990	0.71526E-06	48	1.000003	-0.19148E-05
9	1.000020	-0.15199E-05	49	0.999998	0.46492E-05
10	1.000002	-0.10729E-05	50	0.999996	-0.10282E-05
11	0.999992	-0.51260E-05	51	1.000016	-0.50664E-06
12	1.000010	0.26822E-05	52	0.999998	-0.41723E-06
13	1.000011	-0.15795E-05	53	0.999993	-0.13411E-06
14	1.000029	-0.10878E-05	54	0.999998	0.13113E-05
15	1.000016	0.65565E-06	55	1.000006	-0.29802E-05
16	0.999991	0.41723E-06	56	1.000013	-0.55432E-05
17	1.000001	0.30994E-05	57	1.000006	0.26822E-06
18	1.000007	0.28610E-05	58	0.999994	0.24252E-05
19	0.999998	-0.39935E-05	59	0.999993	-0.23842E-06
20	0.999994	-0.26226E-05	60	1.000010	0.11325E-05
21	0.999994	-0.67055E-06	61	0.999991	0.25332E-06
22	1.000004	-0.41723E-06	62	1.000011	0.51036E-05
23	1.000009	-0.32634E-05	63	0.999992	-0.29802E-05
24	0.999993	-0.15497E-05	64	1.000002	0.49472E-05
25	0.999999	0.16093E-05	65	0.999993	-0.67949E-05
26	1.000010	0.13560E-05	66	1.000003	0.10133E-05
27	0.999991	0.15497E-05	67	0.999992	0.27418E-05
28	0.999997	0.25034E-05	68	1.000006	-0.28461E-05
29	0.999997	-0.15423E-05	69	0.999997	-0.95367E-06
30	0.999993	-0.21160E-05	70	0.999979	-0.42617E-05
31	0.999993	0.39935E-05	71	0.999996	-0.16391E-05
32	0.999982	0.10431E-05	72	1.000000	-0.23842E-05
33	1.000011	0.23842E-06	73	1.000011	-0.16391E-05
34	1.000016	-0.14305E-05	74	0.999993	0.59605E-07
35	1.000003	0.24438E-05	75	0.999998	-0.25034E-05
36	1.000000	-0.19073E-05	76	0.999994	-0.81658E-05
37	0.999994	0.71526E-06	77	0.999980	-0.83447E-06
38	0.999992	0.17881E-06	78	0.999986	0.47088E-05
39	0.999984	-0.46492E-05	79	1.000005	0.21383E-05
40	1.000001	-0.20862E-05	80	0.999985	-0.39339E-05

———— **EXAMPLE 4.7** ————————————————————————

Recall that Newton–Cotes integration (Section 3.3) and polynomial interpolation with equally spaced abscissas x_i both entail solution of linear equation with coefficient matrix of the form

$$\mathbf{A} = \begin{bmatrix} 1 & 1 & \cdots & 1 \\ x_1 & x_2 & \cdots & x_n \\ \vdots & \vdots & & \vdots \\ x_1^{n-1} & x_2^{n-1} & \cdots & x_n^{n-1} \end{bmatrix}, \qquad (4.14)$$

in which for some fixed h, $x_{i+1} - x_i = h$. In Example 3.10 it was seen that this matrix engenders unseemly numerical error. Actually, this matrix (4.14) shows up in many engineering and physical contexts, and is known as the *Vandermonde matrix*.

The Vandermonde matrix (4.14) is famous not only for the regularity of its occurrence in applications, but also for its tendency toward instability. This instability, or ill-conditioning, as we will call it in Section 4.4.2, gives us an opportunity to investigate the relative effectiveness of different pivoting strategies in Gauss elimination. We took

$$x_i = ih, \qquad 1 \leq i \leq n,$$

with $h = 0.1$, and took n to be 8. The coefficients of the vector $\mathbf{b}$ were chosen randomly. We compared:

1. No pivoting (naive elimination).
2. Partial pivoting (subroutine GAUSS).
3. Maximal pivoting.
4. Partial pivoting in double precision.

The error was assessed by the root-mean-square (rms) residual and rms difference defined, respectively, as

$$\text{rms residual} = \left(\frac{1}{n} \sum_{i=1}^{n} r_i^2 \right)^{1/2}$$

with $\mathbf{r} = (r_1, r_2, \cdots, r_n)^T = \mathbf{A}\tilde{\mathbf{x}} - \mathbf{b}$ and

$$\text{rms difference} = \left[\frac{1}{n} \sum_{i=1}^{n} (1 - \tilde{y}_i)^2 \right]^{1/2}.$$

Here $\tilde{\mathbf{x}}$ is the computed solution to $\mathbf{A}\mathbf{x} = \mathbf{b}$ and $\tilde{\mathbf{y}}$ the computed solution to $\mathbf{A}\mathbf{y} = \mathbf{c}$, where, as in (4.12),

$$c_i = a_{i1} + a_{i2} + \cdots + a_{in}$$

(so the exact answer is $y_i = 1$, $1 \leq i \leq n$). Furthermore, r_i is the ith component of the residual vector $\mathbf{r}$. We display the results of our study in Table 4.8. With the exception of the obvious benefits of partial pivoting in double precision, the

TABLE 4.8   Observed Root-Mean-Square Errors Under Different Pivoting Strategies

rms Errors	No Pivoting	Partial Pivoting	Maximal Pivoting	Partial Pivoting in Double Precision
rms residual	4.3×10^{-4}	1.3×10^{-3}	1.1×10^{-3}	1.7×10^{-14}
rms difference	1.0×10^{-3}	4.0×10^{-3}	1.1×10^{-3}	7.11×10^{-15}

relative merits of the pivoting strategies are inconclusive. The calculations were done on a DEC 10.   ∎

*4.2.5   Analysis of Computational Effort

Here we calculate the total number of operations needed for the solution of the linear system (4.4) by Gauss elimination. Forward elimination [equation (4.9)] for fixed i, j, and k requires one multiplication and one addition. (We are regarding subtraction as addition with possibly negative summands.) If k is fixed, there are $n - k$ values for i and $n + 1 - k$ values for j, so the number of additions and multiplications is $(n - k)(n + 1 - k)$. There are $(n - k)$ divisions, one for each value of i. The total number of additions and multiplications are both given by summing over the range of k:

$$\sum_{k=1}^{n-1} (n - k)(n + 1 - k) = \sum_{k=1}^{n-1} [(n^2 + n) - k(2n + 1) + k^2]$$

$$= (n - 1)(n^2 + n)$$

$$- (2n + 1) \sum_{k=1}^{n-1} k + \sum_{k=1}^{n-1} k^2$$

$$= (n - 1)(n^2 + n)$$

$$- (2n + 1) \frac{n(n - 1)}{2} + \frac{(n - 1)n(2n - 1)}{6}$$

$$= \frac{n^3}{3} - \frac{n}{3}. \tag{4.15}$$

We have used the facts that for $n \geq 2$,

$$1 + 2 + \cdots + (n - 1) = \frac{n(n - 1)}{2}$$

and

$$1^2 + 2^2 + \cdots + (n - 1)^2 = \frac{(n - 1)n(2n - 1)}{6}.$$

TABLE 4.9   Computational Effort for Gauss Elimination*

	Forward Elimination	Back Substitution
Additions	$\dfrac{n^3}{3} - \dfrac{n}{3}$	$\dfrac{n(n-1)}{2}$
Multiplications	$\dfrac{n^3}{3} - \dfrac{n}{3}$	$\dfrac{n(n-1)}{2}$
Divisions	$\dfrac{n(n-1)}{2}$	n

*n, order of the system.

These identities can be proven by induction (Problem 8). The number of divisions equals

$$\sum_{k=1}^{n-1} (n - k) = (n - 1) + (n - 2) + \cdots + 1 = \sum_{k=1}^{n-1} k = \frac{n(n - 1)}{2}.$$

In back substitution as defined by (4.10), for each value of k we require one division, $n - k$ additions, and $n - k$ multiplications. By summing over the range of k, one can conclude that the number of additions and multiplications equals

$$\sum_{k=1}^{n} (n - k) = \frac{n(n - 1)}{2}$$

and only n divisions are needed. Thus the effort required for forward elimination dominates, and we can conclude that the number of multiplications and additions grows proportionally to $n^3/3$. Thus, as the order n becomes large, the relative cost of back substitution becomes negligible. For that reason, the sensitivity check of recovering the known solution $\mathbf{y}$ described in connection with (4.12) and (4.13) is economical. We summarize the findings of this section in Table 4.9.

4.3
MATRIX INVERSION, DETERMINANTS, AND BAND SYSTEMS

4.3.1   Gauss Elimination for Matrix Inversion

The elimination method can be used to find inverses of matrices. Let $\mathbf{A}$ be an nth-order nonsingular matrix, and let the n columns of the inverse matrix be denoted by $\mathbf{x}^{(1)}, \ldots, \mathbf{x}^{(n)}$. That is, $\mathbf{A}^{-1} = (\mathbf{x}^{(1)}, \mathbf{x}^{(2)}, \ldots, \mathbf{x}^{(n)})$. Then (Appendix A, property 21)

$$\mathbf{A}\mathbf{A}^{-1} = \mathbf{A}(\mathbf{x}^{(1)}, \ldots, \mathbf{x}^{(n)}) = (\mathbf{A}\mathbf{x}^{(1)}, \ldots, \mathbf{A}\mathbf{x}^{(n)}). \qquad (4.16)$$

On the right of (4.16), $\mathbf{A}\mathbf{x}^{(j)}$ $(j = 1, \ldots, n)$ denotes the column vector obtained after the matrix $\mathbf{A}$ has been multiplied by vector $\mathbf{x}^{(j)}$. The columns of the identity matrix will be represented by $\mathbf{e}^{(1)}, \ldots, \mathbf{e}^{(n)}$. That is, the jth coordinate of $\mathbf{e}^{(j)}$ is 1 and the other coordinates are 0, for $j = 1, \ldots, n$. Now, by definition of matrix inverse (Appendix A, property 17),

$$\mathbf{A}\mathbf{A}^{-1} = \mathbf{I} = (\mathbf{e}^{(1)}, \ldots, \mathbf{e}^{(n)}).$$

That is,

$$(\mathbf{A}\mathbf{x}^{(1)}, \ldots, \mathbf{A}\mathbf{x}^{(n)}) = (\mathbf{e}^{(1)}, \ldots, \mathbf{e}^{(n)}).$$

By matching the columns of the left- and right-hand sides of the equation above, we assemble the set of linear equations

$$\mathbf{A}\mathbf{x}^{(1)} = \mathbf{e}^{(1)}, \quad \mathbf{A}\mathbf{x}^{(2)} = \mathbf{e}^{(2)}, \quad \ldots, \quad \mathbf{A}\mathbf{x}^{(n)} = \mathbf{e}^{(n)}. \tag{4.17}$$

Observe that the coefficient matrix $\mathbf{A}$ of each of the systems above is the same. The forward elimination process, therefore, needs to be done only once. Back substitution has to be repeated n times, once for each of the vectors $\mathbf{x}^{(j)}$. Specifically, the first derived system is obtained by operating on the array

$$\begin{bmatrix} a_{11} & a_{12} & \cdots & a_{1n} & 1 & 0 & \cdots & 0 \\ a_{21} & a_{22} & \cdots & a_{2n} & 0 & 1 & \cdots & 0 \\ \vdots & \vdots & & \vdots & \vdots & \vdots & & \vdots \\ a_{n1} & a_{n2} & \cdots & a_{nn} & 0 & 0 & \cdots & 1 \end{bmatrix}$$

according to the rule

$$a_{ij}^{(1)} = a_{ij} - \frac{a_{i1}}{a_{11}} a_{1j} \qquad (2 \le i \le n,\, 2 \le j \le 2n),$$

where the elements of the right-hand vectors (i.e., the elements of the identity matrix) are now placed in the positions $a_{ij}(i = 1, \ldots, n$ and $j = n + 1, \ldots, 2n)$. Further derived systems are obtained by (4.9), with the modification that j ranges from $k + 1$ to $2n$, in constructing the elements $a_{ij}^{(k)}$ for each value of k.

In the notation of (4.17), one obtains the column vectors $\mathbf{x}^{(j)}$ of the inverse matrix by performing back substitution but with the values $a_{k,n+j}^{(k-1)}$ replacing $a_{k,n+1}^{(k-1)}$ in (4.10).

──────── **EXAMPLE 4.8** ────────

Let $\mathbf{A}$ be the matrix of coefficients of the system given in Example 4.1:

$$\mathbf{A} = \begin{bmatrix} 2 & 1 & 1 \\ 1 & 3 & 2 \\ 1 & 2 & 2 \end{bmatrix}.$$

The inverse of $\mathbf{A}$ will be now determined. The calculations of the derived systems are given in Table 4.10.

TABLE 4.10   Computations of Example 4.8

Original System	2	1	1	1	0	0
	1	3	2	0	1	0
	1	2	2	0	0	1
First Derived System	2	1	1	1	0	0
	0	$\frac{5}{2}$	$\frac{3}{2}$	$-\frac{1}{2}$	1	0
	0	$\frac{3}{2}$	$\frac{3}{2}$	$-\frac{1}{2}$	0	1
Second Derived System	2	1	1	1	0	0
	0	$\frac{5}{2}$	$\frac{3}{2}$	$-\frac{1}{2}$	1	0
	0	0	$\frac{3}{5}$	$-\frac{1}{5}$	$-\frac{3}{5}$	1

By the successive use of back substitution we find that

$$
\mathbf{A}^{-1} = \begin{bmatrix} \frac{2}{3} & 0 & -\frac{2}{3} \\ 0 & 1 & -1 \\ -\frac{1}{3} & -1 & \frac{5}{3} \end{bmatrix}.
$$

For example, the first column is obtained by using back substitution on the $(n + 1)$st column of the second derived system. Thus according to (4.10), with $a_{ij}^{(-1)}$ denoting the (i, j) coordinate of $\mathbf{A}^{-1}$,

$$
a_{31}^{(-1)} = -\frac{1/5}{3/5} = -\frac{1}{3},
$$

and continuing recursively,

$$
a_{21}^{(-1)} = \frac{-1/2 + 3/2 \times 1/3}{5/2} = 0
$$

$$
a_{11}^{(-1)} = \frac{1 + 1/3}{2} = \frac{2}{3}.
$$

The second and third columns of $\mathbf{A}^{-1}$ are obtained in similar fashion from the $(n + 2)$nd and $(n + 3)$rd columns of the second derived system. ∎

Note that if $\mathbf{A}^{-1}$ is known, the solution $\mathbf{x}$ of equation (4.2) can be obtained according to

$$
\mathbf{x} = \mathbf{A}^{-1}\mathbf{b} \tag{4.18}
$$

which is verified as property 18 in Appendix A. However, solution of (4.2) by computing the inverse of the matrix of coefficients and the use of (4.18) is relatively expensive. The computation of an inverse requires about three times as many operations as the Gauss elimination method given in Section 4.2.4 for solving (4.2). Moreover, the effects of roundoff error in inverting **A** and multiplying as in (4.18) tend to be more pronounced. In short, the technique given here should be reserved for cases in which the matrix inverse is actually needed.

Subroutine INVG for matrix inversion is given in Table 4.11. A professional-level code would require about half the memory but would be more difficult to understand.

We remark that if **B** is an $n \times n$ matrix with columns $[\mathbf{b}^{(1)}, \mathbf{b}^{(2)}, \ldots, \mathbf{b}^{(n)}]$ and **x** now is a matrix of vectors $\mathbf{x} = (x^{(1)}, x^{(2)}, \ldots, x^{(n)})$, the technique of this section can be used to find the matrix **x** satisfying

$$\mathbf{A}\mathbf{x} = \mathbf{B}. \tag{4.19}$$

The idea is to perform the preceding elimination technique on the $n \times (2n)$ tableau

$$[\mathbf{a}^{(1)}, \mathbf{a}^{(2)}, \ldots, \mathbf{a}^{(n)} \vdots \mathbf{b}^{(1)}, \mathbf{b}^{(2)}, \ldots, \mathbf{b}^{(n)}].$$

TABLE 4.11 Subroutine INVG for Matrix Inversion

```
      SUBROUTINE INVG(N,A,B,DELT,C)
C
C     ********************************************************
C     *  FUNCTION: THIS SUBROUTINE COMPUTES THE INVERSE MATRIX  *
C     *            BY GAUSS ELIMINATION                        *
C     *  USAGE:                                                *
C     *        CALL SEQUENCE: CALL INVG(N,A,B,DELT,C)          *
C     *        EXTERNAL FUNCTIONS/SUBROUTINES: SUBROUTINE GAUSS *
C     *  PARAMETERS:                                           *
C     *        INPUT:                                          *
C     *            N=ORDER OF THE MATRIX                       *
C     *            A=N BY N ARRAY OF MATRIX VALUES             *
C     *        DELT=MACHINE ZERO (TOLERANCE)                   *
C     *        OUTPUT:                                         *
C     *            B=N BY N ARRAY, THE COMPUTED INVERSE OF A   *
C     *            C=N BY 2*N ARRAY, A WORK MATRIX             *
C     ********************************************************
C
      DIMENSION A(N,N),B(N,N),C(N,2*N)
C     *** COMPUTE WORKING MATRIX C FROM MATRIX A ***
C     ***     AND THE APPENDED IDENTITY MATRIX    ***
C     *** NUMBER OF COLUMNS OF MATRIX C IS 2*N   ***
      DO 1 I=1,N
        DO 1 J=1,N
          C(I,J)=A(I,J)
          C(I,N+J)=0.0
          IF(I.EQ.J) C(I,N+J)=1.0
    1 CONTINUE
C     *** COMPUTE MATRIX INVERSE BY GAUSSIAN ELIMINATION ***
      M=N
      CALL GAUSS(N,M,C,DELT)
      DO 2 I=1,N
        DO 2 J=1,N
          B(I,J)=C(I,J+N)
    2 CONTINUE
      RETURN
      END
```

Then perform Gaussian elimination as we have just described, the difference being that the identity matrix in the initial tableau is to be replaced by matrix **B**.

*4.3.2 Evaluation of Determinants

The concept of determinants is firmly embedded in classical matrix theory books, and determinants are used in many application areas. On the other hand, they play a minor role in computational methodology, and they are not simple to define. For that reason we elected not to discuss them in Appendix A. Instead, we refer the reader to standard linear algebra texts such as that by Nering (1963, Chap. 3) for the definition and elementary properties of determinants. Determinant evaluation is a relatively simple extension of the Gaussian elimination method, as we now demonstrate.

Let det (**A**) denote the determinant of matrix **A**. The operations (4.9) do not change the value of the determinant of a matrix and the interchange of two rows (as required in pivoting) has the effect of multiplying the determinant by the factor (-1). The reader may verify by expansion along the diagonal that because it is an upper triangular matrix, the determinant of the coefficient matrix of the $(n-1)$st derived system is equal to the product of its diagonal elements a_{11}, $a_{22}^{(1)}, a_{33}^{(2)}, \ldots, a_{nn}^{(n-1)}$. Therefore, the determinant of **A** is given by

$$\det (\mathbf{A}) = (-1)^p a_{11} a_{22}^{(1)} a_{33}^{(2)} \cdots a_{nn}^{(n-1)}, \qquad (4.20)$$

p being the number of times the row-interchange operation has been used during the entire computation of the derived systems. Table 4.12 gives subroutine DETC for determinant calculation.

TABLE 4.12   Subroutine DETC for Calculating Determinants

```
      SUBROUTINE DETC(N,A,D,DELT)
C
C     ****************************************************************
C     *  FUNCTION: THIS SUBROUTINE COMPUTES THE DETERMINANT OF A  *
C     *            MATRIX USING GAUSSIAN ELIMINATION              *
C     *  USAGE:                                                    *
C     *        CALL SEQUENCE: CALL DETC(N,A,D,DELT)               *
C     *  PARAMETERS:                                               *
C     *        INPUT:                                              *
C     *            N=ORDER OF THE MATRIX                          *
C     *            A=N BY N ARRAY OF MATRIX ELEMENTS             *
C     *          DELT=MACHINE ZERO (TOLERANCE)                   *
C     *        OUTPUT:                                             *
C     *            D=DETERMINANT OF THE MATRIX                    *
C     ****************************************************************
C
      DIMENSION A(N,N)
C     *** TEST IF MATRIX A CONTAINS A SINGLE ELEMENT ***
      IF(N.LE.1) THEN
         D=A(1,1)
         RETURN
C     *** IF NOT PERFORM INITIALIZATION ***
      ELSE
         D=1.0
      END IF
      DO 1 K=1,N-1
```

TABLE 4.12   (Continued)

```
C       *** PERFORM PARTIAL PIVOTING ***
        U=ABS(A(K,K))
        IN=K
        DO 2 I=K+1,N
            IF (ABS(A(I,K)).GT.U) THEN
                U=ABS(A(I,K))
                IN=I
            END IF
2       CONTINUE
C       *** EXCHANGE ROWS K AND IN IF NEEDED ***
        IF (K.NE.IN) THEN
            D=-D
            DO 3 J=K,N
                X=A(K,J)
                A(K,J)=A(IN,J)
                A(IN,J)=X
3           CONTINUE
        END IF
        IF (U.LT.DELT) THEN
            D=0.0
            RETURN
        END IF
        D=D*A(K,K)
C       *** PERFORM GAUSSIAN ELIMINATION ON MATRIX ELEMENTS ***
        DO 4 I=K+1,N
            DO 4 J=K+1,N
                A(I,J)=A(I,J)-A(I,K)*A(K,J)/A(K,K)
4       CONTINUE
1   CONTINUE
C       *** COMPUTE MATRIX DETERMINANT ***
        D=D*A(N,N)
        RETURN
        END
```

EXAMPLE 4.9

To obtain the determinant of the matrix in Example 4.1, observe that

$$\det \begin{bmatrix} 2 & 1 & 1 \\ 1 & 3 & 2 \\ 1 & 2 & 2 \end{bmatrix} = (-1)^0 \det \begin{bmatrix} 2 & 1 & 1 \\ 0 & \frac{5}{2} & \frac{3}{2} \\ 0 & 0 & \frac{3}{5} \end{bmatrix} = (-1)^0 \cdot 2 \cdot \frac{5}{2} \cdot \frac{3}{5} = 3.$$

---

*4.3.3   Band Systems

In many numerical procedures, especially those for computation of cubic spline functions [equation (2.31)] and solution of boundary-value problems of differential equations by difference equations (Szidarovszky and Yakowitz, 1978, Sec. 8.2.3), linear systems having a *band structure* arise. The special case of tridiagonal matrices has special significance. For such matrices the nonzero elements must be on the main diagonal or on the two diagonals just above and below the main diagonal. The tridiagonal matrices are therefore characterized by the property that $a_{ij} = 0$ if $|i - j| \geq 2$. In general, a square matrix $\mathbf{A} = (a_{ij})$ is called a *band matrix* if there is a positive integer k substantially smaller than the order n of the matrix such that $a_{ij} = 0$ if $|i - j| \geq k$. In the case of tridiagonal matrices, $k = 2$. The number k is said to be the *width* of the band matrix.

It may be seen that the storage requirements for band matrices are much less than for general matrices of the same order. For instance, an $n \times n$ tridiagonal matrix has at most $n + 2(n - 1) = 3n - 2$ nonzero elements. This property is very important if the band matrices are associated with large order n.

To conserve storage capacity, represent the tridiagonal coefficient matrix equation as follows, the elements outside the band shown being zero:

$$
\begin{aligned}
d_1 x_1 + c_1 x_2 &= b_1 \\
a_1 x_1 + d_2 x_2 + c_2 x_3 &= b_2 \\
a_2 x_2 + d_3 x_3 + c_3 x_4 &= b_3 \\
&\cdot \\
&\cdot \\
&\cdot \\
a_{n-2} x_{n-2} + d_{n-1} x_{n-1} + c_{n-1} x_n &= b_{n-1} \\
a_{n-1} x_{n-1} + d_n x_n &= b_n.
\end{aligned}
\tag{4.21}
$$

Now, instead of storing an $n \times n$ matrix, we need only to store the vectors $\mathbf{a} = (a_i)$, $\mathbf{d} = (d_i)$, $\mathbf{c} = (c_i)$ with dimensions $n - 1$, n, and $n - 1$, respectively. By choosing the diagonal element d_1 as a pivot for the first derived system, we need to eliminate x_1 from the second equation only, and all the other equations will remain the same. The first derived system is therefore the following:

$$
\begin{aligned}
d_1 x_1 + c_1 x_2 &= b_1 \\
d_2^{(1)} x_2 + c_2 x_3 &= b_2^{(1)} \\
a_2 x_2 + d_3 x_3 + c_3 x_4 &= b_3 \\
&\cdot \\
&\cdot \\
&\cdot \\
a_{n-2} x_{n-2} + d_{n-1} x_{n-1} + c_{n-1} x_n &= b_{n-1} \\
a_{n-1} x_{n-1} + d_n x_n &= b_n,
\end{aligned}
$$

where

$$
d_2^{(1)} = d_2 - \frac{a_1}{d_1} c_1, \qquad b_2^{(1)} = b_2 - \frac{a_1}{d_1} b_1.
\tag{4.22}
$$

By choosing the diagonal element $d_2^{(1)}$ for computing the second derived system, we need to eliminate x_2 from only the third equation. Now the second derived system has the form

$$d_1 x_1 + c_1 x_2 \qquad\qquad = b_1$$

$$d_2^{(1)} x_2 + c_2 x_3 \qquad\qquad = b_2^{(1)}$$

$$d_3^{(2)} x_3 + c_3 x_4 \qquad\qquad = b_3^{(2)}$$

$$a_3 x_3 + d_4 x_4 + c_3 x_5 \qquad\qquad = b_4$$

$$\vdots \qquad\qquad\qquad \vdots$$

$$a_{n-2} x_{n-2} + d_{n-1} x_{n-1} + c_{n-1} x_n = b_{n-1}$$

$$a_{n-1} x_{n-1} + d_n x_n = b_n.$$

We continue in this fashion, proceeding from one equation to the next according to

$$d_{k+1}^{(k)} = d_{k+1} - \frac{a_k}{d_k^{(k-1)}} c_k, \qquad b_{k+1}^{(k)} = b_{k+1} - \frac{a_k}{d_k^{(k-1)}} b_k^{(k-1)} \qquad (4.23)$$

for $k = 1, 2, \ldots, n - 1$. By the procedure above, we obtain the upper triangular system

$$d_1 x_1 + c_1 x_2 \qquad\qquad = b_1$$

$$d_2^{(1)} x_2 + c_2 x_3 \qquad\qquad = b_2^{(1)}$$

$$d_3^{(2)} x_3 + c_3 x_4 \qquad\qquad = b_3^{(2)}$$

$$\vdots \qquad\qquad\qquad \vdots$$

$$d_{n-1}^{(n-2)} x_{n-1} + c_{n-1} x_n = b_{n-1}^{(n-2)}$$

$$d_n^{(n-1)} x_n = b_n^{(n-1)}.$$

Back substitution also has an efficient form, since from the last equation above we get

$$x_n = \frac{b_n^{(n-1)}}{d_n^{(n-1)}},$$

and the preceding equations imply that

$$x_k = \frac{b_k^{(k-1)} - c_k x_{k+1}}{d_k^{(k-1)}} \qquad (k = n - 1, \ldots, 1). \qquad (4.24)$$

TABLE 4.13   Subroutine BAND for Band Matrices

```
        SUBROUTINE BAND(N,A,B,C,D)
C
C      **************************************************************
C      *   FUNCTION: THIS SUBROUTINE COMPUTES THE SOLUTION OF A     *
C      *            BAND SYSTEM OF WIDTH 2 OF LINEAR EQUATIONS      *
C      *            USING A MODIFIED GAUSSIAN ELIMINATION           *
C      *   USAGE:                                                   *
C      *        CALL SEQUENCE: CALL BAND(N,A,B,C,D)                 *
C      *   PARAMETERS:                                              *
C      *       INPUT:                                               *
C      *           A=N-1 BY 1 ARRAY OF SYSTEM 1-ST SUBDIAGONAL      *
C      *               COEFFICIENTS                                 *
C      *           C=N-1 BY 1 ARRAY OF SYSTEM 1-ST SUPERDIAGONAL    *
C      *               COEFFICIENTS                                 *
C      *           D=N BY 1 ARRAY OF SYSTEM DIAGONAL COEFFICIENTS   *
C      *           B=N BY 1 ARRAY OF RIGHT HAND SIDE VALUES         *
C      *       OUTPUT:                                              *
C      *           B=N BY 1 ARRAY OF THE COMPUTED SOLUTION VALUES   *
C      **************************************************************
C
        DIMENSION A(N-1),B(N),C(N-1),D(N)
C      *** COMPUTE GAUSSIAN ELIMINATION FACTORS ***
        DO 1 I=2,N
          X=A(I-1)/D(I-1)
          D(I)=D(I)-X*C(I-1)
          B(I)=B(I)-X*B(I-1)
      1 CONTINUE
C      *** COMPUTE SYSTEM SOLUTION AND STORE IN VECTOR B ***
        B(N)=B(N)/D(N)
        DO 2 I=1,N-1
          B(N-I)=(B(N-I)-C(N-I)*B(N-I+1))/D(N-I)
      2 CONTINUE
        RETURN
        END
```

By the algorithm above, the total number of additions and multiplications equals

$$2(n - 1) + n - 1 = 3n - 2,$$

which for larger n is essentially nothing compared to the $n^3/3$ additions and multiplications required by conventional Gaussian elimination. The number of divisions is

$$(n - 1) + n = 2n - 1.$$

Table 4.13 presents subroutine BAND for implementing this method.

EXAMPLE 4.10

We now solve the system of equations

$$
\begin{aligned}
x_1 + 2x_2 \qquad\qquad\quad &= 2 \\
x_1 + 3x_2 + 2x_3 \qquad &= 7 \\
x_2 + 4x_3 + 2x_4 &= 15 \\
4x_3 + \;\; x_4 &= 11.
\end{aligned}
$$

TABLE 4.14   Computations of Example 4.10

Original System	1	2				2
	1	3	2			7
		1	4	2		15
			4	1		11
First Derived System	1	2				2
		1	2			5
		1	4	2		15
			4	1		11
Second Derived System	1	2				2
		1	2			5
			2	2		10
			4	1		11
Third Derived System	1	2				2
		1	2			5
			2	2		10
				-3		-9

Then Table 4.14 shows the computational results. The method of back substitution leads to the solution:

$$x_4 = \frac{-9}{-3} = 3,$$

$$x_3 = \frac{10 - 2 \cdot 3}{2} = 2,$$

$$x_2 = \frac{5 - 2 \cdot 2}{1} = 1,$$

$$x_1 = \frac{2 - 2 \cdot 1}{1} = 0.$$

4.4
MATRIX CONDITIONING

4.4.1   A Qualitative Notion of Conditioning

In solving linear equations by computers, the effect of the accumulated errors on the computed solution should be estimated to check the accuracy of the results. As we have noted in Section 4.2.4, one common test involves finding the *residual vector,* **r**, which is defined to be

$$\mathbf{r} = \mathbf{A}\tilde{\mathbf{x}} - \mathbf{b}, \tag{4.25}$$

where $\tilde{\mathbf{x}}$ is a computed approximating solution. Clearly, if $\tilde{\mathbf{x}}$ is actually the exact solution to (4.2), the residual is the zero vector. On the other hand, if the residual is relatively large, one can be confident that $\tilde{\mathbf{x}}$ is not a very good approximation to the solution. Unfortunately, the test of seeing whether the residual vector is acceptably small is not without some potential peril. In Example 4.11 the computation of the residual vector will be illustrated and following that, in Example 4.12, we will have an instance in which the residual is small but the proposed solution is nevertheless far from the correct solution.

───── **EXAMPLE 4.11** ───

As an illustration of a residual computation, consider the equation

$$x_1 + x_2 = 2 \tag{4.26}$$
$$3x_1 - x_2 = 2.$$

Suppose, furthermore, that we computed the solution to be

$$\tilde{x}_1 = 0.999, \qquad \tilde{x}_2 = 1.002.$$

By substituting these calculated values into equation (4.25), we get the components of the residual vector:

$$r_1 = 0.999 + 1.002 - 2 = 0.001$$
$$r_2 = 3 \cdot 0.999 - 1.002 - 2 = -0.005.$$

Since these residuals 0.001 and -0.005 are small, we are tempted to think that the approximating solution is also very close to the exact solution, which is actually the case, for the exact solution of (4.26) is $x_1 = x_2 = 1$. ∎

───── **EXAMPLE 4.12** ───

Consider next the system of equations

$$x_1 + x_2 = 2 \tag{4.27}$$
$$1.001x_1 + x_2 = 2.001,$$

where the exact solution is again given by $x_1 = x_2 = 1$. Assume that the calculated solution is $\tilde{x}_1 = 0, \tilde{x}_2 = 2$. In computing the residual, we find that

$$r_1 = 0 + 2 - 2 = 0$$
$$r_2 = 1.001 \cdot 0 + 2 - 2.001 = -0.001.$$

The first equation is satisfied exactly and the second residual coordinate is small (-0.001). Note that the error in each term of the computed solution is 1; that is, their relative errors are 100%. ∎

By way of the example above, we see that a system of simultaneous linear equations can almost be satisfied by completely erroneous answers. Equations for which this is possible are spoken of as being *ill-conditioned*. As a guide to intuition

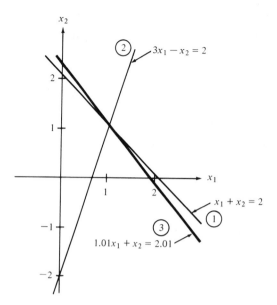

FIGURE 4–3 Graphs of Solutions to Linear Equations

concerning the ill-conditioned property, Figure 4.3 shows the lines associated with the equations of the previous examples.

The solution of the system is the value of x_1 and x_2 at the point of the intersection of the two straight lines that represent the equations. Thus the solution of system (4.26) is the point of intersection of lines ① and ②, and the solution of system (4.27) is given by the point of intersection of lines ① and ③. In the latter case we can see that the two lines are almost parallel, so it is hard to tell just where they really cross or, equivalently, just where the solution lies.

From this vantage point, one can sense why a slight change in coefficients can, when the lines are nearly parallel (or, what amounts to the same thing, when the linear system is nearly singular), shift the solution by an enormous amount in the case of (4.27). For if either the slope or the intercept between two nearly parallel lines such as ① and ③ is altered, it is clear that there will be a dramatic shift in their intersection point.

*4.4.2   Quantification of Ill-Conditioning

To definitively relate the magnitudes of the coordinates of the residual vector to errors in the computed solution, it is necessary to introduce vector and matrix norms. The *norm* of a vector $\mathbf{x}$ is a nonnegative number, usually denoted as $\|\mathbf{x}\|$, which is intuitively a measure of vector length. There are a number of different vector norms, but they are characterized by the following properties:

(a) $\|\mathbf{x}\| \geq 0$.

(b) $\|\mathbf{x}\| = 0$ if and only if $\mathbf{x} = \mathbf{0}$.

(c) $\|\alpha\mathbf{x}\| = |\alpha| \cdot \|\mathbf{x}\|$ for all vectors $\mathbf{x}$ and real or complex numbers α.

(d) $\|\mathbf{x} + \mathbf{y}\| \leq \|\mathbf{x}\| + \|\mathbf{y}\|$ for all vectors $\mathbf{x}$, $\mathbf{y}$.

Similarly, the *norm* $\|\mathbf{A}\|$ of a square matrix $\mathbf{A}$ is a nonnegative number. Matrix norms satisfy properties analogous to (a) through (d). Also, for two $n \times n$ matrices $\mathbf{A}$ and $\mathbf{B}$,

(e) $\|\mathbf{AB}\| \leq \|\mathbf{A}\| \cdot \|\mathbf{B}\|.$ (4.28)

For additional information about norms, see, for example, Lancaster (1969, Chap. 6). Regarding the norm $\|\mathbf{A}\|$ as a function of the matrix $\mathbf{A}$, and the vector norm $\|\mathbf{x}\|$ as a function defined over vectors $\mathbf{x}$, $\|\mathbf{A}\|$ and $\|\mathbf{x}\|$ are *consistent* if for all $\mathbf{A}$ and $\mathbf{x}$,

$$\|\mathbf{Ax}\| \leq \|\mathbf{A}\| \cdot \|\mathbf{x}\|.$$

───── **EXAMPLE 4.13** ─────────────────────────

For any column vector $\mathbf{x} = (x_1, \ldots, x_n)^T$ define

$$\|\mathbf{x}\| = \max_{1 \leq i \leq n} |x_i|.$$

This is called the *uniform norm*. The uniform matrix norm is defined for any $n \times n$ matrix $\mathbf{A} = (a_{ij})$ by

$$\|\mathbf{A}\| = \max_{1 \leq i \leq n} \sum_{j=1}^{n} |a_{ij}|$$

and it is consistent with the uniform vector norm. For

$$\|\mathbf{Ax}\| = \max_{1 \leq i \leq n} \left| \sum_{j=1}^{n} a_{ij} x_j \right| \leq \max_{1 \leq i \leq n} \sum_{j=1}^{n} |a_{ij}| \cdot |x_j|$$

$$\leq \left(\max_{1 \leq i \leq n} \sum_{j=1}^{n} |a_{ij}| \right) \max_{1 \leq i \leq n} |x_i| = \|\mathbf{A}\| \cdot \|\mathbf{x}\|.$$

■

───── **EXAMPLE 4.14** ─────────────────────────

The vector and matrix norms defined by

$$\|\mathbf{x}\| = \left(\sum_{i=1}^{n} x_i^2 \right)^{1/2}$$

and

$$\|\mathbf{A}\| = \left(\sum_{i=1}^{n} \sum_{j=1}^{n} a_{ij}^2 \right)^{1/2}$$

are consistent. The vector norm $\|\mathbf{x}\|$ is called the *Euclidean norm*, and $\|\mathbf{A}\|$ the *Frobenius norm*.

■

An obvious but highly significant consequence of the definition of norm is that for $\tilde{x}$ the computed approximation to the solution of $Ax = b$, and r the residual $r = A\tilde{x} - b$,

$$\|\tilde{x} - x\| \leq \|A^{-1}\| \cdot \|r\| \tag{4.29}$$

for any consistent vector and matrix norms.

This inequality can be verified as follows. The definition of the residual vector implies that

$$r = A\tilde{x} - b = A\tilde{x} - Ax = A(\tilde{x} - x).$$

By premultiplying this relation by the inverse A^{-1}, we get

$$\tilde{x} - x = A^{-1}r,$$

and by taking the norm of both sides and using the definition of consistent norm, we obtain

$$\|\tilde{x} - x\| = \|A^{-1}r\| \leq \|A^{-1}\| \cdot \|r\|.$$

Formula (4.29) allows us to bound the error of the computed solution vector $\tilde{x}$ in terms of the residual vector and the norm of the inverse of the coefficient matrix.

With the notion of matrix norm in hand, we define the *condition number* of a nonsingular matrix A by

$$\text{Cond } (A) = \|A\| \cdot \|A^{-1}\|.$$

One says that a matrix is *ill-conditioned* if Cond (A) is a large number. Now ill-conditionedness is not precisely defined, since it will depend on the matrix norm used as well as whatever one takes as "large." As a rule of thumb, for the norms above, if Cond (A) is greater than 1000, one should be alert for trouble. Always, Cond $(A) \geq 1$, which can be proven in the following way. Observe first that

$$\text{Cond } (A) = \|A\| \cdot \|A^{-1}\| \geq \|AA^{-1}\| = \|I\|.$$

On the other hand, with any nonzero vector x,

$$\|x\| = \|Ix\| \leq \|I\| \cdot \|x\|,$$

and by dividing this inequality by $\|x\| \neq 0$, we get

$$\|I\| \geq 1.$$

The condition number allows compact expressions relating error in the solution of a linear equation to errors or perturbations in the coefficient matrix A or vector b. Assume that the right-hand-side vector b is slightly modified to be $\tilde{b}$. Let x^*

be the solution of $\mathbf{Ax} = \mathbf{b}$, and $\tilde{\mathbf{x}}$ the solution of $\mathbf{Ax} = \tilde{\mathbf{b}}$. Then since $\|\tilde{\mathbf{x}} - \mathbf{x}^*\|$ $\leq \|\mathbf{A}^{-1}\| \|\tilde{\mathbf{b}} - \mathbf{b}\|$ and $\|\mathbf{x}^*\| \geq \|\mathbf{b}\|/\|\mathbf{A}\|$, we see that

$$\frac{\|\tilde{\mathbf{x}} - \mathbf{x}^*\|}{\|\mathbf{x}^*\|} \leq (\|\mathbf{A}^{-1}\| \|\mathbf{A}\|) \frac{\|\tilde{\mathbf{b}} - \mathbf{b}\|}{\|\mathbf{b}\|}.$$

That is,

$$\frac{\|\tilde{\mathbf{x}} - \mathbf{x}^*\|}{\|\mathbf{x}^*\|} \leq \text{Cond } (\mathbf{A}) \frac{\|\tilde{\mathbf{b}} - \mathbf{b}\|}{\|\mathbf{b}\|}. \tag{4.30}$$

Formulas related to (4.30) are derived in Szidarovszky and Yakowitz (1978, Sec. 6.3). They are useful in bounding error in the solution resulting from roundoff in the coefficients. Golub and Van Loan (1983, p. 72) offer the rule of thumb that under the uniform norm, and with error in Gaussian elimination arising from roundoff only, for $\mathbf{x}^*$ the exact and $\tilde{\mathbf{x}}$ the computed solution of $\mathbf{Ax} = \mathbf{b}$,

$$\boxed{\frac{\|\mathbf{x}^* - \tilde{\mathbf{x}}\|}{\|\mathbf{x}^*\|} \approx \text{Cond } (\mathbf{A}) \cdot \text{(machine epsilon)}.} \tag{4.31}$$

EXAMPLE 4.15

In Example 4.6, we found that if we chose the coefficients of the matrix $\mathbf{A}$ to be random numbers, then the matrix solution seemed stable, even with order $n = 80$. For the particular case of $n = 6$, and the coefficients picked randomly according to the program in Table 4.6, we computed that under the uniform norm, Cond $(\mathbf{A}) = 51.9$. Next we took $\mathbf{A} = (a_{ij})$ to be the Vandermonde matrix with elements $a_{ij} = (0.2j)^{i-1}$, which was seen in Examples 4.7 and 3.10 to have unstable behavior. The order n was taken to be 6, and after an obvious modification of the program in Table 4.15, we found the condition number to be about 79,000.

TABLE 4.15 Program to Calculate the Condition Number of a Matrix

```
C       PROGRAM CONDNO
C
C       ************************************************************
C       GENERATES A 6 BY 6 MATRIX,INVERTS THE MATRIX,COMPUTES THE
C       UNIFORM NORMS OF A AND A INVERSE AND COMPUTES THE CONDITION
C       NUMBER OF MATRIX A
C       CALLS:  INVG,GAUSS
C       OUTPUT: A=6 BY 6 ORIGINAL MATRIX
C               ANORM =UNIFORM NORM OF MATRIX A
C               AINORM=UNIFORM NORM OF MATRIX A INVERSE
C               CONDNO=CONDITION NUMBER OF MATRIX A
C       ************************************************************
C
```

TABLE 4.15   (Continued)

```
        DIMENSION C(6,12),B(6,6),A(6,6)
C
C       *** GENERATE THE ELEMENTS OF MATRIX A              ***
C
        KSEED=12345
        DO 1 I=1,6
        DO 1 J=1,6
          A(I,J)=RAN(KSEED)
      1 CONTINUE
        WRITE(10,4)((A(I,J),J=1,6),I=1,6)
      4 FORMAT(2X,6F9.6,/)
        EPS=1.E-7
C
C       *** SUBROUTINE INVG COMPUTES THE INVERSE OF MATRIX A   ***
C
        CALL INVG(6,A,B,EPS,C)
C
C       *** COMPUTES THE UNIFORM NORM OF MATRIX A AND A INVERSE ***
C
        DO 3 J=1,6
          AN=0.
          AIN=0.
          DO 2 K=1,6
            AN=AN+ABS(A(J,K))
            AIN=AIN+ABS(B(J,K))
      2   CONTINUE
        ANORM=AMAX1(AN,ANORM)
        AINORM=AMAX1(AIN,AINORM)
      3 CONTINUE
C
C       *** COMPUTES THE CONDITION NUMBER OF MATRIX A         ***
C
        CONDNO=ANORM*AINORM
        WRITE(10,7)ANORM,AINORM,CONDNO
      7 FORMAT(//' NORMS AND CONDITION NUM.',//3E15.7)
        STOP
        END
```

Direct calculation of condition number by a technique such as in Table 4.15 is worthwhile if the order n of the linear system of equations is not large and if we feel any uneasiness at all about the computed solution. However, the expense of the computation is about three times that of solving the linear system itself, and for large systems, therefore, one would like to sidestep this expense. Golub and Van Loan (1983, Sec. 4.5), and Rice (1983, Sec. 6.6B) offer numerical approximations of the condition number, the computational effort of which only grows as n^2, instead of n^3, as exact calculation requires.

*4.5
LU DECOMPOSITION AND CHOLESKY'S METHOD

*4.5.1   Dolittle Reduction

First we show the relation of the Gaussian elimination method to a special triangular factorization of square matrices. Assume that matrix **A** is square and nonsingular. Let $\mathbf{A}_k$ denote the coefficient matrix determined by the kth derived system. That is, $\mathbf{A}_k$ is the same as $\mathbf{A}^{(k)} = (a_{ij}^{(k)})$, as determined by (4.9), except that the $(n + 1)$st column is omitted. One may then verify that $\mathbf{A}_k = \mathbf{L}_k \mathbf{A}_{k-1}$, where

$$
\mathbf{L}_k =
\begin{bmatrix}
1 & & & & & & \\
 & \cdot & & & & & \\
 & & \cdot & & & & \\
 & & & \cdot & & & \\
 & & & 1 & & & \\
 & & & -\dfrac{a_{k+1,k}^{(k-1)}}{a_{kk}^{(k-1)}} & 1 & & \\
 & & & \cdot & & \cdot & \\
 & & & \cdot & & & \cdot \\
 & & & -\dfrac{a_{nk}^{(k-1)}}{a_{kk}^{(k-1)}} & & & 1
\end{bmatrix}
\tag{4.32}
$$

(all elements not shown here are equal to zero).

Repeated application of this principle implies that

$$
\mathbf{A}_{n-1} = \mathbf{L}_{n-1}\mathbf{A}_{n-2} = \cdots = \mathbf{L}_{n-1}\mathbf{L}_{n-2}\cdots\mathbf{L}_1\mathbf{A}.
\tag{4.33}
$$

Since the inverse matrix

$$
\mathbf{L}_k^{-1} =
\begin{bmatrix}
1 & & & & & & \\
 & \cdot & & & & & \\
 & & \cdot & & & & \\
 & & & \cdot & & & \\
 & & & 1 & & & \\
 & & & \dfrac{a_{k+1,k}^{(k-1)}}{a_{kk}^{(k-1)}} & 1 & & \\
 & & & \cdot & & \cdot & \\
 & & & \cdot & & & \cdot \\
 & & & \dfrac{a_{nk}^{(k-1)}}{a_{kk}^{(k-1)}} & & & 1
\end{bmatrix}
\tag{4.34}
$$

is lower triangular with unit diagonal elements, the matrix

$$
\mathbf{L} = (\mathbf{L}_{n-1}\mathbf{L}_{n-2}\cdots\mathbf{L}_1)^{-1} = \mathbf{L}_1^{-1}\cdots\mathbf{L}_{n-2}^{-1}\mathbf{L}_{n-1}^{-1}
$$

is also lower triangular with unit diagonal elements, and (4.33) implies that

$$
\mathbf{A} = \mathbf{LU},
\tag{4.35}
$$

where $\mathbf{U} = \mathbf{A}_{n-1}$. In this decomposition $\mathbf{U}$ is the upper triangular matrix produced in the final derived system of Gaussian elimination, and as we have seen, $\mathbf{L}$ is a lower triangular matrix with unit diagonal elements. Representation (4.35) is called the *Dolittle LU decomposition* of $\mathbf{A}$. (The Dolittle decomposition is very close to the well-known *Crout decomposition*. The only difference between the two is that in the Crout case, $\mathbf{U}$ has unit diagonal values, and the diagonal values of the lower-diagonal matrix $\mathbf{L}$ are unrestricted.) We will soon show that a linear system may quickly be solved once the LU decomposition is known. But first, let us see how to decompose efficiently.

Decomposition (4.35) can be computed directly, without performing Gaussian elimination. Let $\mathbf{A} = (a_{ij})$, $\mathbf{L} = (l_{ij})$, and $\mathbf{U} = (u_{ij})$. Then (4.35) implies that for all i and j,

$$a_{ij} = \sum_{k=1}^{\min\{i,j\}} l_{ik}u_{kj}, \tag{4.36}$$

since for $k > j$, $u_{kj} = 0$ and for $i < k$, $l_{ik} = 0$.

Now we may successively determine the elements of $\mathbf{U}$ and $\mathbf{L}$ as follows. If $i \leq j$, then by recalling that $l_{ii} = 1$ and by rearranging (4.36), we get

$$u_{ij} = a_{ij} - \sum_{k=1}^{i-1} l_{ik}u_{kj} \qquad (j = i, \ldots, n). \tag{4.37}$$

Observe that the right-hand side depends only on the first $i - 1$ rows of $\mathbf{U}$ and $i - 1$ columns of $\mathbf{L}$. If $i > j$, then, similarly,

$$l_{ij} = \frac{1}{u_{jj}}\left(a_{ij} - \sum_{k=1}^{j-1} l_{ik}u_{kj}\right) \qquad (j = 1, \ldots, i - 1). \tag{4.38}$$

In our subroutine, the actual computation begins by defining $u_{11} = a_{11}$, and $l_{ii} = 1$, $l_{i1} = a_{i1}/u_{11}$, $i = 2, \ldots, n$. Then for each $j = 2, \ldots, n$, we compute u_{ij} and l_{ij} by (4.37) and (4.38) for increasing i. By this procedure, each coefficient is available by the time it is needed. In brief, $\mathbf{U}$ and $\mathbf{L}$ are found, column by column. This rule for constructing the LU decomposition is called *Dolittle reduction*. Observe that the method can be used if and only if all diagonal elements $u_{11}, u_{22}, \ldots, u_{nn}$ differ from zero. As in Gaussian elimination, partial pivoting for avoiding small values in the diagonal of $\mathbf{U}$ may be accomplished by row interchanges. We refer the reader to Stewart (1973, pp. 137–139) for details on

TABLE 4.16   Subroutine COMPLU for LU Decomposition

```
        SUBROUTINE COMPLU(N,A,L,U,DELT)
C
C     ****************************************************************
C     *   FUNCTION: THIS SUBROUTINE COMPUTES THE LU DECOMPOSITION    *
C     *             OF A GIVEN MATRIX A USING DOLITTLE REDUCTION      *
C     *   USAGE:                                                      *
C     *       CALL SEQUENCE: CALL COMPLU(N,A,L,U,DELT)                *
C     *   PARAMETERS:                                                 *
C     *       INPUT:                                                  *
C     *           N=ORDER OF THE MATRIX                               *
C     *           A=N BY N ARRAY OF MATRIX ELEMENTS                   *
C     *         DELT=MACHINE ZERO (TOLERANCE)                         *
C     *       OUTPUT:                                                 *
C     *           L=N BY N ARRAY OF LOWER TRIANGULAR MATRIX           *
C     *             ELEMENTS (ALL ZERO ELEMENTS ABOVE DIAGONAL)       *
C     *           U=N BY N ARRAY OF UPPER TRIANGULAR MATRIX           *
C     *             ELEMENTS (ALL ZERO ELEMENTS BELOW DIAGONAL)       *
C     ****************************************************************
C
```

TABLE 4.16   (Continued)

```
      DIMENSION A(N,N),U(N,N)
      REAL L(N,N)
      U(1,1)=A(1,1)
C     *** TEST IF MATRIX A CONTAINS A SINGLE ELEMENT ***
      IF(N.LE.1) THEN
         L(1,1)=1.0
         RETURN
      END IF
C     *** TEST IF MATRIX U'S (1,1) ELEMENT IS VERY SMALL ***
      IF(ABS(U(1,1)).LT.DELT) THEN
         WRITE(6,6)
         STOP
      END IF
      DO 1 I=2,N
         L(I,1)=A(I,1)/U(1,1)
    1 CONTINUE
      DO 2 J=2,N
         U(1,J)=A(1,J)
C     *** COMPUTE JTH COLUMN OF U ***
         DO 3 I=2,J
            SUM=A(I,J)
            DO 9 K=1,I-1
               SUM=SUM-L(I,K)*U(K,J)
    9       CONTINUE
      U(I,J)=SUM
    3 CONTINUE
         IF(J.NE.N) THEN
C     *** COMPUTE JTH COLUMN OF L ***
         DO 4 I=J+1,N
            SUM=A(I,J)
            DO 5 K=1,J-1
               SUM=SUM-L(I,K)*U(K,J)
    5       CONTINUE
            L(I,J)=SUM
            IF(ABS(U(J,J)).LT. DELT) THEN
               WRITE(6,6)
    6          FORMAT(2X,' THE METHOD CANNOT BE USED ')
               STOP
            END IF
            L(I,J)=L(I,J)/U(J,J)
    4    CONTINUE
         END IF
    2 CONTINUE
      DO 7 J=1,N
         L(J,J)=1.0
    7 CONTINUE
      DO 8 I=2,N
         DO 8 J=1,I-1
            U(I,J)=0.0
            L(J,I)=0.0
    8 CONTINUE
      RETURN
      END
```

pivoting and overwriting.

Subroutine COMPLU (Table 4.16) implements LU decomposition by (4.37) and (4.38). A more sophisticated code would save memory by "overwriting" **A** with **L** and **U**, instead of requiring separate arrays.

The computational effort required by the Dolittle reduction approach is essentially the same as that of Gauss elimination. One advantage of Dolittle reduction is that whereas by forward elimination, for fixed i, j, $a_{ij}^{(k)}$ is pulled in and out of storage as k goes from 1 to $n - 1$, with Dolittle reduction, the values u_{ij} and l_{ij} are computed once and for all according to (4.37) and (4.38), respectively. By this "one-pass" feature, we have the capability of accumulating sums in double precision and thereby reducing the effect of roundoff error, while avoiding the memory burden of declaring all storage locations to be double precision.

━━━━━ **EXAMPLE 4.16** ━━━

Consider again the matrix given in Examples 4.1 and 4.2. The triangular decomposition of the form

$$\begin{bmatrix} 2 & 1 & 1 \\ 1 & 3 & 2 \\ 1 & 2 & 2 \end{bmatrix} = \begin{bmatrix} 1 & 0 & 0 \\ l_{21} & 1 & 0 \\ l_{31} & l_{32} & 1 \end{bmatrix} \begin{bmatrix} u_{11} & u_{12} & u_{13} \\ 0 & u_{22} & u_{23} \\ 0 & 0 & u_{33} \end{bmatrix}$$

will now be calculated.

For $i = 1$, (4.37) implies that

$$u_{11} = a_{11} = 2$$

$$u_{12} = a_{12} = 1$$

$$u_{13} = a_{13} = 1,$$

since $l_{11} = 1$. Then equation (4.38) gives

$$l_{21} = \frac{a_{21}}{u_{11}} = \frac{1}{2}, \qquad l_{31} = \frac{a_{31}}{u_{11}} = \frac{1}{2}.$$

For $i = 2$, we find that

$$u_{22} = a_{22} - l_{21}u_{12} = 3 - \frac{1}{2} = \frac{5}{2}$$

$$u_{23} = a_{23} - l_{21}u_{13} = 2 - \frac{1}{2} = \frac{3}{2}$$

$$l_{32} = \frac{a_{32} - l_{31}u_{12}}{u_{22}} = \frac{2 - 1/2}{5/2} = \frac{3}{5},$$

and finally, for $i = 3$,

$$u_{33} = a_{33} - l_{31}u_{13} - l_{32}u_{23} = 2 - \tfrac{1}{2} - \tfrac{3}{5} \cdot \tfrac{3}{2} = \tfrac{3}{5}.$$

Thus the LU decomposition of **A** is

$$\begin{bmatrix} 2 & 1 & 1 \\ 1 & 3 & 2 \\ 1 & 2 & 2 \end{bmatrix} = \begin{bmatrix} 1 & 0 & 0 \\ \frac{1}{2} & 1 & 0 \\ \frac{1}{2} & \frac{3}{5} & 1 \end{bmatrix} \begin{bmatrix} 2 & 1 & 1 \\ 0 & \frac{5}{2} & \frac{3}{2} \\ 0 & 0 & \frac{3}{5} \end{bmatrix}.$$

Note that matrix **U** and the coefficient matrix of the second derived system given in Example 4.1 are the same.   ■

Once the triangular decomposition $\mathbf{A} = \mathbf{LU}$ is known, the solution of $\mathbf{Ax} = \mathbf{b}$ can be found in two steps. First, solve the equation

$$\boxed{\mathbf{Lz} = \mathbf{b},} \qquad (4.39)$$

where $\mathbf{L}$ is the lower triangular matrix and $\mathbf{z}$ is the unknown vector. An obvious modification of back substitution can be used, where z_1 is calculated from the first (rather than the last) equation, then z_2 is determined from the second equation, and so on. Next solve the equation

$$\boxed{\mathbf{Ux} = \mathbf{z},} \qquad (4.40)$$

where $\mathbf{x}$ is the unknown vector, and the result $\mathbf{z}$ of the previous step constitutes the coefficient vector. The solution $\mathbf{x}$ of this equation can be obtained by use of the original version (4.10) of back substitution. Then $\mathbf{x}$ is the solution of the original equations, since, by substitution, we have

$$\mathbf{Ax} = (\mathbf{LU})\mathbf{x} = \mathbf{L}(\mathbf{Ux}) = \mathbf{Lz} = \mathbf{b}. \qquad (4.41)$$

───── **EXAMPLE 4.17** ─────

By using the result of Example 4.16, we shall now solve the equations given in Example 4.1. In this case (4.39) has the particular form

$$z_1 \qquad\qquad = 4$$
$$\tfrac{1}{2}z_1 + z_2 \qquad = 6$$
$$\tfrac{1}{2}z_1 + \tfrac{3}{5}z_2 + z_3 = 5,$$

which implies that $z_1 = 4$, $z_2 = 4$, and $z_3 = \tfrac{3}{5}$. Now (4.40) can be written as

$$2x_1 + x_2 + x_3 = 4$$
$$\tfrac{5}{2}x_2 + \tfrac{3}{2}x_3 = 4$$
$$\tfrac{3}{5}x_3 = \tfrac{3}{5},$$

and after back substitution $x_1 = x_2 = x_3 = 1$. ∎

*4.5.2 Cholesky's Method

In the special case that $\mathbf{A}$ is real and symmetric, the LU decomposition can be modified so that the two factors are the transpose of each other. That is, the factorization

$$\mathbf{A} = \mathbf{U}^T\mathbf{U} \qquad (4.42)$$

can be obtained, in which $\mathbf{U}$ is upper triangular. Specifically, by arguments leading to the Dolittle reduction (4.36), one can verify that the elements of $\mathbf{U}$ can be obtained by the recurrence relations

$$
\begin{aligned}
u_{ii} &= \left(a_{ii} - \sum_{k=1}^{i-1} u_{ki}^2 \right)^{1/2}, \qquad i = 1, 2, \ldots, n, \\
u_{ij} &= \frac{1}{u_{ii}} \left(a_{ij} - \sum_{k=1}^{i-1} u_{ki} u_{kj} \right) \qquad (j > i).
\end{aligned}
\tag{4.43}
$$

This decomposition is called *Cholesky's method* or the *method of square roots*.

──────── **EXAMPLE 4.18** ────────

Cholesky's method will be applied to the symmetric matrix

$$
\mathbf{A} = \begin{bmatrix} 2 & 1 & 1 \\ 1 & 3 & 2 \\ 1 & 2 & 2 \end{bmatrix}.
$$

For $i = 1$, (4.43) yields

$$
u_{11} = \sqrt{a_{11}} = \sqrt{2}
$$

$$
u_{12} = \frac{a_{12}}{u_{11}} = \frac{\sqrt{2}}{2}
$$

$$
u_{13} = \frac{a_{13}}{u_{11}} = \frac{\sqrt{2}}{2}.
$$

For $i = 2$, we have

$$
u_{22} = \sqrt{a_{22} - u_{12}^2} = \sqrt{3 - \frac{1}{2}} = \frac{\sqrt{10}}{2}
$$

$$
u_{23} = \frac{a_{23} - u_{12}u_{13}}{u_{22}} = \frac{2 - (\sqrt{2}/2)(\sqrt{2}/2)}{\sqrt{10}/2} = \frac{3\sqrt{10}}{10},
$$

and finally, for $i = 3$, we find that

$$
u_{33} = \sqrt{a_{33} - u_{13}^2 - u_{23}^2} = \sqrt{2 - \frac{1}{2} - \frac{9}{10}} = \frac{\sqrt{15}}{5}.
$$

Thus

$$\mathbf{U} = \begin{bmatrix} \sqrt{2} & \dfrac{\sqrt{2}}{2} & \dfrac{\sqrt{2}}{2} \\[3mm] 0 & \dfrac{\sqrt{10}}{2} & \dfrac{3\sqrt{10}}{10} \\[3mm] 0 & 0 & \dfrac{\sqrt{15}}{5} \end{bmatrix}.$$

∎

Because the arguments of the square root in (4.43) can be negative, Cholesky's method can require complex arithmetic. In the important special case in which $\mathbf{A}$ is positive definite, that is, $\mathbf{x}^T\mathbf{A}\mathbf{x} > 0$ for all nonzero vectors $\mathbf{x}$, Cholesky's method requires only real numbers and pivoting is never needed for avoiding division by zero.

*4.6
ITERATIVE METHODS

For linear systems of small or moderate size (say, of order not exceeding 200), either Gaussian elimination or LU decomposition is effective and efficient, and these are the methods recommended. For certain classes of high-order linear equations, which arise, for example, in solving differential equations, iterative methods are attractive. In this section some of the better-known iterative methods are described.

*4.6.1   Jacobi Iteration

Consider again (4.1) with $m = n$. Assume that $\mathbf{A}$ is nonsingular and the rows have been exchanged, as necessary, so that the diagonal elements are nonzero. Equations (4.1) can then be rewritten so that the ith equation is explicit for x_i:

$$x_1 = -\frac{1}{a_{11}}(a_{12}x_2 + a_{13}x_3 + \cdots + a_{1n}x_n - b_1)$$

$$x_2 = -\frac{1}{a_{22}}(a_{21}x_1 + a_{23}x_3 + \cdots + a_{2n}x_n - b_2)$$

$$\cdots\cdots$$

$$\tag{4.44}$$

$$x_n = -\frac{1}{a_{nn}}(a_{n1}x_1 + a_{n2}x_2 + \cdots + a_{n,n-1}x_{n-1} - b_n).$$

Assume that an initial approximation of the solution has been given, or simply choose an arbitrary vector. Let $\mathbf{x}^{(0)} = (x_1^{(0)}, \ldots, x_n^{(0)})^T$ denote this initial approximation. Substitute it into the right-hand side of (4.44) and evaluate. The elements of the resulting vector give the next approximations of the unknowns. Let vector $\mathbf{x}^{(1)} = (x_1^{(1)}, \ldots, x_n^{(1)})^T$ denote these approximations, and then substitute the new vector $\mathbf{x}^{(1)}$ into the right side of (4.44) to get a further approximation, $\mathbf{x}^{(2)} = (x_1^{(2)}, \ldots, x_n^{(2)})^T$, and so on. The general step is given by

$$
\begin{aligned}
x_1^{(k+1)} &= -\frac{1}{a_{11}} (a_{12}x_2^{(k)} + a_{13}x_3^{(k)} + \cdots + a_{1n}x_n^{(k)} - b_1) \\[2mm]
x_2^{(k+1)} &= -\frac{1}{a_{22}} (a_{21}x_1^{(k)} + a_{23}x_3^{(k)} + \cdots + a_{2n}x_n^{(k)} - b_2) \\[2mm]
&\quad \cdots \cdots \\[2mm]
x_n^{(k+1)} &= -\frac{1}{a_{nn}} (a_{n1}x_1^{(k)} + a_{n2}x_2^{(k)} + \cdots + a_{n,n-1}x_{n-1}^{(k)} - b_n).
\end{aligned}
\tag{4.45}
$$

Scheme (4.45) is called the *Jacobi iteration method*.

It can be proven that under certain conditions (see Szidarovszky and Yakowitz, 1978, Sec. 6.2), for $k \to \infty$, the sequence of vectors $\mathbf{x}^{(k)}$ converges to the exact solution of equations (4.1). One such condition is that each diagonal element of the matrix of coefficients satisfy the condition:

$$
|a_{ii}| > \sum_{\substack{j=1 \\ j \neq i}}^{n} |a_{ij}|, \qquad i = 1, \ldots, n.
\tag{4.46}
$$

If this condition is satisfied, then $\mathbf{A}$ is said to be *diagonally dominant*.

The criteria for stopping the iteration process are usually either:

1. The number of iterations has exceeded some predetermined maximum K, or
2. The difference between successive values of all x_i's are less than some predetermined tolerance ε.

One iteration step requires at most n divisions, n^2 multiplications, and n^2 additions (or subtractions), which implies that if the number of iteration steps to get the desired accuracy of the solution is less than $n/3$, this method needs fewer arithmetic operations than the Gaussian elimination.

━━━━━ **EXAMPLE 4.19** ━━━━━

We wish to apply the Jacobi method to the following system:

$$
\begin{aligned}
64x_1 - 3x_2 - x_3 &= 14 \\
x_1 + x_2 + 40x_3 &= 20 \\
2x_1 - 90x_2 + x_3 &= -5.
\end{aligned}
$$

The matrix of coefficients is not diagonally dominant, since in the second equation

$$
|1| < |1| + |40|.
$$

TABLE 4.17   Jacobi Iterations for Example 4.19

	k = 0	k = 1	k = 2	k = 3	k = 4	k = 5
$x_1^{(k)}$	0.21875	0.22916	0.22959	0.22955	0.22954	0.22954
$x_2^{(k)}$	0.05556	0.06598	0.06616	0.06613	0.06613	0.06613
$x_3^{(k)}$	0.50000	0.49592	0.49262	0.49261	0.49261	0.49261

But by interchanging the second and third equations, a system with a diagonally dominant matrix of coefficients is obtained:

$$64x_1 - 3x_2 - x_3 = 14$$

$$2x_1 - 90x_2 + x_3 = -5$$

$$x_1 + x_2 + 40x_3 = 20.$$

For this case, the Jacobi iteration scheme (4.45) takes the particular form

$$x_1^{(k+1)} = \qquad\qquad 0.04688x_2^{(k)} + 0.01563x_3^{(k)} + 0.21875$$

$$x_2^{(k+1)} = \quad 0.02222x_1^{(k)} \qquad\qquad + 0.01111x_3^{(k)} + 0.05556$$

$$x_3^{(k+1)} = -0.02500x_1^{(k)} - 0.02500x_2^{(k)} \qquad\qquad + 0.50000,$$

where all coefficients have been rounded to five decimal places. Let the error bound of the successive values of x_i be given as 10^{-5}, and take $\mathbf{x}^{(0)}$ to be the constant terms on the right-hand sides of the iteration scheme. The numerical results are presented in Table 4.17. ∎

*4.6.2   Gauss–Seidel Iteration

Consider again the recursive Jacobi iteration scheme (4.45). Observe that in calculating the "new" value of $x_2^{(k+1)}$, the previous value of $x_1^{(k)}$ is used on the right-hand side although the "new" value, $x_1^{(k+1)}$ is already known. Similarly, for obtaining the new value $x_3^{(k+1)}$, the "old" values $x_1^{(k)}$ and $x_2^{(k)}$ are used, although the new, and presumably more accurate values $x_1^{(k+1)}$ and $x_2^{(k+1)}$ of these variables are already available. A modification typically (but not always) giving faster convergence can be devised if in the calculation of $x_i^{(k+1)}$ ($2 \le i \le n$), the updated new values $x_1^{(k+1)}, \ldots, x_{i-1}^{(k+1)}$ are used in place of the earlier values $x_1^{(k)}, \ldots, x_{i-1}^{(k)}$, in (4.45). This modification results in the Gauss–Seidel iterative method, which is defined by the recursive scheme

$$x_1^{(k+1)} = -\frac{1}{a_{11}}(a_{12}x_2^{(k)} + a_{13}x_3^{(k)} + \cdots + a_{1n}x_n^{(k)} - b_1)$$

$$x_2^{(k+1)} = -\frac{1}{a_{22}}(a_{21}x_1^{(k+1)} + a_{23}x_3^{(k)} + \cdots + a_{2n}x_n^{(k)} - b_2)$$

$$\cdot\quad\cdot\qquad\cdot\quad\cdot\quad\cdot\qquad\cdot\quad\cdot\quad\cdot\quad\cdot$$

$$x_n^{(k+1)} = -\frac{1}{a_{nn}}(a_{n1}x_1^{(k+1)} + a_{n2}x_2^{(k+1)} + \cdots + a_{n,n-1}x_{n-1}^{(k+1)} - b_n).$$

(4.47)

Conditions for the convergence of this method are discussed by Szidarovszky and Yakowitz (1978, Sec. 6.2). Table 4.18 gives a computer subroutine for the Gauss–Seidel method.

TABLE 4.18   Subroutine SEID for Gauss–Seidel Iterations

```
        SUBROUTINE SEID(N,X0,X,A,B,EPS)
C
C     ****************************************************************
C     *  FUNCTION: THIS SUBROUTINE COMPUTES THE SOLUTION FOR A      *
C     *            SYSTEM OF N EQUATIONS IN N UNKNOWNS USING THE    *
C     *            GAUSS SEIDEL ITERATION METHOD                    *
C     *  USAGE:                                                     *
C     *       CALL SEQUENCE: CALL SEID(N,X0,X,A,B,EPS)              *
C     *  PARAMETERS:                                                *
C     *       INPUT:                                                *
C     *          N=NUMBER OF EQUATIONS AND UNKNOWNS (< 500)         *
C     *          X0=N BY 1 ARRAY OF INITIAL SOLUTION VALUES         *
C     *          A=N BY N ARRAY OF MATRIX COEFFICIENTS              *
C     *          B=N BY 1 ARRAY OF RIGHT HAND SIDE VECTOR           *
C     *         EPS=ERROR BOUND                                     *
C     *       OUTPUT:                                               *
C     *          X=N BY 1 ARRAY OF SOLUTION VALUES                  *
C     ****************************************************************
C
        DIMENSION A(N,N),B(N),X0(N),X(N),U(500)
C     *** INITIALIZATION ***
        K=1
        M=1
        DO 1 I=1,N
          U(I)=X0(I)
          B(I)=B(I)/A(I,I)
          DO 5 J=1,N
             IF(I.NE.J) A(I,J)=A(I,J)/A(I,I)
     5    CONTINUE
          A(I,I)=0.0
     1  CONTINUE
        DO WHILE(M.LT.N+1)
C     *** COMPUTE SOLUTION VALUES X ***
          DO 2 I=1,N
            X(I)=0.0
            DO 3 J=1,N
               X(I)=X(I)-A(I,J)*U(J)
     3      CONTINUE
            X(I)=X(I)+B(I)
            U(I)=X(I)
     2    CONTINUE
C     *** TEST IF SOLUTION VALUES X ARE CLOSE ***
          M=1
          DO WHILE(ABS(X(M)-X0(M)).LT.EPS.AND.M.LT.N+1)
            M=M+1
          END DO
C     *** IF NOT THEN RESET THE INITIAL X0 ***
C     ***    AND CONTINUE ITERATION PROCESS   ***
          IF(M.LT.N+1) THEN
            DO 4 I=1,N
               X0(I)=X(I)
     4      CONTINUE
C     *** OTHERWISE SOLUTION VALUES X ARE GOOD ***
C     ***    CEASE ITERATION PROCESS AND RETURN   ***
            K=K+1
          END IF
        END DO
        RETURN
        END
```

─────── **EXAMPLE 4.20** ───────────────────────────────

Application of Gauss–Seidel iteration to the equation in Example 4.19 is summarized in Table 4.19.

TABLE 4.19   Gauss–Seidel Iterations for Example 4.20

	$k = 0$	$k = 1$	$k = 2$
$x_1^{(k)}$	0.21875	0.22916	0.22955
$x_2^{(k)}$	0.05556	0.06621	0.06613
$x_3^{(k)}$	0.50000	0.49262	0.49261

In comparison to the Jacobi iteration calculation (Table 4.17), fewer steps are needed to obtain the same accuracy.   ■

Iterative methods are particularly popular for solution of band systems arising in numerical methods for partial differential equations. For such equations, in many cases the width k (see Section 4.3.3) is relatively wide—often proportional to $n^{1/2}$—but the band itself has relatively few nonzero entries. Let p denote a bound to the number of nonzero elements in each row. The effort required by elimination for solution of banded systems is proportional to $k^2 n$, and for iterative methods, proportional to $2pnM$, where M is the number of iterations necessary for acceptable accuracy. Thus iterative methods are preferable if

$$2pnM < k^2 n,$$

that is

$$M < \frac{k^2}{2p} \sim \frac{n}{p}.$$

Unfortunately, it is typically difficult to assess M until computations have already begun.

─────── **EXAMPLE 4.21** ───────────────────────────────

It is known (Szidarovszky and Yakowitz, 1978, Sec. 6.2) that the Gauss–Seidel method converges whenever the matrix $\mathbf{A}$ is positive definite, that is, whenever $\mathbf{A}$ is symmetric and $\mathbf{x}^T\mathbf{A}\mathbf{x} > 0$ for all $\mathbf{x}$ except the zero vector. In Table 4.20 we have called on subroutine SEID to solve a 100-variable equation with a positive definite coefficient matrix. The lower diagonal coefficients were chosen at random, as was the coefficient vector $\mathbf{b}$. The upper diagonal elements were then determined by the condition that the matrix be symmetric, and the diagonal elements were set to 100, a number large enough to ensure that the matrix is positive definite. The maximum residual value was printed out at each iteration (Table 4.21). The run was made on a VAX.

TABLE 4.20   Calling Program for the Gauss–Seidel Algorithm

```
C       PROGRAM SEIDEL
C
C       ****************************************************************
C       THIS PROGRAM WILL GENERATE A SYSTEM OF RANDOM NUMBERS
C       FOR THE COEFFICIENT MATRIX A AND THE RIGHT HAND SIDE VECTOR B
C       IT THEN CALLS SUBROUTINE SIED TO SOLVE THE SYSTEM
C       CALLS:    SEID
C       OUTPUT(FROM MODIFIED SUBROUTINE):
C               K=NUMBER ITERATION
C               EMAX=MAXIMUM (FOR ALL EQUATIONS)  ABSOLUTE
C                   ERROR
C       ****************************************************************
C
        DIMENSION A(100,100),B(100),X0(100),X(100)
        N=100
        KSEED=12345
C
C       *** FIRST THE SYSTEM IS GENERATED USING THE RANDOM NUMBER ***
C       *** GENERATOR RAN                                         ***
C
        DO 1 I=1,N
           DO 2 J=1,I
              A(I,J)=RAN(KSEED)
              A(J,I)=A(I,J)
    2      CONTINUE
           A(I,I)=N*1.
           B(I)=RAN(KSEED)
    1   CONTINUE
        EPS=1.E-9
C
C       *** SUBROUTINE SEID WILL SOLVE THE SYSTEM USING GAUSS    ***
C       *** SEIDEL ITERATION. PRINTING IS DONE IN THE SUBROUTINE ***
C
        CALL SEID(N,X0,X,A,B,EPS)
        STOP
        END
```

**TABLE 4.21
Application of the
Gauss–Seidel
Algorithm**

k	Absolute Value Maximum Coordinate in Residual After Iteration k
1	9.6686576E-03
2	2.2037476E-03
3	3.6594970E-04
4	2.3348257E-05
5	2.3739412E-06
6	3.8836151E-07
7	1.8626451E-08
8	4.6566129E-09
9	4.6566129E-10

*4.7
EIGENVALUES

Let $\mathbf{A}$ be an $n \times n$ matrix. An *eigenvalue* for $\mathbf{A}$ is a (perhaps complex) number λ which, for some nonzero vector $\mathbf{x}$, satisfies

$$\mathbf{A}\mathbf{x} = \lambda\mathbf{x}. \tag{4.48}$$

The vector $\mathbf{x}$ as above is the *eigenvector* associated with the eigenvalue λ. A matrix $\mathbf{A}$ may have complex eigenvalues and eigenvectors with complex coefficients, even in cases in which all coefficients of $\mathbf{A}$ are real.

Our discussion of eigenvalues and their computation will be devastatingly abridged. It is intended for the reader who has already some introduction to the theory [such as is offered by Bellman (1970) or Nering (1963)]. We merely offer an elementary algorithm, the *Rayleigh quotient iteration* (RQI) method and its subroutine, its analysis and justification being beyond the scope of this text. The present algorithm can be useful for certain simple, stable eigenvalue problems, and does share some foundation with the elegant and effective but, alas, highly complicated QR eigenvalue method. The theory for both the RQI and QR methods is given in detail in Szidarovszky and Yakowitz (1978, Secs. 7.2.3 and 7.3).

The RQI method is an iterative method, and as such, requires the user to provide starting estimates. Let $\mathbf{x}^{(0)}$ be an initial guess of the eigenvector. (For theoretical reasons, it is well to take an initial guess that has a nonzero imaginary part, or the method has no hope of locating complex eigenvalues.) Assume that after k iterations, we have obtained an approximation $\mathbf{x}^{(k)}$ of the eigenvector. A RQI iteration has two steps;

1. Approximate the eigenvalue by the Rayleigh quotient:

$$\lambda^{(k)} = \frac{(\mathbf{x}^{(k)})^T \mathbf{A} \mathbf{x}^{(k)}}{(\mathbf{x}^{(k)})^T \mathbf{x}^{(k)}}. \tag{4.49}$$

2. Update eigenvector approximation and normalize. Compute the solution $\mathbf{q} = (q_1, \ldots, q_n)^T$ of

$$(\mathbf{A} - \lambda^{(k)}\mathbf{I})\mathbf{q} = \mathbf{x}^{(k)} \tag{4.50}$$

and normalize, defining $\|\mathbf{q}\| = \Sigma_{i=1}^n |q_i|$, and define $\mathbf{x}^{(k+1)} = (x_1^{(k+1)}, \ldots, x_n^{(k+1)})^T$, where

$$x_i^{(k+1)} = \frac{q_i}{\|q\|} \qquad (1 \le i \le n).$$

When successive iterations $\mathbf{x}^{(k)}$ and $\mathbf{x}^{(k+1)}$ are within a prescribed tolerance, stop the process and accept the most recent estimates. Subroutine EIGEN (Table 4.22) implements the RQI method. It calls GAUS1, and requires that subroutine to have been appropriately modified for complex variables. In Table 4.22 we give these declaration modifications.

We have found the RQI method to be satisfactory for certain small matrices, and were it not for roundoff error, its properties would make it interesting. But the fundamental trouble is that when the successive approximations $\lambda^{(k)}$ of the eigenvalue λ are in fact getting close to λ, the matrix in (4.50) must become ill-conditioned. In fact, if $\lambda^{(k)}$ in (4.50) equals an eigenvalue, in view of (4.48) the matrix $\mathbf{A} - \lambda^{(k)}\mathbf{I}$ is singular. Another drawback to the RQI and related elementary techniques is that locating all the eigenvalues of a given matrix will either require a good deal of guesswork, or systematic matrix deflation, such as discussed in the preceding reference. In the computer study discussed in Example 4.22, we confront this problem by repeatedly calling EIGEN with randomly chosen initial eigenvector estimates until a complete set of eigenvalues are obtained.

The highly technical QR algorithm mentioned above overcomes both the instability problem and the problem of location of all the eigenvalues, and should be employed if we have a large (order greater than, say, 10) matrix. The eigenvalue subroutines in the IMSL package are founded on the QR approach.

TABLE 4.22   Subroutine EIGEN for Eigenvalue Computation*

```
      SUBROUTINE EIGEN(N,A,EPS,EIGVEC,EIGVAL)
C
C
C   *****************************************************************
C   *  FUNCTION: THIS SUBROUTINE COMPUTES AN EIGENVALUE AND         *
C   *            EIGENVECTOR OF THE MATRIX A BY RAYLEIGH            *
C   *            QUOTIENT ITERATIONS                               *
C   *  USAGE:                                                       *
C   *       CALL SEQUENCE: CALL EIGEN(N,A,EPS,EIGVEC,EIGVAL)        *
C   *       EXTERNAL FUNCTIONS/SUBROUTINES:                         *
C   *                      SUBROUTINE GAUS1(N,1,ND,B,DELT)         *
C   *                      (MODIFIED FOR COMPLEX*16)               *
C   *  PARAMETERS:                                                  *
C   *     INPUT:                                                    *
C   *         N=DIMENSION OF THE SQUARE MATRIX A                    *
C   *         A=N BY N MATRIX FOR WHICH THE EIGENVALUE AND          *
C   *           EIGENVECTOR IS TO BE COMPUTED. N MUST BE            *
C   *           LESS THEN OR EQUAL 20                               *
C   *       EPS=MINIMUM CHANGE TO COMPONENTS OF THE                 *
C   *           EIGENVECTOR OR TO THE EIGENVALUE TO PROCEED         *
C   *           TO THE NEXT ITERATION                               *
C   *     EIGVEC=INITIAL GUESS AT THE VALUE OF THE EIGENVECTOR     *
C   *     OUTPUT:                                                   *
C   *     EIGVEC=THE NORMALIZED EIGENVECTOR CORRESPONDING TO        *
C   *            EIGVAL                                             *
C   *     EIGVAL=THE COMPUTED EIGENVALUE                            *
C   *****************************************************************
C
      IMPLICIT COMPLEX*16 (A-H,O-Z)
      REAL EPS,ANORM,DELT
      DIMENSION A(N,N),EIGVEC(N),B(20,20)
      DELT=1.E-10
      EIGSAV=CMPLX(1.D+10,1.D+10)
      DO 100 WHILE (CDABS(EIGSAV-EIGVAL).GT.EPS)
C   *** UPDATE EIGVAL ***
      ANUM=CMPLX(0.D0,0.D0)
      DENOM=CMPLX(0.D0,0.D0)
      DO 60 I=1,N
        DENOM=DENOM+CONJG(EIGVEC(I))*EIGVEC(I)
        DO 50 J=1,N
          ANUM=ANUM+A(I,J)*CONJG(EIGVEC(I))*EIGVEC(J)
50      CONTINUE
60    CONTINUE
      EIGSAV=EIGVAL
      EIGVAL=ANUM/DENOM
```

TABLE 4.22   (Continued)

```
C      *** CHECK FOR STOPPING CRITERIA ***
          IF(CDABS(EIGSAV-EIGVAL).GT.EPS) THEN
C      *** ASSEMBLE THE LINEAR EQUATIONS ***
          DO 20 I=1,N
             B(I,N+1)=EIGVEC(I)
             DO 10 J=1,N
                B(I,J)=A(I,J)
                IF(I.EQ.J) B(I,J)=B(I,J)-EIGVAL
   10        CONTINUE
   20     CONTINUE
C      *** SOLVE THE SYSTEM OF EQUATIONS ***
          CALL GAUS1(N,1,20,B,DELT)
C      *** NORMALIZE THE EIGENVECTORS ***
          ANORM=0.0
          DO 30 I=1,N
             ANORM=ANORM+CDABS(B(I,N+1))
   30     CONTINUE
          DO 40 I=1,N
             EIGVEC(I)=B(I,N+1)/ANORM
   40     CONTINUE
       ELSE
       ENDIF
  100 CONTINUE
      RETURN
      END
```

*Modifications Needed for GAUS1 are

1. Declaration Statements

```
IMPLICIT COMPLEX*16 (A–H,O–Z)
REAL DELT,U
```

2. Replace ABS function by CDABS, throughout.

────── **EXAMPLE 4.22** ──────────────────────────

Here we find an eigenvalue/eigenvector pair for the matrix

$$\mathbf{A} = \begin{bmatrix} -4 & -3 & -7 \\ 2 & 3 & 2 \\ 4 & 2 & 7 \end{bmatrix}.$$

By means of the calling program in Table 4.23, we applied subroutine EIGEN to this matrix, to obtain the eigenvalue iterates displayed in Table 4.24. In a subsequent computation, the main program was modified to repeatedly obtain random initial eigenvector approximations (by calling the random number generator). Each such random starting eigenvector was passed to EIGEN and an eigenvalue/eigenvector pair was returned. After just a few repetitions, the complete set of eigenvalues, namely 1, 2, and 3, was uncovered, the processing time being negligible. Subroutine EIGEN and its calling program use double-precision complex variables to alleviate the ill-conditioning mentioned above. The variable declarations in GAUS1 must be modified accordingly, as shown in Table 4.22.

TABLE 4.23 Program for Eigenvalue Computation

```
C       PROGRAM EIGENVAL
C
C       ****************************************************************
C       THIS PROGRAM DEMONSTRATES THE COMPUTATION OF EIGENVALUES
C       AND EIGENVECTORS
C       CALLS:          EIGEN,GAUS1(MODIFIED FOR COMPLEX*16)
C       OUTPUT:         EIGVAL=THE COMPUTED EIGENVALUE
C                       EIGVEC=THE CORRESPONDING EIGENVECTOR
C       ****************************************************************
C
        IMPLICIT COMPLEX*16(A-H,O-Z)
        REAL EPS,R1,R2
        DIMENSION A(3,3),X(3)
        KSEED=12345
        N=3
C
C       *** CONSTRUCT THE TEST MATRIX AND GENERATE AN INITIAL   ***
C       *** EIGENVECTOR AT RANDOM                               ***
C
        A(1,1)=CMPLX(-4.D0,0.D0)
        A(1,2)=CMPLX(-3.D0,0.D0)
        A(1,3)=CMPLX(-7.D0,0.D0)
        A(2,1)=CMPLX(2.D0,0.D0)
        A(2,2)=CMPLX(3.D0,0.D0)
        A(2,3)=CMPLX(2.D0,0.D0)
        A(3,1)=CMPLX(4.D0,0.D0)
        A(3,2)=CMPLX(2.D0,0.D0)
        A(3,3)=CMPLX(7.D0,0.D0)
        DO 10 I=1,N
           R1=RAN(KSEED)
           R2=RAN(KSEED)
           X(I)=CMPLX(R1,R2)
     10 CONTINUE
C
C       *** SUBROUTINE EIGEN WILL COMPUTE ONE EIGENVALUE AND THE ***
C       *** NORMALIZED EIGENVECTOR                              ***
C
        EPS=.0001
        CALL EIGEN(N,A,EPS,X,Y)
        WRITE(10,*) Y
        WRITE(10,*)(X(I),I=1,N)
        STOP
        END
```

TABLE 4.24 Output of Eigenvalue Computation

Successive Eigenvalue Estimates from EIGEN

```
(1.509799502841826,1.146020421424286)
(1.509960413519013,1.110282405927048)
(1.538221058711549,0.2790532785529105)
(1.314871445299046,-0.2512895871666240)
(1.151647595028324,0.1986352616137635)
(1.056218466137056,-3.6794216895795270E-02)
(0.9983627632762747,4.4887362292787727E-03)
(1.000016738414568,1.5167001284078572E-05)
(0.9999999999562698,-5.0627799336817185E-10)
```

Output from EIGEN

Computed eigenvalue:

```
(0.9999999999562698,-5.0627799336817185E-10)
```

Computed eigenvector:

```
(-0.3785400561347732,0.3266610597406686)
(0.1892700278666965,-0.1633305300719654)
(0.1892700281338001,-0.1633305298028695)
```

4.8
SUPPLEMENTARY NOTES AND DISCUSSIONS

With the advent of digital computers, linear equations have become crucial and pervasive in all quantitative fields of science. An especially important need for their solution arose in the early days of digital computing, which coincided with the advent of thermonuclear devices. At one stage, deciding whether such devices would work at all or work too well depended on the solution of a large linear system resulting from discretization of a partial differential equation. In brief, there was a period (about 1950–1960) during which some of the very best scholars in computer science and mathematics devoted considerable effort to devising ingenious techniques for computational solution of linear equations. Sophisticated iterative methods related to those discussed in Section 4.6 were invented and analyzed during this period, as were "Monte Carlo" or probabilistic simulation techniques for linear equations. [For an introduction to the latter topic, the reader is referred to Yakowitz (1977), especially Chaps. 3 and 6.] Iterative and Monte Carlo methods initially attracted much interest and hope, and indeed these methods were well suited to the limited memory capacity of early electronic computers.

But at present, the fortunate state of affairs is that for small to moderate-sized linear systems ("moderate" currently being about 200 unknowns, and growing), Gauss elimination and its close cousin, LU decomposition, are clearly the methods of choice. We say "fortunately" because, in comparison to efficient iterative and Monte Carlo methods, their analysis and implementation is relatively simple. Even for larger systems, one often finds that special properties such as bandedness (discussed in Section 4.3.3) make variants of Gauss elimination competitive. However, "supercomputers" effectively employ parallel processing and vector structures, and for these machines, iterative methods may be more competitive.

Standard library codes for solving linear equations are invariably based on Gaussian elimination, or, more often its close cousin, LU decomposition (Section 4.5). As an important case in point, the several major IMSL routines for linear equations employ LU decomposition. They also incorporate equilibration and iterative refinement, which are heuristic features aimed at sharpening the calculations. Equilibration was discussed in Section 4.2.3. Problem 28 sketches the idea behind iterative refinement. Both equilibration and iterative refinement are analyzed in Golub and Van Loan (1983, Sec. 4.5).

In our exposition of pivoting and ill-conditioning, by example we have seen that instabilities can occur in small, innocuous linear systems such as those that arise in engineering and practical applications. We suggested that partial pivoting should be used as a matter of course and that an "instability detector" should be incorporated even if it is no more than an examination of the residual. If ill-conditioning is suspected, in our estimation the first line of defense should be double or multiple precision. In the past this was avoided because of fast memory constraints, but modern machines such as the VAX computer overcome this constraint to a large extent by storing parts of big arrays on disk. Because of quick "swapping" between fast (chip) memory and disk memory, these machines effectively handle huge arrays.

The domain in which iterative methods are currently thought to be superior on present-day standard computers includes solution of linear equations arising from discretization of partial differential equations. The application of iterative methods requires consideration of fairly delicate issues such as whether a given method will be convergent for the desired equation, and if several methods are convergent, which one is the best. Some iterative methods depend on a scaling parameter, the choice of which is somewhat at the user's discretion. In summary, our feeling is that if the direct methods we have discussed are not effective on a linear equation that you need to solve, you had better consult a numerical analyst or be prepared and willing to learn about some fairly deep aspects of numerical analysis. For further study and some basic references in iterative methods for linear equations, we refer the reader to Szidarovszky and Yakowitz (1978, Sec. 6.2).

In optional Section 4.7, we gave a rudimentary scheme and program for finding matrix eigenvalues and eigenvectors. The references in that section will lead the reader to more elegant procedures.

PROBLEMS

Section 4.2

1. Solve the following systems of linear equations by hand calculation, using naive Gaussian elimination.

(a) $x_1 + x_2 + x_3 = 3$

$\quad x_1 - x_2 - x_3 = 3$

$\quad 2x_1 + x_2 + 7x_3 = 6.$

(b) $x_1 + 2x_2 + 3x_3 = 4$

$\quad 2x_1 + 6x_2 + 10x_3 = 14$

$\quad -x_1 - 3x_2 - 6x_3 = 9.$

(c) $x_1 + 2x_2 + 3x_3 = 4$

$\quad 2x_1 + 4x_2 + 9x_3 = 4$

$\quad 4x_1 + 3x_2 + 2x_3 = 1.$

Show the intermediate derived systems and the back-substitution calculations. Check your solution $\mathbf{x}$ by verifying that $\mathbf{Ax} = \mathbf{b}$ in each case.

2. Following the procedures of Example 4.5, use subroutine GAUSS (Table 4.2) to solve $\mathbf{Ax} = \mathbf{b}$ for

$$b_i = \sqrt{i}, \qquad a_{ij} = \left(\frac{i}{n}\right)^{j-1}, \qquad 1 \le i, j \le n.$$

Find the residual vector and the computed solution $\tilde{\mathbf{y}}$ to

$$\mathbf{Ay} = \mathbf{c}$$

with $c_i = \sum_{j=1}^{n} a_{ij}$, $1 \le i \le n$. Compare $\tilde{y}_i$ with 1. Do this for $n = 5, 7, 9, \ldots$, until residuals are hopelessly large. (**HINT:** Do not forget to "float" integers.)

3. Suppose that $\mathbf{U}$ is an upper triangular matrix. That is, $\mathbf{U} = (u_{ij})$ and $u_{ij} = 0$ if $i > j$. Show that if for some i, $u_{ii} = 0$, there will either be no solutions, or many, to $\mathbf{Ux} = \mathbf{b}$. What is the criterion that a solution exist in this case?

4. Use your computer's random number generator to construct coefficients for the $n \times n$ matrix $\mathbf{A} = (a_{ij})$ and vector $\mathbf{b} = (b_i)$. Use GAUSS (Table 4.2) to solve the system, and calculate the residual $\mathbf{r}$. Make n as large as you can, subject to the maximum magnitude of a coordinate of $\mathbf{r}$ being less than 10^{-5}. Try to get the biggest n in your class.

5. In some cases Gauss elimination might run faster on a virtual memory machine if it were column rather than row oriented. A trivial change in subroutine GAUSS (Table 4.2) assures that in forward elimination, coordinates of successive derived systems are computed column by column rather than row by row. What is that trivial change?

6. Modify subroutine GAUSS (Table 4.2) so that no pivoting is performed. Redo Problem 2 with your modified routine.

*7. Modify subroutine GAUSS so that it performs maximal pivoting. Check it by solving the standard example problem (Example 4.1). By examining residuals, compare its performance with that of unmodified GAUSS on the linear equation given in Problem 2. (**HINT:** In maximal pivoting, we must keep track of column interchanges.)

8. Prove by induction that

$$1 + 2 + \cdots + (n - 1) = \frac{n(n - 1)}{2}$$

and

$$1^2 + 2^2 + \cdots + (n - 1)^2 = \frac{(n - 1)n(2n - 1)}{6}.$$

These formulas were needed in our analysis of operation count in Gaussian elimination.

9. In Example 4.15 and Problem 2 we have dealt with a form of the notoriously ill-conditioned coefficient matrix with (i, j) element

$$a_{ij} = i^{j-1}, \qquad 1 \le i, j \le n.$$

Another familiar ill-conditioned matrix is the *Hilbert segment matrix,* defined as

$$a_{ij} = \frac{1}{i + j - 1}, \qquad 1 \le i, j \le n. \tag{4.51}$$

Redo Problem 2 with the same $\mathbf{b}$ vector, but use the matrix (4.51) as the matrix of coefficients.

10. Apply a commercial code, such as a subroutine in the IMSL library, if available, for solving Problem 9. Compare the outputs.

11. Convert GAUSS (Table 4.2) so that it operates in double precision. (Convert ABS to DABS.) Redo Problem 2, comparing the biggest order n with negligible residuals to that obtained in single precision.

Section 4.3

12. By hand, and showing the intermediate derived systems, obtain the inverse of the matrix

$$A = \begin{bmatrix} 1 & 1 & 1 \\ 1 & -1 & -1 \\ 2 & 1 & 7 \end{bmatrix}.$$

Verify that for your computation of A^{-1} and coefficient vector $\mathbf{b} = (3, 3, 6)^T$, $A^{-1}\mathbf{b}$ agrees with the solution you obtained for Problem 1(a).

13. Use subroutine INVG (Table 4.11) to find the inverse of the matrix

$$A = (a_{ij}) \quad \text{with} \quad a_{ij} = \left(\frac{i}{5}\right)^{j-1}, \quad 1 \leq i, j \leq 5.$$

Check to verify that for A^{-1} your computed inverse, and I the fifth-order identity matrix,

$$A^{-1}A = I.$$

Try to recover the identity for higher-order n, with $a_{ij} = \left(\frac{i}{n}\right)^{j-1}$.

*14. Show that the number of additions and multiplications required to invert an nth-order matrix by the elimination method is $n^3 +$ terms of order n and n^2. (**Hint:** Look carefully at developments in Section 4.2.5.)

*15. Find the determinant of

$$\begin{bmatrix} 1 & 2 & 3 \\ 1 & 3 & 1 \\ 2 & 1 & 1 \end{bmatrix}$$

by subdeterminant expansion. Verify that this value is also given by (4.20).

*16. Use subroutine DETC (Table 4.12) to find the determinant of the matrix in Problem 15. Verify that this agrees with your hand computation.

*17. Use the fact that interchange of two rows of a matrix has the effect of changing the determinant by a factor of (-1), and subtraction of a multiple of one row from another does not change the determinant at all to verify (4.20).

*18. Solve the following band systems by Gauss elimination. Show the derived systems.

(a) $2x_1 + 3x_2 \qquad = 1$

$\quad x_1 + 2x_2 + x_3 = 1$

$\qquad\quad x_2 - x_3 = 1.$

(b)
$$x_1 + 2x_2 \qquad\qquad = 5$$
$$x_1 - x_2 + x_3 \qquad = -1$$
$$2x_2 - x_3 + x_4 = 4$$
$$x_3 - x_4 = 0.$$

(c)
$$x_1 + x_2 \qquad\qquad = 4$$
$$2x_1 + x_2 + x_3 \qquad = 6$$
$$x_2 + x_3 - 3x_4 = 2$$
$$x_3 + x_4 = 0.$$

*19. Use subroutine BAND (Table 4.13) to solve the linear equations in Problem 18.

Section 4.4

*20. In Example 4.15 we saw that the condition numbers of the matrix classes of Problems 2 and 4 are high and low, respectively. Find the condition number of order $n = 6$ of the Hilbert segment matrix, defined in Problem 9.

Section 4.5

*21. Actually specify the matrices L_1 and L_2, as in the decomposition (4.33), for the matrix **A** in Example 4.1. Verify that for your specified matrices, $\mathbf{A}_2 = L_2L_1\mathbf{A}$ is upper triangular.

*22. Solve the linear equation of Problem 1(a) by Dolittle reduction. Show your work.

*23. Redo Problem 2 by LU decomposition, using subroutine COMPLU. Record the residuals.

*24. Declare the variable SUM in COMPLU to be double precision. By this means, we perform summations (4.37) and (4.38) in double precision (but store the summand in single precision). Redo Problem 2, and compare with Problem 23.

Section 4.6

*25. Compute two Jacobi iteration steps for the linear equation:
$$500x_1 + x_2 - 2x_3 = 400$$
$$2x_1 - 250x_2 + x_3 = 120$$
$$-x_1 + x_2 + 400x_3 = 300.$$
Take $x_1^{(0)} = x_2^{(0)} = x_3^{(0)} = 0$ as initial values.

*26. Redo Problem 25 by the Gauss–Seidel method. Do one iteration by hand, showing your work, and then apply subroutine SEID (Table 4.18) to do several more.

*27. In *iterative refinement,* if $\mathbf{r}$ is the residual error in a computational solution $\mathbf{x}$ of $\mathbf{Ax} = \mathbf{b}$, we solve $\mathbf{Au} = \mathbf{r}$, and take $\tilde{\tilde{\mathbf{x}}} = \tilde{\mathbf{x}} + \mathbf{u}$ as an "update" of the solution. We compute the residual associated with $\tilde{\tilde{\mathbf{x}}}$ and repeat the process. To hope for success, the residual must be computed in double precision. Redo Problem 2, with $n = 7$, using iterative refinement. Use LU decomposition once (by COM-PLU).

Section 4.7

*28. Find at least one eigenvalue and eigenvector for the following matrix:

$$\begin{bmatrix} -4 & -3 & -7 & 8 \\ 2 & 3 & 2 & 7 \\ 4 & 2 & 7 & -3 \\ 3 & 4 & -1 & 1 \end{bmatrix}.$$

Your solution should satisfy (4.48).

CHAPTER 5

Nonlinear Equations

5.1
PRELIMINARIES

The concern of this chapter is equations of the form

$$f(x) = 0, \tag{5.1}$$

where x and $f(x)$ are real, complex, or vector valued.

Our task will be to find a value or values x^* of the independent variable or vector x that satisfies nonlinear equation (5.1). Such values are technically known as *roots* of (5.1) or *zeros* of the function $f(x)$. On the other hand, in accordance with common usage, we will sometimes speak of roots of a function, when we really mean zeros. If $f(x)$ is linear [i.e., $f(x) = \mathbf{A}x - \mathbf{b}$, for some matrix $\mathbf{A}$ and vector $\mathbf{b}$], the methods discussed in Chapter 4 are more appropriate for solving the equation.

As an illustration of how nonlinear equations arise in the simplest engineering situations, consider the electric circuit shown in Figure 5.1. Assume that $R = 20$ ohms, $L = 4$ henrys, and the voltage, as a function of the time t, is $V(t) = 100\sqrt{2} \sin (5t)$ for $t \geq 0$, and $V(t) = 0$ for $t < 0$. If we assume that $i(0) = 0$, the current can be expressed as

$$i(t) = 5e^{-5t} \sin \left(\frac{\pi}{4} \right) + 5 \sin \left(5t - \frac{\pi}{4} \right).$$

In order to compute the earliest positive time at which the current is zero, we must solve the nonlinear equation

$$i(t) = 5e^{-5t} \sin \left(\frac{\pi}{4} \right) + 5 \sin \left(5t - \frac{\pi}{4} \right) = 0.$$

Any of the three major methods described in this chapter are adequate for finding that to the accuracy shown, the solution is $t^* = 0.78815$. However, one cannot

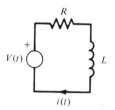

FIGURE 5–1
Electric Circuit

solve this sort of problem by using only a fixed sequence of operations involving FORTRAN library functions.

Nonlinear equation techniques are standard fare in computer library and calculator implementations of functional inverses. For example, once some technique (perhaps such as described in Chapter 2) has been found for computing sin (x), methods to be described here provide a convenient implementation of $\sin^{-1}(y)$ by solving the equation $f(x) = \sin(x) - y = 0$. In this regard it is informative to note that the path of finding a polynomial approximation to $\sin^{-1}(y)$ itself is not practical. Hart et al. (1968, p. 51) mention that to get a polynomial approximation with error less than 10^{-8} on $[-1, 1]$, one would need a polynomial of degree at least 10,000, no matter how the polynomial is constructed.

In discussing methods for linear equations (Chapter 4), we drew the distinction between direct methods (e.g., Gauss elimination), which require only a predetermined number of operations, and iterative methods (e.g., the Jacobi method), which for $k = 1, 2, \ldots$ iteratively construct x_k on the basis of x_{k-1}, and possibly earlier values, until some stopping condition is met. Whereas in the case of linear equations, iterative methods are merely an option, in the case of nonlinear equations, they are mandatory. Developments in mathematics imply that direct methods for nonlinear equations do not exist except for certain restrictive cases.

5.2
THE GRAPHICAL APPROACH

We have mentioned that all prominent nonlinear equation methods are iterative. The user must supply the initial "guess," x_0, of the root location. Whether or not the successive iterates $x_1, x_2, \ldots$ converge will, in many cases, depend on the initial guess being in the "vicinity" of a root. Ideas offered in the present section are useful for choosing a sensible initial guess.

An adequate approximation for a root can often be found by strictly graphical (hand calculator and plotting paper) techniques. Such graphical techniques amount to simply calculating a few functional values and perhaps interpolating between arguments for which the function "straddles" zero. In fact, before the computer age, this was the standard approach in many instances. By strategically choosing successive "sample" points, we could indeed get very accurate answers. Now, with computers so readily available, once we resolve to use a strategy, for the sake of convenience and avoidance of error, the strategy should be encoded and run. The "computer methods" offered in this chapter predate the computer age—in fact, they predate the twentieth century. But they are still appropriate for commonplace problems and the obvious gateway to recent techniques.

We offer a brief discussion of graphical methods without strategy. A first step is to seek a finite ''search'' interval that we can count on to contain a root. In the special case of polynomial equations, many theoretical results relating roots to coefficients are available. Below we give one sample of such a result and then provide an example of its use. In the case of polynomial equations, all roots can readily be localized into a disk in the complex plane. Define reverse-order terms

$$f(x) = a_n x^n + a_{n-1} x^{n-1} + \cdots + a_0. \tag{5.2}$$

Then all (real or complex) roots x^* of $f(x)$ satisfy the inequality

$$|x^*| \le 1 + \frac{1}{|a_n|} \max \{|a_0|, |a_1|, \ldots, |a_{n-1}|\}, \tag{5.3}$$

where the ''max'' operation denotes the maximum of the values $|a_0|, |a_1|, \ldots, |a_{n-1}|$. This and other expressions relating polynomial coefficients to roots are given by Wilf (1960).

━━━ **EXAMPLE 5.1** ━━━

Consider the polynomial

$$f(x) = x^3 - 3x^2 - x + 9.$$

In our case $n = 3$, $a_0 = 9$, $a_1 = -1$, $a_2 = -3$, and $a_3 = 1$. Then by (5.3), all roots x^* of $f(x)$ must satisfy the relation

$$|x^*| \le 1 + \max \{3, 1, 9\} = 10.$$

Thus the real roots, if any, lie within the interval $[-10, 10]$. Table 5.1 gives the values of $f(x)$ between -10 and $+10$ at integer arguments. From this, and the fact that polynomials are continuous, we see that the interval $[-2, -1]$ contains a root. In Figure 5.2 we have plotted 11 evenly spaced points in this interval.

TABLE 5.1   Values of $f(x)$

x	$f(x)$	x	$f(x)$	x	$f(x)$
-10	-1281	-3	-42	4	21
-9	-954	-2	-9	5	54
-8	-687	-1	6	6	111
-7	-474	0	9	7	198
-6	-309	1	6	8	321
-5	-186	2	3	9	486
-4	-99	3	6	10	699

It would take luck to see that to three significant places, $x^* = -1.52$. Greater accuracy requires a graph with greater resolution. The graphical approach is infeasible here if we require that the relative error of the root be less than 0.01. We should push ahead to the purely algorithmic methods described in the following sections.

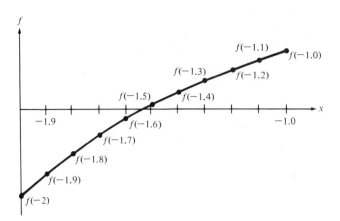

FIGURE 5–2   Plot of Function Points for Graphical Approach

5.3
THE BISECTION METHOD

Assume that the real function $f(x)$ is continuous on the interval $[a, b]$ and that $f(a)$ and $f(b)$ have different signs. Then there is at least one root of $f(x)$ between a and b.

Let $x_L = a$, $x_R = b$, and define x_M to be the midpoint of $[a, b]$; that is, $x_M = (x_L + x_R)/2$. Thus view L, R, and M as abbreviations for "left," "right," and "middle." If $f(x_M) = 0$, we are done. If $f(x_M) \neq 0$, either $f(x_L)f(x_M) < 0$ or $f(x_M)f(x_R) < 0$. In other words, either $f(x_L)$ and $f(x_M)$, or $f(x_M)$ and $f(x_R)$ have different signs. In the first case, interval (x_L, x_M) contains a root, and in the second case, a root is in (x_M, x_R). Observe that by the use of this step, we can locate a root of $f(x)$ within an interval of length $(b - a)/2$, which is the half of the length of the original interval. By repeating the foregoing step successively with $[x_L, x_M]$ or $[x_M, x_R]$ now playing the role of $[a, b]$, we may locate a root of $f(x)$ in an interval half again as small. The general step of the algorithm is given as follows. Assume that x_L and x_R, $x_L < x_R$, satisfying $f(x_L)f(x_R) < 0$, are available from the preceding step. Then

Let

$$x_M = \frac{x_L + x_R}{2}$$

If $f(x_M)f(x_L) < 0$, set

$$x_R = x_M;$$

otherwise, set

$$x_L = x_M.$$

(5.4)

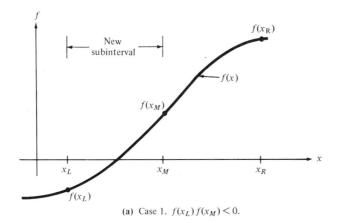

(a) Case 1. $f(x_L)f(x_M) < 0$.

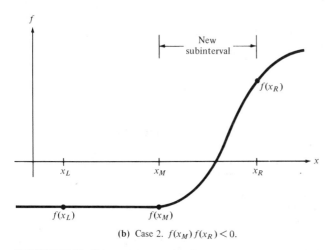

(b) Case 2. $f(x_M)f(x_R) < 0$.

FIGURE 5–3   One Bisection Step

In Figure 5.3 we illustrate this single bisection step.

For purposes of discussing a succession of bisection steps, we resubscript $x_0 = x_L$ and $x_1 = x_R$, and x_2, x_3, and x_4 are the successive x_M values. This process, illustrated in Figure 5.4, is called the *bisection method*. After K bisection steps, the root is localized within an interval of length

$$\frac{b - a}{2^K}.$$

Table 5.2 provides subroutine BISE for implementing the bisection method. The calling parameter EPS must be larger than the product of the root magnitude and machine epsilon (this product being a bound for the error of approximating the root by a machine word), or else roundoff error may prevent program termination. The user is expected to supply a function subprogram giving the function whose zero is sought. The example that follows gives details on calling BISE.

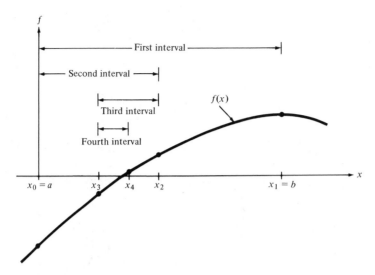

FIGURE 5–4 Successive Bisection Steps

TABLE 5.2 Subroutine BISE for the Bisection Method

```
      SUBROUTINE BISE(A,B,X,EPS)
C
C   ***************************************************************
C   *   FUNCTION: THIS SUBROUTINE APPROXIMATES THE ROOT           *
C   *             OF F(X)=0 OVER THE INTERVAL [A,B] USING THE      *
C   *             BISECTION METHOD                                 *
C   *   USAGE:                                                     *
C   *       CALL SEQUENCE: CALL BISE(A,B,X,EPS)                    *
C   *       EXTERNAL FUNCTIONS/SUBROUTINES: FUNCTION F(X)          *
C   *   PARAMETERS:                                                *
C   *       INPUT:                                                 *
C   *           A=INTERVAL LEFT ENDPOINT                           *
C   *           B=INTERVAL RIGHT ENDPOINT                          *
C   *         EPS=ERROR BOUND                                      *
C   *       OUTPUT:                                                *
C   *           X=BISECTION APPROXIMATION OF THE ROOT             *
C   ***************************************************************
C
      XL=A
      XR=B
      U=F(XL)
      XM=XR
      DO WHILE(ABS(XM-XL).GE.EPS)
         W=F(XM)
         IF(U*W.LT.0.0) THEN
            XR=XM
         ELSE
            XL=XM
         END IF
         XM=(XL+XR)/2.0
      END DO
      X=XM
      RETURN
      END
```

—————— **EXAMPLE 5.2** ——————————————————————————

From Example 5.1 we know that the function

$$f(x) = x^3 - 3x^2 - x + 9$$

has a root in the interval $(-2, -1)$. Thus we take $a = -2$ and $b = -1$, or in the notation of (5.4), $x_L = -2$, $x_R = -1$, and $x_M = (-2 - 1)/2 = -\frac{3}{2}$.

$$f(x_L) = f(-2) = -8 - 12 + 2 + 9 < 0$$

$$f(x_M) = f\left(-\frac{3}{2}\right) = \frac{-27 - 27 \cdot 2 + 3 \cdot 4 + 9 \cdot 8}{8}$$

$$= \frac{3}{8} > 0.$$

Since $f(x_L)f(x_M) < 0$, the root must be in $[-2, -\frac{3}{2}]$, and for the next bisection step,

$$x_L = -2 \quad \text{and} \quad x_R = -\frac{3}{2}.$$

A computer program that uses BISE to continue these steps is given in Table 5.3, and successive values of x_k and $f(x_k)$ ($k \geq 0$) are given in Table 5.4. Double

TABLE 5.3   Program for Bisection Example

```
C      PROGRAM BISECT
C
C      ****************************************************************
C      THIS PROGRAM ILLUSTRATES THE BISECTION METHOD BY FINDING
C      A ROOT OF THE FUNCTION F(X)=X**3-3*X**2-X+9 IN THE INTERVAL
C      [-2.,-1.]
C      CALLS:    BISE (MODIFIED FOR DOUBLE PRECISION)
C      OUTPUT(FROM FUNCTION F):
C                 X=VALUE OF X AT CURRENT ITERATION
C                 F=FUNCTIONAL VALUE AT X
C      ****************************************************************
C
       IMPLICIT DOUBLE PRECISION(A-H,O-Z)
       A=-2.0D0
       B=-1.0D0
       EPS=0.5D-6
C
C      *** SUBROUTINE BISE WILL USE FUNCTION F (BELOW) TO FIND   ***
C      *** A ROOT IN THIS INTERVAL                               ***
C
       CALL BISE(A,B,X,EPS)
       STOP
       END
C
C      *** FUNCTION F IS CALLED BY BISE TO CALCULATE FUNCTIONAL  ***
C      *** VALUES FOR THE PASSED POINT X                         ***
C
       FUNCTION F(X)
       IMPLICIT DOUBLE PRECISION(A-H,O-Z)
       F=X**3-3.0D0*X**2-X+9.0D0
       WRITE(10,1)X,F
     1 FORMAT(22X,D16.7,4X,D16.7)
       RETURN
       END
```

TABLE 5.4   **Successive Bisection Iterates**

k	x_k	$f(x_k)$	Number of Correct Significant Decimals in x_k
0	-0.2000000D+01	-0.9000000D+01	0
1	-0.1000000D+01	0.6000000D+01	1
2	-0.1500000D+01	0.3750000D+02	2
3	-0.1750000D+01	-0.3796875D+01	1
4	-0.1625000D+01	-0.1587891D+01	1
5	-0.1562500D+01	-0.5764160D+00	2
6	-0.1531250D+01	-0.9329224D-01	2
7	-0.1515625D+01	0.1426964D+00	2
8	-0.1523438D+01	0.2516413D-01	3
9	-0.1527344D+01	-0.3394836D-01	3
10	-0.1525391D+01	-0.4363216D-02	4
11	-0.1524414D+01	0.1040768D-01	3
12	-0.1524902D+01	0.3024037D-02	4
13	-0.1525146D+01	-0.6691384D-03	5
14	-0.1525024D+01	0.1177562D-02	5
15	-0.1525085D+01	0.2542400D-03	6
16	-0.1525116D+01	-0.2074421D-03	5
17	-0.1525101D+01	0.2340073D-04	6
18	-0.1525108D+01	-0.9202025D-04	6
19	-0.1525105D+01	-0.3430965D-04	6
20	-0.1525103D+01	-0.5454435D-05	6
21	-0.1525102D+01	0.8973153D-05	7

precision was used in this driver program, and several to follow, for more exacting answers. Of course, BISE was accordingly modified for double precision. Table 5.4 was obtained by inserting a WRITE statement into BISE. ■

If we regard the most recently computed midpoint x_k as an approximation of the root, then from discussion at the beginning of this subsection, one may say that the approximation error is "about" halved at each bisection iteration. Another way of viewing this is to say that the approximation x_k picks up approximately a significant bit at each iteration, or one significant decimal place in three or four iterations, since $2^{-3} > \frac{1}{10} > 2^{-4}$. One can see this phenomenon exhibited clearly in Table 5.4. After every three to four iterations, the x_k value becomes one significant decimal closer to the final entry.

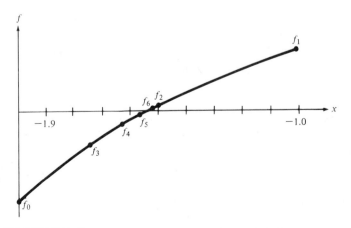

FIGURE 5–5   Bisection Iteration of Example 5.2

In Figure 5.5 we have plotted successive bisection midpoints x_k and values $f_k = f(x_k)$. It is instructive to compare the way the bisection method quickly concentrates about the root, with the corresponding Figure 5.2 for the "graphical" approach. Accuracy indicated in Table 5.4 is infeasible by graphical means.

5.4
THE SECANT METHOD

Let x_0 and x_1 be two real numbers. In contrast to the bisection method, we do not insist that $x_0 < x_1$ or that $f(x_0)$ and $f(x_1)$ have different signs. Let us approximate the function $f(x)$ by a linear interpolating polynomial with interpolating points x_0 and x_1. From the Lagrange representation (Section 2.4.2)

$$l_0(x) = \frac{x - x_1}{x_0 - x_1} \quad \text{and} \quad l_1(x) = \frac{x - x_0}{x_1 - x_0} \tag{5.5}$$

and so

$$p(x) = f(x_0) \frac{x - x_1}{x_0 - x_1} + f(x_1) \frac{x - x_0}{x_1 - x_0}. \tag{5.6}$$

The line $p(x)$ is the secant line for $f(x)$ at x_0 and x_1. The *secant method* has us obtain a new approximation x_2 of the root by solving the linear equation $p(x) = 0$. After some simple algebra, we confirm that x_2 is given by

$$x_2 = x_1 - \frac{f(x_1)(x_1 - x_0)}{f(x_1) - f(x_0)}.$$

One step of the secant method is shown in Figure 5.6. We see in this figure that x_2 is the x-intercept of the secant line connecting $(x_0, f(x_0))$ with $(x_1, f(x_1))$. This secant line itself is the linear polynomial $p(x)$ given by (5.6).

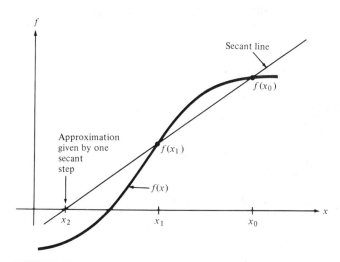

FIGURE 5–6 One Secant Step

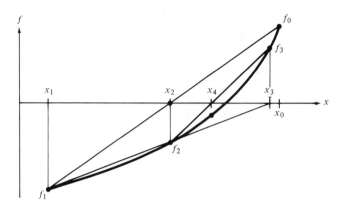

FIGURE 5–7 The Secant Method

By repeating this procedure, and letting x_1 and x_2 play the roles of x_0 and x_1, we have

$$x_3 = x_2 - \frac{f(x_2)(x_2 - x_1)}{f(x_2) - f(x_1)},$$

and so on. The general secant method iteration step is

$$x_{k+1} = x_k - \frac{f(x_k)(x_k - x_{k-1})}{f(x_k) - f(x_{k-1})} \qquad (k \geq 1). \tag{5.7}$$

Note that the value of the new root approximation depends on the previous *two* approximations and corresponding functional values. The successive steps of the secant method are illustrated in Figure 5.7.

The computer subroutine SECA for the secant method is given in Table 5.5.

TABLE 5.5 Subroutine SECA for the Secant Method

```
      SUBROUTINE SECA(A,B,X,EPS)
C
C  ****************************************************************
C  *  FUNCTION: THIS SUBROUTINE COMPUTES THE APPROXIMATE ROOT  *
C  *            OF F(X)=0 USING THE SECANT METHOD              *
C  *  USAGE:                                                   *
C  *      CALL SEQUENCE: CALL SECA(A,B,X,EPS)                  *
C  *      EXTERNAL FUNCTIONS/SUBROUTINES: FUNCTION F(X)        *
C  *  PARAMETERS:                                              *
C  *      INPUT:                                               *
C  *         A=INITIAL APPROXIMATION X(0)                      *
C  *         B=INITIAL APPROXIMATION X(1)                      *
C  *        EPS=ERROR BOUND                                    *
C  *      OUTPUT:                                              *
C  *         X=SECANT APPROXIMATION OF THE ROOT               *
C  ****************************************************************
C
C      *** INITIALIZATION ***
      V=F(B)
      U=F(A)
      X=B
```

TABLE 5.5   (Continued)

```
C       *** COMPUTE APPROXIMATE ROOT ITERATIVELY ***
        DO WHILE(ABS(X-A).GT.EPS)
            X=B-V*(B-A)/(V-U)
            A=B
            B=X
            U=V
            V=F(X)
        END DO
        RETURN
        END
```

EXAMPLE 5.3

Here the secant method is applied to the function $f(x) = x^3 - 3x^2 - x + 9$ of Example 5.2. Again we take $x_0 = -2$ and $x_1 = -1$. Thus according to (5.7) with $k = 1$, we calculate that

$$f(x_0) = f(-2) = -9, \qquad f(x_1) = f(-1) = 6,$$

and

$$x_2 = -1 - \frac{6[-1 - (-2)]}{6 - (-9)} = -1 - \frac{6}{15} = -1.4.$$

Continuing in this fashion, but using the calling program in Table 5.6 to avail

TABLE 5.6   Program for Secant Example

```
C       PROGRAM SECANT
C
C       ****************************************************************
C       THIS PROGRAM ILLUSTRATES THE SECANT METHOD BY FINDING
C       A ROOT OF THE FUNCTION F(X)=X**3-3*X**2-X+9 IN THE INTERVAL
C       [-2.,-1.]
C       CALLS:   SECA (MODIFIED FOR DOUBLE PRECISION)
C       OUTPUT(FROM FUNCTION F):
C               X=VALUE OF X AT CURRENT ITERATION
C               F=FUNCTIONAL VALUE AT X
C       ****************************************************************
C
        IMPLICIT DOUBLE PRECISION(A-H,O-Z)
        A=-2.0D0
        B=-1.0D0
        EPS=0.1D-6
        CALL SECA(A,B,X,EPS)
        STOP
        END
C
C       *** FUNCTION F IS CALLED BY SECA. THIS IS THE FUNCTION TO ***
C       *** BE ZEROED                                            ***
C
        FUNCTION F(X)
        IMPLICIT DOUBLE PRECISION(A-H,O-Z)
        F=X**3-3.0D0*X**2-X+9.0D0
        WRITE(10,1)X,F
    1   FORMAT(22X,D16.8,4X,D16.8)
        RETURN
        END
```

TABLE 5.7   Successive Secant Iterations

k	x_k	$f(x_k)$	Number of Correct Significant Decimals in x_k
0	-0.10000000D+01	0.60000000D+01	0
1	-0.20000000D+01	-0.90000000D+01	1
2	-0.14000000D+01	0.17760000D+01	1
3	-0.15681818D+01	-0.66586448D+00	2
4	-0.15223208D+01	0.42019874D-01	3
5	-0.15250431D+01	0.89434209D-03	4
6	-0.15251023D+01	-0.12469362D-05	8
7	-0.15251023D+01	0.36912695D-10	12

ourselves of subroutine SECA (Table 5.5), we calculate Table 5.7. The root approximations are presented only to eight digits, but the last column shows the accuracy of the stored numbers. Successive secant points are displayed in Figure 5.8.

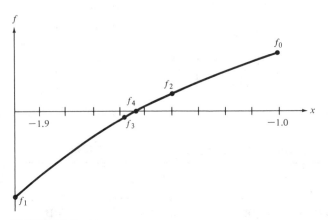

FIGURE 5–8   Secant Iterations of Example 5.3

It is known (Szidarovszky and Yakowitz, 1978, Sec. 5.1.3) that in a neighborhood of a root of a smooth function, the secant method converges far faster than the bisection rule. Recall that successive estimates obtained by the latter gain a new significant decimal every three iterations. Successive iterates of the secant method, close to the solution, increase the number of correct significant decimals by about 50%. The more rapid convergence of the secant method on our test problem is evident in Table 5.7, compared with Table 5.4. Much fewer steps were used to obtain the same accuracy, and in the last step the number of correct decimal places jumps from 8 to 12. On the other hand, the bisection method, provided that one can find points on either side of a root, is fail-safe: As long as the magnitude $|f(x_k)|$ is large enough that the computer can correctly distinguish the sign, the bisection method always converges. The secant method may diverge unless the starting points are sufficiently close to the root, as shown in Example 5.4.

──────── **EXAMPLE 5.4** ────────────────────────────────

Here we see a case in which successive approximations x_k actually get farther from the root. If we do so much as multiply the polynomial of preceding examples by exp (x), thus obtaining

$$f(x) = (x^3 - 3x^2 - x + 9) \exp(x),$$

then the roots are unchanged. But the secant method, starting, as before, with $x_0 = -1, x_1 = -2$, diverges. A glance at the graph of $f(x)$ (Figure 5.9) suggests why; Once consecutive points are located to the left of the hump, the algorithm will thereafter "think" that the root is to the left because the secant line connecting the most recent pair of estimates x_{k-1} and x_k has negative slope and intersects the axis to the left of these points. Locations of the first four secant function points are shown in Figure 5.9.

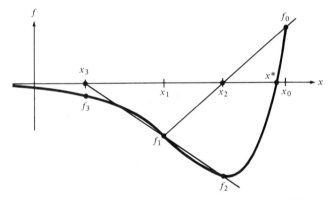

FIGURE 5–9   Function Demonstrating Secant Divergence

─── ■

5.5
NEWTON'S METHOD

The secant method has us choose the successive point as the zero of the linear interpolation polynomial based on the most recent two points. Newton's method, analogously, has us choose the successive point as the zero to the linear Taylor's polynomial, the expansion being about the current point. Whereas two points, x_0 and x_1, were needed to "start" the secant iterations, only one point, x_0, determines the Newton iterations. But a single Newton step requires evaluation of the derivative $f'(x)$ as well as $f(x)$ itself.

Let us denote the linear Taylor's polynomial about x_0 by

$$p(x) = f(x_0) + f'(x_0)(x - x_0).$$

Then the new approximation x_1 of the root is the solution of the linear equation $p(x) = 0$. This implies that

$$x_1 = x_0 - \frac{f(x_0)}{f'(x_0)}.$$

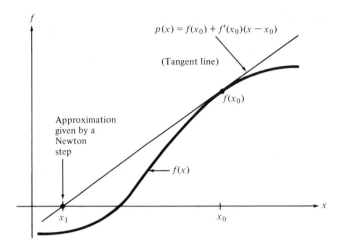

FIGURE 5–10   One Step of Newton's Method

The linear Taylor's polynomial is the tangent line at x_0. A single Newton's step is illustrated in Figure 5.10. The successor x_1 is the x-intercept of this tangent line.

By repeating this procedure, a sequence $\{x_k\}$ is determined by the recursive rule

$$x_{k+1} = x_k - \frac{f(x_k)}{f'(x_k)}. \tag{5.8}$$

This method (5.8) is called *Newton's method* (or, alternatively, the *Newton–Raphson method*).

The computer subroutine NEIT for Newton's method is given in Table 5.8. It requires the user to supply function subprograms F(X) and DF(X) for $f(x)$ and $f'(x)$, respectively.

TABLE 5.8   Subroutine NEIT for Newton's Method

```
      SUBROUTINE NEIT(X0,X,EPS)
C
C     ****************************************************************
C     *  FUNCTION: THE SUBROUTINE APPROXIMATES THE ROOT OF          *
C     *            F(X)=0 GIVEN THE INITIAL POINT X0 AND THE         *
C     *            DERIVATIVE FUNCTION DF(X) USING THE NEWTON        *
C     *            METHOD                                            *
C     *  USAGE:                                                      *
C     *      CALL SEQUENCE: CALL NEIT(X0,X,EPS)                      *
C     *      EXTERNAL FUNCTIONS/SUBROUTINES: FUNCTION F(X)           *
C     *                                      FUNCTION DF(X)          *
C     *  PARAMETERS:                                                 *
C     *      INPUT:                                                  *
C     *          X0=INITIAL ROOT APPROXIMATION                       *
C     *          EPS=ERROR BOUND                                     *
C     *      OUTPUT:                                                 *
C     *          X=NEWTON APPROXIMATION OF THE ROOT                  *
C     ****************************************************************
C
C     *** INITIALIZATION ***
      X=X0-(F(X0)/DF(X0))
```

TABLE 5.8 (Continued)

```
C      *** COMPUTE APPROXIMATE ROOT ITERATIVELY ***
       DO WHILE(ABS(X-X0).GT.EPS)
          X0=X
          X=X0-(F(X0)/DF(X0))
       END DO
       RETURN
       END
```

──── EXAMPLE 5.5 ────

Newton's method is now applied to our familiar polynomial equation

$$f(x) = x^3 - 3x^2 - x + 9 = 0.$$

Let the initial approximation be given by $x_0 = -2$. Then, according to (5.8) with $k = 0$ and $f'(x) = 3x^2 - 6x - 1$,

$$x_1 = x_0 - \frac{f(x_0)}{f'(x_0)} = -2 - \frac{-9}{23} = -1.608696.$$

The successive Newton iterations are obtained according to the program in Table 5.9 and are given in Table 5.10. The faster convergence of Newton's method compared to the secant method is obvious. For purposes of comparison with Figures 5.5 and 5.8, the Newton iterations are plotted in Figure 5.11.

TABLE 5.9 Program for Illustration of Newton's Method

```
C      PROGRAM NEWTON
C
C      ****************************************************************
C      THIS PROGRAM ILLUSTRATES THE NEWTONS METHOD ON THE FUNCTION
C      F(X)=X**3-3*X**2-X+9 STARTING AT X=-2.0
C      CALLS:   NEIT (MODIFIED FOR DOUBLE PRECISION)
C      OUTPUT(FROM FUNCTION F):
C               X=VALUE OF X AT CURRENT ITERATION
C               F=FUNCTIONAL VALUE AT X
C      ****************************************************************
C
       IMPLICIT DOUBLE PRECISION(A-H,O-Z)
       X0=-2.0D0
       EPS=0.5D-6
C
C      *** SUBROUTINE NEIT IS USED TO FIND A ROOT STARTING WITH   ***
C      *** THE INITIAL GUESS X0                                   ***
C
       CALL NEIT(X0,X,EPS)
       STOP
       END
C
C      *** FUNCTION F IS CALLED BY NEIT TO CALCULATE FUNCTIONAL   ***
C      *** VALUES FOR THE PASSED POINT X                         ***
C
       FUNCTION F(X)
       IMPLICIT DOUBLE PRECISION(A-H,O-Z)
       F=X**3-3.0D0*X**2-X+9.0D0
       WRITE(10,1)X,F
    1  FORMAT(22X,D16.7,4X,D16.7)
       RETURN
       END
```

TABLE 5.9   (Continued)

```
C
C      *** FUNCTION DF(X) IS CALLED BY NEIT TO CALCULATE THE        ***
C      *** DERIVATIVE AT X                                          ***
C

       FUNCTION DF(X)
       IMPLICIT DOUBLE PRECISION(A-H,O-Z)
       DF=3.0D0*X**2-6.0D0*X-1.0D0
       RETURN
       END
```

TABLE 5.10   Successive Newton Iterates

k	x_k	$f(x_k)$	Number of Correct Significant Decimals in x_k
0	-0.2000000D+01	-0.9000000D+01	1
1	-0.1608696D+01	-0.1318156D+01	1
2	-0.1528398D+01	-0.4994256D-01	3
3	-0.1525108D+01	-0.8208588D-04	6
4	-0.1525102D+01	-0.2230209D-09	11

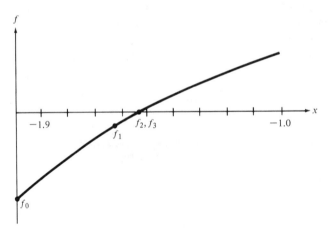

FIGURE 5–11   Newton Iterations for Example 5.5

In the following optional section (Section 5.6) we give theoretical reasons why Newton's method converges even faster than the bisection and secant methods. In the neighborhood of a root of a smooth function, the number of correct significant figures approximately doubles at each Newton iteration. This phenomenon is exhibited in Table 5.10. (Not all the correct digits of x_4 are shown.) Because of its typical efficiency, Newton's method is popular when the derivative of $f(x)$ is available. The price paid for the increased convergence rate of Newton's method over the bisection and secant methods is that it requires two function calls [$f(x_k)$ and its derivative, $f'(x_k)$] per iteration rather than one.

The derivative requirement of Newton's method can be sidestepped by the suitable numerical differentiation techniques given in Section 3.2. By making the step size h in the derivative formula decrease proportionally to $|f(x_k)|$, one can preserve the rapid convergence property of Newton's method (Dennis and Schnabel, 1983). Details of this approach are given in Problem 21. Note that if the derivative $f'(x)$ is approximated by

$$\frac{f(x_k) - f(x_{k-1})}{x_k - x_{k-1}},$$

the secant method is obtained. [But in the secant case, the step size h is proportional to $f(x_{k-1})$ rather than $f(x_k)$.]

──────── **EXAMPLE 5.6** ──

Let us examine how square roots may be extracted by elementary arithmetical operations alone through use of Newton's method. It is evident that $\sqrt{A}$ is the positive root of the equation

$$f(x) = x^2 - A = 0.$$

Then, from (5.8), Newton's method chooses

$$x_{k+1} = x_k - \frac{f(x_k)}{f'(x_k)} = x_k - \frac{x_k^2 - A}{2x_k} = \frac{1}{2}\left(x_k + \frac{A}{x_k}\right). \tag{5.9}$$

For example, if $A = x_0 = 5$, then

$$x_1 = \tfrac{1}{2}(5 + 1) = 3$$

$$x_2 = \tfrac{1}{2}(3 + \tfrac{5}{3}) = \tfrac{7}{3}$$

and continuing, we get the list below. The estimate x_5 of $\sqrt{5}$ is correct to the accuracy indicated:

$$x_3 = 2.238095238$$

$$x_4 = 2.236068896$$

$$x_5 = 2.236067977.$$

This example is particularly significant because we observed in Chapter 2 that the square-root function has no close low-degree polynomial approximators.

Other integer roots $\sqrt[n]{A}$ can be obtained arithmetically by this procedure, by finding the root of $f(x) = x^n - A = 0$. Then the Newton method gives the following generalization of (5.9):

$$x_{k+1} = x_k - \frac{x_k^n - A}{n x_k^{n-1}} = \frac{1}{n}\left[(n - 1)x_k + \frac{A}{x_k^{n-1}}\right]. \tag{5.10}$$

EXAMPLE 5.7

The bothersome function of Example 5.4 can cause divergence of Newton as well as secant iterations. Here we illustrate a different type of difficulty. Take

$$f(x) = 1 - 2 \exp(-|x|).$$

The function is shown in Figure 5.12. It has no derivative at 0, but this could be trivially corrected. The real problem is that at a distance from the root, the slope is so close to 0 that the next Newton iteration seriously overshoots and ends up even farther from the roots. This, in turn, results in even a worse gyration in the opposite direction at the next iteration. If Newton iterations are started at $x_0 = 2.5$, then as shown in Figure 5.12, Newton iterates oscillate about the origin with ever-increasing amplitude until overflow results. On the other hand, if $0 < x_0 < x^*$, rapid convergence will be observed.

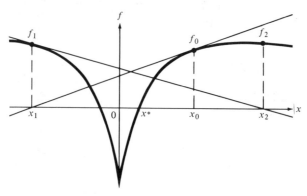

FIGURE 5–12   Function Demonstrating Newton Divergence

ENGINEERING EXAMPLE

A train was traveling in the positive x direction when the locomotive lost power. The engineer radioed to the preceding town for a replacement, and let the train coast under its own momentum. At time $t = 0$, the replacement locomotive left location $x = 0$ and proceeded at a constant velocity of $dx_2(t)/dt = 50$ km/h toward the train. At that instant the train was located at $x_1(0) = 30$ km, and it was traveling at a velocity $dx_1(t)/dt = 40$ km/h. It was decelerating, due to viscous friction so that its position was

$$x_1(t) = 110 - 80 \exp(-0.5t).$$

The question is: At what time did the replacement locomotive catch up to the train? That is, we wish to find T such that

$$f(T) = x_2(T) - x_1(T) = 0.$$

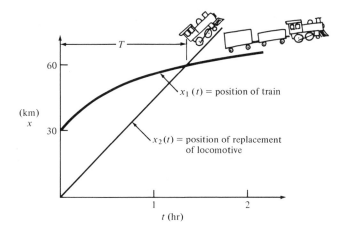

FIGURE 5–13 Position of Replacement Locomotive and Train

The locomotive's location at any time t is $x_2(t) = 50t$. In Figure 5.13 we have attempted to illustrate the functions involved. Thus we seek the root of

$$f(t) = x_2(t) - x_1(t) = 50t - 110 + 80 \exp(-0.5t).$$

The secant method was used to find the zero of $f(t)$, and initial points x_0 and x_1 were chosen to be 1 and 2, respectively. Tables 5.11 and 5.12 give a trivial calling program and output, from which we see that the time T (in hours) that the replacement engine catches the train is $T = 1.409051$.

TABLE 5.11 Program for Engineering Example

```
C       PROGRAM TRAIN
C
C       ************************************************************
C       THIS PROGRAM COMPUTES THE SOLUTION FOR THE TRAIN EXAMPLE
C       BY THE SECANT METHOD APPLIED TO THE FUNCTION
C       F(T)=50*T-110+80*EXP(-.5*T)
C       CALLS:    SECA
C       OUTPUT:
C                 T=SOLUTION TIME
C       OUTPUT(FROM FUNCTION F):
C                 X=VALUE OF X AT CURRENT ITERATION
C                 F=FUNCTIONAL VALUE AT X
C       ************************************************************
C
        T0=1.
        T1=2.
        EPS=1.E-7
C
C       *** SUBROUTINE SECA WILL FIND A ROOT IN THIS INTERVAL      ***
C
        CALL SECA(T0,T1,T,EPS)
        WRITE (10,*)'THE SOLUTION IS ',T
        STOP
        END
```

TABLE 5.11   (Continued)

```
C
C    *** FUNCTION F IS CALLED BY SECA. THIS IS THE FUNCTION TO ***
C    *** BE ZEROED                                             ***
C
     FUNCTION F(X)
     F=50.*X-110.+80.*EXP(-.5*X)
     WRITE(10,1)X,F
   1 FORMAT(22X,E16.8,4X,E16.8)
     RETURN
     END
```

TABLE 5.12   Output for Engineering Example

j	Time, t_j	Distance Between Trains, $x_2(t_j) - x_1(t_j)$
0	0.20000000E+01	0.19430357E+02
1	0.10000000E+01	-0.11477547E+02
2	0.13713467E+01	-0.11325874E+01
3	0.14059724E+01	-0.93017578E-01
4	0.14090706E+01	0.58746338E-03
5	0.14090512E+01	0.38146973E-05
6	0.14090511E+01	0.00000000E+00
7	0.14090511E+01	0.00000000E+00

The Solution: 1.409051

*5.6
CONVERGENCE AND ERROR PROPERTIES OF THE BISECTION, SECANT, AND NEWTON'S METHODS

Let x^* denote the root of the equation $f(x) = 0$, where $f(x)$ is a real-valued function. After k bisection steps, the root is located inside an interval of length $(b - a)/2^k$. Consequently, either endpoint approximates the root within the error bound $(b - a)/2^k$.

In applying the secant or the Newton method, a sequence x_0, x_1, x_2, . . . is computed. It is known (e.g., Szidarovszky and Yakowitz, 1978, Secs. 5.1.3 and 5.1.4) that if $f''(x)$ exists and is bounded and furthermore if $f'(x)$ differs from zero in the neighborhood of the root x^*, then for the starting point x_0 sufficiently close to x^*, both of these methods are convergent and there exist constants C and r such that

$$|x_{k+1} - x^*| \le C|x_k - x^*|^r. \tag{5.11}$$

In the case of the secant method $r = (1 + \sqrt{5})/2 \approx 1.615$, and for Newton's method $r = 2$. The exponent r is known as the *convergence rate*.

It is not difficult to confirm that methods having larger r ultimately have smaller error, no matter what C is. Thus, all things being equal, we would like to achieve as large a convergence rate r as possible without entailing excessive additional computational cost or numerical instability.

The Newton and secant methods were motivated by approximating the function $f(x)$ by linear polynomials (Taylor's and interpolation polynomials, respectively) and choosing the successive estimate to be the root of the linear polynomial. This suggests that one could perhaps get a faster convergence rate by increasing the degree of the approximating polynomial, thereby getting a better "fit" to $f(x)$. The Taylor expansion avenue does generalize (Problem 22): If we approximate $f(x)$ by a Taylor's polynomial of degree D, for well-behaved $f(x)$, the rate of convergence will be $D + 1$. (Thus Newton's method is the special case that $D = 1$.) However, this avenue has not become popular. Recall from Chapters 2 and 3 that Taylor's expansions have not led to convenient numerical schemes because of the bother of evaluating higher derivatives. Moreover, other troubles arise from Taylor's polynomial approximations. For example, one must solve the polynomial nonlinear equation problem in order to find the roots of the approximating Taylor's polynomial. Then the question arises as to which of the roots should be then selected as the next approximation. To summarize, the secant and Newton's methods, despite their age and simplicity, are still the methods of choice for routine nonlinear equation problems.

5.7
POLYNOMIAL ROOTS

5.7.1 Background Information

Because of their pervasive role in science and mathematics, polynomial equations, that is, equations of the form

$$f(z) = a_n z^n + a_{n-1} z^{n-1} + \cdots + a_0 = 0 \tag{5.12}$$

warrant special attention. The coefficients a_k ($0 \leq k \leq n$) may be either real or complex numbers. For such polynomials, the roots can be either real or complex. (Recall that a polynomial with real coefficients may have complex roots. In particular, the imaginary number $i = \sqrt{-1}$ is defined as a root of the real polynomial $z^2 + 1$.) In this section, we use z, rather than x, to denote the independent variable. This is to emphasize that here we presume polynomials to be evaluated on the set of complex, rather than real, numbers.

The bisection method is not applicable to finding complex roots, because the implementation of the method requires ordering the values of $f(x)$, which must therefore be real. It is known (e.g., Ortega and Rheinboldt, 1970) that the Newton and secant methods are convergent to simple roots. Because derivatives of polynomials are easy to obtain, and because of its fast convergence rate and stable behavior (e.g., Ralston and Rabinowitz, 1978, Sec. 8.6) near multiple roots, Newton's method is appropriate for polynomial equations.

We will use the following facts (see Herstein, 1964, Chap. 5) regarding the roots of the nth-degree polynomial (5.12) with real or complex coefficients:

1. There are n (not necessarily distinct) real or complex roots.
2. If n is odd and all coefficients are real, there is at least one real root.
3. If all coefficients are real and complex roots exist, they occur in conjugate pairs.
4. If z^* is a root of (5.12), then necessarily

$$f(z) = (z - z^*)g(z), \tag{5.13}$$

where $g(z)$ is a polynomial of degree $n - 1$.

This last fact can be utilized for reducing the degree of a polynomial, as successive roots are computed, by *synthetic division*.

Specifically, let

$$f(z) = \sum_{j=0}^{n} a_j z^j \quad \text{and} \quad g(z) = \sum_{j=0}^{n-1} b_j z^j$$

be related as in property 4. Then

$$f(z) = a_n z^n + a_{n-1} z^{n-1} + \cdots + a_1 z + a_0 \qquad (5.14)$$
$$= (z - z^*)(b_{n-1} z^{n-1} + b_{n-2} z^{n-2} + \cdots + b_0).$$

By equating coefficients of like powers of z in (5.14), we can solve for the b_j's. Thus, collecting the coefficients of z^i, $i = n, n - 1, \ldots, 1$, we conclude that

$$a_i = b_{i-1} - z^* b_i$$

or

$$\boxed{b_{i-1} = a_i + z^* b_i, \qquad i = n, n - 1, \ldots, 1.} \qquad (5.15)$$

In (5.15) we define $b_n = 0$ and recursively use that formula to get b_{n-1}, b_{n-2}, $\ldots$, b_0. The activity of obtaining lower-degree factor polynomials $g(z)$ is referred to as polynomial *deflation*.

─────── **EXAMPLE 5.8** ───────

Observe that $z^* = 2$ is a root of the polynomial equation

$$f(z) = z^4 - 2z^3 + 3z^2 - 7z + 2 = 0.$$

In the nomenclature of (5.14) and (5.15), $n = 4$, and

$$a_4 = 1, \quad a_3 = -2, \quad a_2 = 3, \quad a_1 = -7, \quad \text{and} \quad a_0 = 2.$$

From (5.15),

$$b_4 = 0$$
$$b_3 = a_4 + 2 \cdot 0 = 1$$
$$b_2 = a_3 + 2 \cdot 1 = 0$$
$$b_1 = a_2 + 2 \cdot 0 = 3$$
$$b_0 = a_1 + 2 \cdot 3 = -1.$$

The factor polynomial $g(z)$ is, then, $z^3 + 3z - 1$. One may check that $g(z)$ really is a factor by multiplying $(z - 2)(z^3 + 3z - 1)$ to recover the original polynomial $f(z)$.

■

5.7.2 Polynomial Roots by Newton's Method

In FORTRAN, arithmetic in complex variables may be undertaken by simply declaring variables of interest to be complex. Recall that Newton's method for finding a root z^* of $f(z)$ is determined by the recursive equation

$$z_{k+1} = z_k - \frac{f(z_k)}{f'(z_k)}.$$

If $f(z)$ is a polynomial, it is clear that its derivative $f'(z)$ exists and is easy to evaluate. In the polynomial case, the root z^* may well be a complex number. We must therefore presume that the approximations z_k and perhaps the coefficients of the polynomial $f(z)$ are complex. Even if we start with a polynomial having real coefficients, after synthetic division by a linear polynomial $(z - z^*)$, with z^* complex, the ''deflated'' polynomial will have complex coefficients.

The reader may confirm that if $f(z)$ has real coefficients and the imaginary part of z_0 is zero, successive Newton iterates will thereafter be constrained to be real and therefore cannot possibly locate complex roots. Thus the initial estimate should be complex.

The subroutine NTPOL given in Table 5.13 will attempt to locate all roots of a given polynomial having real or complex coefficients. It proceeds by successively applying Newton's method and polynomial deflation to obtain polynomials of lower and lower degree. In this routine, Horner's rule (Section 2.2) is used for evaluating $f(z)$ and its derivative. The program makes use of the observation that (5.15) for the coefficients of the deflated polynomial is identical to the intermediate terms b_i in the formula (2.2) for Horner's method. In specifying the coefficients $(A(1),. . .,A(N+1))$ to the subroutine NTPOL, it is presumed that $f(z)$ is represented according to

$$F(Z) = A(N+1)Z^N + \cdots + A(2)Z + A(1).$$

Note that this differs from polynomial representations such as (5.12) used earlier in this chapter. This modified representation is needed because in FORTRAN, array arguments must be positive. If the subroutine is called according to the main program statement

TABLE 5.13   Subroutine NTPOL for Newton Polynomial Root Computation

```
      SUBROUTINE NTPOL(N,A,ROOT,EPS)
C
C     ****************************************************************
C     *   FUNCTION: THIS SUBROUTINE COMPUTES ALL ROOTS FOR THE      *
C     *             POLYNOMIAL                                       *
C     *                  F(Z)=A(1)+A(2)*Z+...+A(N+1)*Z**N           *
C     *             USING NEWTON'S METHOD                           *
C     *   USAGE:                                                     *
C     *        CALL SEQUENCE: CALL NTPOL(N,A,ROOT,EPS)             *
C     *   PARAMETERS:                                                *
C     *        INPUT:                                                *
C     *            N=DEGREE OF THE COMPLEX POLYNOMIAL               *
C     *            A=N+1 BY 1 ARRAY OF COMPLEX POLYNOMIAL           *
C     *               COEFFICIENTS                                   *
C     *            EPS=ERROR BOUND                                   *
C     *        OUTPUT:                                               *
C     *        ROOT=N BY 1 ARRAY OF COMPLEX ROOT APPROXIMATIONS    *
C     ****************************************************************
C
C
```

TABLE 5.13 (Continued)

```
        IMPLICIT COMPLEX(A-H,O-Z)
        REAL EPS
        DIMENSION A(N+1),B(100),ROOT(N)
C       *** COMPUTE POLYNOMIAL ROOT ***
        ZSTAR=(1.0,1.0)
        DO WHILE(N.GT.1)
          X=ZSTAR+2.0*EPS
          DO WHILE(CABS(X-ZSTAR).GE.EPS)
            ZSTAR=X
C           *** COMPUTE U=F(ZSTAR) AND ITS DERIVATIVE ***
C           *** U=DF(ZSTAR) BY HORNERS RULE           ***
            DO 1 I=N+1,1,-1
              U=ZSTAR*U+A(I)
              IF(I.NE.1) THEN
                V=ZSTAR*V+A(I)*(I-1)
              END IF
   1        CONTINUE
            X=ZSTAR-U/V
            U=(0.0,0.0)
            V=(0.0,0.0)
          END DO
          ZSTAR=X
          ROOT(N)=ZSTAR
C         *** DEFLATE POLYNOMIAL BY DIVISION (Z-ZSTAR) ***
          N=N-1
          DO 2 I=N+1,1,-1
            B(I)=A(I+1)+ZSTAR*B(I+1)
   2      CONTINUE
          DO 3 I=1,N+1
            A(I)=B(I)
            B(I)=0.
   3      CONTINUE
          IF(ABS(AIMAG(ZSTAR)).GT.EPS) THEN
            ZSTAR=CONJG(ZSTAR)
          ELSE
            ZSTAR=(1.0,1.0)
          END IF
        END DO
        ROOT(1)=-A(1)/A(2)
        RETURN
        END
```

$$\text{CALL NTPOL(N,A,ROOT,EPS),}$$

it is to be interpreted that N is the degree of the polynomial and arrays A and ROOT must be declared complex in the calling program and dimensioned appropriately. We will illustrate the use of subroutine NTPOL in the example to follow. The N roots of $f(z)$ will be returned in the array ROOT.

EXAMPLE 5.9

The complex Newton's method and subroutine NTPOL (Table 5.13) for Newton extraction of polynomial roots and deflation will be applied to the function

$$f(z) = z^3 - 3z^2 - z + 9,$$

which has been subject of earlier examples.

Let the initial approximation of the root be selected as $z = 1 + i$. Then

$$f(z_0) = (1 + i)^3 - 3(1 + i)^2 - (1 + i) + 9$$

$$= 1 + 3i - 3 - i - 3 - 6i + 3 - 1 - i + 9 = 6 - 5i$$

$$f'(z_0) = 3(1 + i)^2 - 6(1 + i) - 1$$

$$= 3 + 6i - 3 - 6 - 6i - 1 = -7,$$

so the first Newton step gives the approximating root:

$$z_1 = z_0 - \frac{f(z_0)}{f'(z_0)} = 1 + i - \frac{6 - 5i}{-7}$$

$$= \frac{13}{7} + \frac{2}{7} i \approx 1.857142 + 0.285714i.$$

Further Newton iterates can be obtained analogously by replacing z_0 by z_1 and by repeatng the calculation above. The computer program performing the entire iteration method is given in Table 5.14, together with the output of the calculation (Table 5.15). We see that the computation uncovered the real root obtained by previous methods as well as a complex conjugate pair of roots. Thus the roots of $f(z)$ are as follows:

$$z_1^* = -1.5251020$$

$$z_2^* = 2.2625510 - 0.8843676i$$

$$z_3^* = 2.2625510 + 0.8843676i.$$

By evaluating $f(z)$ at these points, we may confirm that, indeed, they are roots.

TABLE 5.14   Program for Root Extraction

```
C       PROGRAM COMPLEX
C
C       ****************************************************************
C       THIS PROGRAM ILLUSTRATES NEWTONS METHOD IN THE COMPLEX CASE
C       BY FINDING ALL ROOTS OF THE FUNCTION F(Z)=Z**3-3*Z**2-Z+9
C       CALLS:    NTPOL
C       OUTPUT:
C               ROOT(I)=ROOTS OF THE FUNCTION I=1,2,3
C       ****************************************************************
C
        COMPLEX A(4),ROOT(3)
        EPS=1.E-6
        N=3
        A(1)=9.
        A(2)=-1.
        A(3)=-3.
        A(4)=1.
C
C       *** SUBROUTINE NTPOL WILL CALCULATE THE ROOTS          ***
C
        CALL NTPOL(N,A,ROOT,EPS)
        WRITE(10,*)(ROOT(I),I=1,3)
        STOP
        END
```

TABLE 5.15   Output of Computation

Polynomial Roots

```
(-1.525102,-5.9604645E-08) (2.262551,-0.8843675) (2.262551,0.8843676)
```

Some authors (e.g., Ralston and Rabinowitz, 1978, Sec. 8.13) solemnly warn about problems arising because of ill-conditioning in the polynomial equation problem. By this, they mean that either slight changes in the coefficients or round-off errors in the computations can result in large shifts of computed roots. A famous polynomial subject to this effect is the polynomial having as its roots the integers -1 to -20. Our impression is that with polynomials of degree 10 or less, ill-conditioning is not usually a difficulty, and the methods of this section are adequate. For higher-degree polynomials, precaution is advisable. Deflation can be undertaken more accurately if it is done in order of increasing root magnitudes. A procedure known as *purification* achieves this as follows:

1. Use some root extraction procedure such as NTPOL to get approximations $z_1, \ldots, z_n$ of the roots.
2. Reindex so that $|z_1| \leq |z_2| \leq \cdots \leq |z_n|$.
3. Run the procedure again, using z_1 as starting point in NTPOL, and after each deflation, take next largest z_j as initial approximation.

The IMSL mathematical library uses the Jenkins–Traub technique (Ralston and Rabinowitz, 1978, pp. 383–391), which can be viewed as a Newton method applied to a rational function (ratio of polynomials) approximation of the given polynomial.

In Section 5.2 we discussed root localization. In the case of polynomials, there exist methods (Szidarovszky and Yakowitz, 1978, Sec. 5.2.2), which, for any given polynomial $p(x)$ of degree n and positive number ε, provide estimates z_i, $1 \leq i \leq n$, such that if z^* is a root of $p(z)$, then for some j, $|z_j - z^*| < \varepsilon$. These localization methods accomplish the task within a predetermined number of computer operations, the number depending on the polynomial coefficients and the error tolerance bound ε. The computations proceed by first bounding the roots into some finite disk in the plane, by (5.3), for example. Then this disk is subdivided into a finite collection of regions having dimensions less than ε, and each of these regions is checked for interior roots. It turns out that in comparison to iterative methods, these localization methods are cumbersome to program and require excessive running time. In particular, the number of regions to be checked grows as the square of $1/\varepsilon$.

*5.7.3   Bairstow's Method

In FORTRAN it is sensible and convenient to locate roots of polynomials by performing Newton's method in complex arithmetic. On the other hand, some popular programming languages (e.g., BASIC and Pascal) do not readily admit complex arithmetic. Moreover, a weakness to the approach of using Newton's method in complex arithmetic, as proposed in the preceding section, is that on many computers, this limits us to single-precision computation. In some cases the root location is extremely sensitive to polynomial coefficient values and avoidance of roundoff is essential. Bairstow's method, in essence, provides the capability of using Newton's method in real arithmetic to locate complex polynomial roots. This enables the FORTRAN user to employ multiple precision and thereby greatly increase the accuracy of the computations. In addition, since all variables are real, Bairstow's method can easily be coded in Pascal or BASIC. Bairstow's method avoids requiring complex arithmetic for polynomials with real coefficients because it obtains quadratic rather than linear factors; these quadratics also have real coefficients. Of course, one can alternatively sidestep complex arithmetic by doing real arithmetic on the real and imaginary parts of each number of Newton's method. Indeed, this is what the computer itself ultimately does. Through this device, the technique of the preceding section may be implemented, but the coding is complicated.

Assume now that all coefficients of $f(z)$ are real. Then, as we have noted, if a complex number $\alpha + i\beta$ is a root of $f(z)$, its complex conjugate, $\alpha - i\beta$, is also a root. As a consequence, the real quadratic

$$h(z) = [z - (\alpha + i\beta)][z - (\alpha - i\beta)] = z^2 - 2\alpha z + (\alpha^2 + \beta^2) \quad (5.16)$$

must be a factor of $f(z)$.

Bairstow's method obtains real quadratic factors of real polynomials. All the complex roots of $f(z)$ are necessarily among the roots of the quadratic factors.

Let $h(z) = z^2 - pz - q$, where p and q are coefficients to be determined. In order that the notation of our discussion of Bairstow's method conform with the standard literature, we will index the coefficients of polynomials in reverse order to preceding sections. Thus we write, instead of (5.12),

$$f(z) = a_0 z^n + a_1 z^{n-1} + \cdots + a_{n-1} z + a_n. \quad (5.17)$$

According to polynomial factorization theory (e.g., Herstein, 1964, Chap. 5), regardless of p and q, we may always write

$$\begin{aligned} f(z) &= a_0 z^n + a_1 z^{n-1} + \cdots + a_n \\ &= (z^2 - pz - q)(b_0 z^{n-2} + b_1 z^{n-3} + \cdots + b_{n-2}) \quad (5.18) \\ &\quad + b_{n-1}(z - p) + b_n. \end{aligned}$$

This equation has the form

$$f(z) = h(z)g(z) + r(z),$$

with $r(z) = b_{n-1}(z - p) + b_n$ being the *remainder polynomial*. (The remainder polynomial is written as a linear polynomial of the difference $z - p$ in order to obtain more convenient formulas later.) The object in Bairstow's method is to calculate coefficients p and q of $h(z)$ so that $r(z) = 0$; that is, so that $h(z)$ is a factor of $f(z)$. Toward this end, we need to represent the coefficients b_i in (5.18) in terms of the coefficients a_i of $f(z)$ and p and q. The coefficients b_i ($i \geq 0$) are determined by matching the coefficients of various powers of z on the left- and right-hand sides of (5.18), which implies that

$$a_0 = b_0$$

$$a_1 = b_1 - pb_0$$

$$a_2 = b_2 - pb_1 - qb_0$$

$$\vdots$$

$$a_{n-1} = b_{n-1} - pb_{n-2} - qb_{n-3}$$

$$a_n = b_n - pb_{n-1} - qb_{n-2}.$$

After some elementary algebraic manipulations, the preceding equation yields an explicit formula for the coefficients b_i:

$$
\begin{aligned}
b_0 &= a_0 \\
b_1 &= a_1 \quad + pb_0 \\
b_2 &= a_2 \quad + pb_1 \quad + qb_0 \\
&\vdots \\
b_{n-1} &= a_{n-1} + pb_{n-2} + qb_{n-3} \\
b_n &= a_n \quad + pb_{n-1} + qb_{n-2}.
\end{aligned}
\tag{5.19}
$$

In view of (5.18), one can verify that $h(z)$ factors $f(z)$ if and only if

$$
b_{n-1} = b_{n-1}(p, q) = 0 \quad \text{and} \quad b_n = b_n(p, q) = 0.
\tag{5.20}
$$

This is a system of two nonlinear equations in the two variables p and q. It can be solved by the multivariable Newton's method to be discussed in Section 5.8. If this is done, the algorithm is called *Bairstow's method*.

As we shall see in the next section, the application of the multivariable Newton's method computes sequences $\{p_k\}$, $\{q_k\}$ which, under suitable conditions, converge to the exact solution of (5.20). The multivariable Newton's method requires the first partial derivatives of $b_{n-1}(p, q)$ and $b_n(p, q)$. As derived in (Johnson and Riess, 1977, pp. 150–151), these derivatives may be obtained by simple recursive relations. Starting from initial values $c_{-2} = c_{-1} = 0$, define the sequence $\{c_i\}$ by the following relations:

$$c_{-2} = c_{-1} = 0; \quad c_i = b_i + pc_{i-1} + qc_{i-2}, \quad i = 0, 1, \ldots, n - 1;$$

that is, let

$$
\begin{aligned}
c_0 &= b_0 \\
c_1 &= b_1 \quad + pc_0 \\
c_2 &= b_2 \quad + pc_1 \quad + qc_0 \\
&\vdots \\
c_{n-1} &= b_{n-1} + pc_{n-2} + qc_{n-3}.
\end{aligned}
\tag{5.21}
$$

Then the first partial derivatives are given by

$$\frac{\partial b_{n-1}(p,\,q)}{\partial q} = c_{n-3}, \qquad \frac{\partial b_n(p,\,q)}{\partial p} = c_{n-1},$$

$$\frac{\partial b_{n-1}(p,\,q)}{\partial p} = \frac{\partial b_n(p,\,q)}{\partial q} = c_{n-2}. \tag{5.22}$$

────── **EXAMPLE 5.10** ──

In the case of our usual polynomial $f(z) = z^3 - 3z^2 - z + 9$, $n = 3$ and (5.19) takes the specific values

$$b_0 = 1$$
$$b_1 = -3 + p$$
$$b_2 = -1 + p(p - 3) + q = p^2 - 3p + q - 1$$
$$b_3 = 9 + p(p^2 - 3p + q - 1) + q(p - 3)$$
$$ = p^3 - 3p^2 + 2pq - p - 3q + 9.$$

In accordance with (5.21), the terms c_i are

$$c_0 = 1$$
$$c_1 = (-3 + p) + p = 2p - 3$$
$$c_2 = (p^2 - 3p + q - 1) + p(2p - 3) + q = 3p^2 - 6p + 2q - 1.$$

Now we see that for our example, the system of nonlinear equations (5.20) becomes

$$b_2(p,\,q) = p^2 - 3p + q - 1 = 0$$

and

$$b_3(p,\,q) = p^3 - 3p^2 + 2pq - p - 3q + 9 = 0.$$

We may simplify the foregoing expression of $b_3(p,\,q)$ by noting that $b_2 = b_{n-1}(p,\,q)$ must be 0. Then, from (5.19), and ignoring the pb_{n-1} addend in b_n, we have $b_3(p,\,q) = a_3 + qb_1 = 9 + q(p - 3)$.

Then the partial derivatives of $b_2(p,\,q)$ and $b_3(p,\,q)$ can be obtained by relations (5.22) to get

$$\frac{\partial b_2}{\partial q} = c_0 = 1, \qquad \frac{\partial b_2}{\partial p} = \frac{\partial b_3}{\partial q} = c_1 = 2p - 3,$$

$$\frac{\partial b_3}{\partial p} = c_2 = 3p^2 - 6p + 2q - 1.$$

In Example 5.11 we shall see the solution of (5.20) is,

$$p \approx 4.525102255, \qquad q \approx -5.901243652,$$

from which we can conclude that

$$z_1^*, z_2^* \approx 2.262551128 \pm 0.8843675975i,$$

since z_1^* and z_2^* are the roots of the quadratic polynomial $z^2 - pz - q$.   ∎

*5.8
SYSTEMS OF NONLINEAR EQUATIONS

Attention now focuses on simultaneous nonlinear equations. Such equations can be expressed by

$$f_i(x_1, \ldots, x_n) = 0 \qquad (i = 1, 2, \ldots, n), \tag{5.23}$$

where the f_i's represent real-valued functions. A *root* is any vector $\mathbf{x}^* = (x_1^*, \ldots, x_n^*)$ of real numbers for which (5.23) is satisfied simultaneously for all i. The multivariable generalization of the Newton's method, which is introduced next, is one of the most popular and effective techniques for solving nonlinear equations.

The single-variable Newton's method—as we have seen—can be interpreted as follows. If an approximation x_k of the root is given, $f(x)$ is approximated by the linear Taylor's polynomial

$$p(x) = f(x_k) + f'(x_k)(x - x_k), \tag{5.24}$$

and the next approximation x_{k+1} is defined to be the root of this linear approximating function. This idea can be generalized to the multivariable case in the following way. Let $\mathbf{x}^{(k)} = (x_1^{(k)}, \ldots, x_n^{(k)})$ be an approximation of the root. Then the univariable linear approximating polynomial (5.24) is replaced by its multivariable counterpart,

$$p_i(x_1, \ldots, x_n) = f_i(x_1^{(k)}, \ldots, x_n^{(k)})$$
$$+ \sum_{j=1}^{n} \frac{\partial f_i(x_1^{(k)}, \ldots, k_n^{(k)})}{\partial x_j} (x_j - x_j^{(k)}) \qquad (i = 1, \ldots, n).$$

The next approximation $\mathbf{x}^{(k+1)}$ is defined to be the solution of the corresponding linear equations:

$$p_i(x_1, \ldots, x_n) = 0 \qquad (i = 1, \ldots, n). \tag{5.25}$$

The multivariable Newton's algorithm becomes more transparent through introduction of matrix notation. The higher-dimensional generalization of a derivative $f'(x)$ is the *Jacobian matrix* $\mathbf{J}(\mathbf{x})$, the (i, j)th element of which is defined to be

$$J_{ij}(\mathbf{x}) = \frac{\partial}{\partial x_j} f_i(x_1, x_2, \ldots, x_n).$$

Define $\mathbf{J}^{(k)} = \mathbf{J}(\mathbf{x}^{(k)})$ to be the Jacobian matrix of $(f_1(\mathbf{x}), \ldots, f_n(\mathbf{x}))$, evaluated at the kth iteration estimate $\mathbf{x}^{(k)}$, and introduce the vector

$$\mathbf{f}^{(k)} = \begin{bmatrix} f_1(\mathbf{x}^{(k)}) \\ \vdots \\ f_n(\mathbf{x}^{(k)}) \end{bmatrix}.$$

Then the *multivariable Newton method* is expressible, analogously to its real counterpart (5.8), as

$$\mathbf{x}^{(k+1)} = \mathbf{x}^{(k)} - (\mathbf{J}^{(k)})^{-1}\mathbf{f}^{(k)}. \tag{5.26}$$

Of course, for computational reasons offered in Chapter 4, evaluation of the inverse matrix is not as efficient as solution of the equivalent linear equation

$$\mathbf{J}^{(k)}(\mathbf{x}^{(k+1)} - \mathbf{x}^{(k)}) = -\mathbf{f}^{(k)}$$

by Gauss elimination methods discussed in Chapter 4. Under certain smoothness assumptions concerning the second partial derivatives of the f_i's, the successive estimates can be shown to have the same convergence rate as in the real case. That is, if $(x_1^*, \ldots, x_n^*)$ is the exact root, then for some constant C and all i, $1 \le i \le n$,

$$|x_i^{(k+1)} - x_i^*| \le C|x_i^{(k)} - x_i^*|^2, \qquad k = 1, 2, \ldots.$$

A precise statement and proof of this assertion is given, for example by Szidarovszky and Yakowitz (1978, Sec. 5.3.1). Subroutine MVNE, which implements the multivariable Newton's method, constitutes Table 5.16. It presumes that the user provides program subfunctions $F(X, I)$, and $DF(X, I, J)$, which returns the values $f_i(\mathbf{x})$ and $J_{ij}(\mathbf{x})$, if $I = i$, $J = j$, and $X = \mathbf{x}$.

TABLE 5.16   Subroutine MVNE for the Multivariable Newton Method

```
      SUBROUTINE MVNE(N,X0,X,EPS)
C
C     *****************************************************************
C     *   FUNCTION: THIS SUBROUTINE COMPUTES THE ROOT VECTOR          *
C     *             OF A SYSTEM OF EQUATIONS F(I,X)=0 I=1,...,N        *
C     *             USING THE MULTIVARIATE NEWTON METHOD              *
C     *   USAGE:                                                      *
C     *       CALL SEQUENCE: CALL MVNE(N,X0,X,EPS)                    *
C     *       EXTERNAL FUNCTIONS/SUBROUTINES:                         *
C     *                      FUNCTION F(I,X)                          *
C     *                      FUNCTION DF(I,J,X0)                      *
C     *                      SUBROUTINE GAUS1(N,M,ND,A,DELT)          *
C     *   PARAMETERS:                                                 *
C     *       INPUT:                                                  *
C     *           N=NUMBER OF SIMULTANEOUS LINEAR EQUATIONS(LESS      *
C     *             THEN 30)                                          *
C     *           X0=N BY 1 ARRAY OF INITIAL APPROXIMATE ROOT         *
C     *              VECTOR VALUES                                    *
C     *           EPS=ERROR BOUND (TOLERANCE)                         *
C     *       OUTPUT:                                                 *
C     *           X=N BY 1 ARRAY OF APPROXIMATE ROOTS                 *
C     *****************************************************************
C
C
```

TABLE 5.16   (Continued)

```
      DIMENSION X0(N),X(N),A(30,31)
C     *** INITIALIZATION ***
      K=1
C     *** COMPUTE VECTOR X AS THE ROOT ***
      DO 5 WHILE(K.LE.N)
         DO 1 I=1,N
            DO 2 J=1,N
               A(I,J)=DF(I,J,X0)
2           CONTINUE
            A(I,N+1)=-F(I,X0)
1        CONTINUE
C        *** PERFORM GAUSSIAN ELIMINATION ***
         M=1
         ND=30
         DELT=0.000001
         CALL GAUS1(N,M,ND,A,DELT)
         DO 3 I=1,N
            X(I)=X0(I)+A(I,N+1)
3        CONTINUE
C        *** IF SOLUTION CHANGE IS SMALL, STOP ***
         K=1
         DO WHILE(ABS(X(K)-X0(K)).LT.EPS.AND.K.LE.N)
            K=K+1
         END DO
         IF(K.LE.N) THEN
            DO 4 I=1,N
               X0(I)=X(I)
4           CONTINUE
         END IF
5     CONTINUE
      RETURN
      END
```

EXAMPLE 5.11

In Example 5.10 we have seen that the application of the Bairstow's method for solving the polynomial equation

$$f(z) = z^3 - 3z^2 - z + 9 = 0$$

requires the solution of the nonlinear equations

$$b_2(p, q) = -1 - 3p + p^2 + q = 0$$

$$b_3(p, q) = 9 - 3q + pq = 0.$$

The elements of the Jacobian matrix are

$$\frac{\partial b_2(p, q)}{\partial p} = 2p - 3, \qquad \frac{\partial b_2(p, q)}{\delta q} = 1$$

$$\frac{\partial b_3(p, q)}{\partial p} = q, \qquad \frac{\partial b_3(p, q)}{\partial q} = -3 + p.$$

Then

$$\mathbf{J}^{(k)} = \begin{bmatrix} 2p^{(k)} - 3 & 1 \\ q^{(k)} & -3 + p^{(k)} \end{bmatrix}$$

and

$$\mathbf{f}^{(k)} = \begin{bmatrix} -1 - 3p^{(k)} + (p^{(k)})^2 + q^{(k)} \\ 9 - 3q^{(k)} + p^{(k)}q^{(k)} \end{bmatrix}.$$

Let us start the application of the multivariable Newton's method with the initial approximates $p^{(0)} = 2$, $q^{(0)} = -3$. Then the corresponding Jacobian matrix $\mathbf{J}^{(0)}$ and vector $\mathbf{f}^{(0)}$ are as follows:

$$\mathbf{J}^{(0)} = \begin{bmatrix} 2 \cdot 2 - 3 & 1 \\ -3 & -3 + 2 \end{bmatrix} = \begin{bmatrix} 1 & 1 \\ -3 & -1 \end{bmatrix}$$

$$\mathbf{f}^{(0)} = \begin{bmatrix} -1 - 3 \cdot 2 + 2^2 + (-3) \\ 9 - 3 \cdot (-3) + 2 \cdot (-3) \end{bmatrix} = \begin{bmatrix} -6 \\ 12 \end{bmatrix}.$$

Since in this case

$$\mathbf{x}^{(0)} = \begin{bmatrix} p^{(0)} \\ q^{(0)} \end{bmatrix} = \begin{bmatrix} 2 \\ -3 \end{bmatrix},$$

one step of the multivariable Newton's method (5.26) requires solution of the linear equation

$$\mathbf{J}^{(0)}(\mathbf{x}^{(1)} - \mathbf{x}^{(0)}) = -\mathbf{f}^{(0)};$$

or in the notation of (5.26)

$$\mathbf{x}^{(1)} = \mathbf{x}^{(0)} - (\mathbf{J}^{(0)})^{-1}\mathbf{f}^{(0)}.$$

In our case we obtain the equation

$$\begin{bmatrix} 1 & 1 \\ -3 & -1 \end{bmatrix} \begin{bmatrix} p^{(1)} - 2 \\ q^{(1)} + 3 \end{bmatrix} = \begin{bmatrix} 6 \\ -12 \end{bmatrix},$$

which has the unique solution $p^{(1)} = 5$, $q^{(1)} = 0$. The step above is then repeated by replacing $p^{(0)}$, $q^{(0)}$ by $p^{(1)}$, $q^{(1)}$, respectively. Successive Newton iterates and the corresponding functional values are displayed in Table 5.17. The roots associated with the quadratic factor were computed by solving the quadratic equation, with p and q as in the last entry of Table 5.17,

$$z^2 - pz - q = 0,$$

by the quadratic formula

$$z_1, z_2 = \frac{p \pm \sqrt{p^2 + 4q}}{2}.$$

By this means we conclude that the complex roots are

$$2.262551128 \pm i0.8843675975.$$

This agrees with our computation in Section 5.7.2.

TABLE 5.17 Application of the Multivariable Newton Method

$p^{(k)}$	$q^{(k)}$	$b_2(p^{(k)}, q^{(k)})$	$b_3(p^{(k)}, q^{(k)})$
2.000000	-3.000000	-6.000000	12.00000
5.000000	0.0000000E+00	9.000000	9.000000
4.357143	-4.500000	0.4132643	2.892857
4.547431	-6.000624	3.6211491E-02	-0.2855511
4.525296	-5.901927	4.9066544E-04	-2.1858215E-03
4.525102	-5.901244	-1.4305115E-06	1.9073486E-06

5.9
SUPPLEMENTARY NOTES AND DISCUSSIONS

The techniques offered in this chapter for solving nonlinear equations (5.1) for $f(x)$ a real function are usually effective and simple to implement. For such functions, prior reflection on the general shape of $f(x)$ is often rewarding: It is helpful to have some idea how many (if any) roots exist, and their approximate location on the real line. Such "localization" is useful, within limits. But once we have some confidence that the computation does not suffer from pathological problems such as shown in Examples 5.4 and 5.7, it is well to implement some nonlinear equation algorithm and be done with the job. A safety feature is that in the vast majority of cases either successive iterations will not be converging at all, or they will be converging to the correct answer. We often find that programmable calculators (especially those programmable in BASIC) are adequate and convenient for solving real nonlinear equations, using any of the methods described in this chapter.

On the other hand, the subject of nonlinear equations, like several other areas of numerical analysis, becomes murky when the problem involves functions $f(x)$ of several variables. To begin with, there is no evident multivariable version of the bisection method, and the task of localizing roots and inferring the nature of the function from a manageable number of evaluations becomes complicated or even hopeless, especially in light of the fact that graphs are not available to guide the intuition. Whereas the secant method has several multivariable interpretations, the extension of Newton's method as described in Section 5.5 is unambiguous in light of the Taylor's series motivation. We will assert that many convergence properties of both the Newton's methods and standard extensions of the secant method remain valid in a suitable multivariable interpretation, but their justification requires much heavier mathematical machinery than is needed in the univariate case. The reader is referred to Szidarovszky and Yakowitz (1978, Chaps 4 and 5) for further analytical developments and references to the advanced literature on nonlinear equations of several variables.

There are strong links between nonlinear equation problems and function optimization. As a simple but instructive example, recall from calculus that an interior point x^* is a local maximum or minimum point of a differentiable function $f(x)$ only if it is a root of $f'(x)$. Thus the task of finding the optimum of a function $f(x)$ defined on an interval $[a, b]$ can be rephrased as the problem of finding all the roots of $f'(x)$ on that interval. Then the optimum point can be found by comparing the values of $f(x)$ at all these points and the endpoints of the interval.

TABLE 5.18   Match of Nonlinear Equation Problem Characteristics and Methods

Nonlinear Equation Characteristics	Method*
Real Function $f(x)$ in (5.1)	
Points a, b, $a < b$, and root known to be in $[a, b]$; $f(x)$ not expensive to evaluate	B, S
Same as above, but speed of convergence a concern	N, S
No points a, b bracketing roots are known	N, S
Same as above, but derivative not available, or expensive to evaluate	S
$f(x)$ a polynomial	
Complex computer arithmetic available, roundoff not troublesome	CN
Complex arithmetic not available, or accuracy important	BA, perhaps in double precision; CN, in double precision
Multivariable function $f(x)$	
Derivatives available	MN
Derivatives unavailable, or MN does not converge	Assistance or further study needed

*B, bisection; BA, Bairstow; CN, complex Newton; S, secant; MN, multivariable Newton; N, Newton.

Conversely, it is sometimes useful to recognize that $\mathbf{x}^* = (x_1^*, \ldots, x_n^*)$ is a solution of the system of nonlinear equations

$$f_i(x_1, \ldots, x_n) = 0 \qquad (1 \leq i \leq n)$$

if and only if the function

$$F(\mathbf{x}) = f_1^2(\mathbf{x}) + \cdots + f_n^2(\mathbf{x})$$

has minimal value at $\mathbf{x} = \mathbf{x}^*$, and that value is zero.

In Table 5.18 we have tried to relate nonlinear equation problem characteristics to what we regard as the most appropriate of the methods of the present chapter, for those characteristics.

PROBLEMS

Section 5.2

1. Find the intervals on which $p(x) = x^3 - 6x^2 + 10x - 5$ is monotonically increasing. How many real roots does $p(x)$ possess?

2. Find an interval of at most unit length that you can assure contains a root of $p(x) = x^3 - 6x^2 + 10x - 5$.

3. Show conclusively (i.e., mathematically, not graphically) that the function $f(x) = e^x - \sin(\pi x/3)$ has a unique root in the interval $[-3.1, -2.8]$.

4. Give a graphical demonstration that the following functions have infinitely many real roots, and localize the smallest positive roots of each of these functions in an interval of unit length.
   (a) $x - \tan(x)$.
   (b) $\cos(x - \cos(3x))$.
   (c) $e^x - \tan(x)$.

5. Show that equation $e^x - 2x - 2 = 0$ has two real roots and that they are inside the intervals $(-1, 0)$ and $(1, 2)$.

6. (a) Show that any root of the polynomial $p(x) = x^5 + 3x^4 - x^3 + 2x^2 - 1$ satisfies the inequality $|x| \leq 4$.
   (b) Show that if x is a real root, $|x| \leq 3.5$.

Section 5.3

7. The polynomial $p(x) = x^3 - 6x^2 + 10x - 5$ has a root in the interval $[3, 4]$. Perform two iterations of the bisection method, using the starting points $x_0 = 3$ and $x_1 = 4$. Write down intermediate steps, giving the values $f(x_1)$, $f(x_2)$, and so on.

8. Find a root of

$$f(x) = e^x - \sin\left(\frac{\pi x}{3}\right)$$

to four significant decimal places by the bisection method. (**HINT:** See Problem 3 for starting points.)

9. Apply the bisection method to find the roots of $f(x) = e^x - 2x - 2$ as accurately as your computer will allow. (**HINT:** See Problem 5.)

10. If $|x_L - x_R| = 1$, how many bisection iterations are required to estimate a root with error less than 10^{-12}?

*11. Consider the "trisection" method. Divide $[a, b]$ into three equal parts. If $f(a)f(b) < 0$, at least one of the subintervals must have a root.
   (a) Modify subroutine BISE (Table 5.2) to trisect. Apply it to Problem 8.
   (b) Give the number of iterations required in the trisection case to obtain the root within a given bound ϵ.
   (c) Compare the number of times function $f(x)$ must be evaluated to assure a certain accuracy, by the bisection and trisection methods.

Section 5.4

12. Do Problem 7 by the secant method. Show your work.

13. Do Problem 9, using the secant method (subroutine SECA, Table 5.5). Record the actual error $x^* - x_k$ at each iteration, x^* being the root.

14. State the problem of finding $\sqrt{30}$ as a nonlinear equation problem with $f(x)$ a quadratic. Solve the problem by the secant method as accurately as your computer permits, with $x_0 = 5$, $x_1 = 6$. What limits the accuracy?

*15. Apply both the secant and bisection methods to find that root x of

$$f(x) = e^x - \sin\left(\frac{\pi x}{3}\right)$$

which is in the interval $[-3.1, -2.8]$. For each method, plot the graphs of $\log |x^* - x_k|$ as a function of k. Visually estimate the slope of the graphs. Try to connect your estimates to discussion in Section 5.6.

Section 5.5

16. Apply two iterations of Newton's method to the polynomial

$$p(x) = x^3 - 6x^2 + 10x - 5,$$

starting at $x = 3$. Show your calculations.

17. Apply Newton's method [via subroutine NEIT (Table 5.8)] to obtain approximations of $\sin^{-1}(x)$, $\cos^{-1}(x)$, $\tan^{-1}(x)$ and $\log(x)$ from $\sin(x)$, $\cos(x)$, $\tan(x)$, and e^x at $x = 0.1, 0.5, 1.0, 2.0$. Compare your estimates against those given by your computer's library function.

18. Apply the result of Example 5.6 [relation (5.10)] to find $\sqrt[3]{123.0}$, $\sqrt[4]{12}$. Compare with your computer's calculation as given by $(123.0)**(1.0/3.0)$ and $(12.)**(1.0/4.0)$.

19. Find $\sqrt{1234.0}$ as accurately as your computer permits by Newton's method. Use the idea in Example 5.6. What limits the accuracy?

*20. (Continuation of Problem 15.) Prepare a graph for Newton's method errors corresponding to those of the bisection and secant calculations. Approximate the slope, and comment in light of Section 5.6.

21. Modify subroutine NEIT so that it replaces $f'(x)$ with the finite difference approximation

$$g(x) = \frac{1}{2\Delta_i} [f(x + \Delta_i) - f(x - \Delta_i)],$$

where $\Delta_i \equiv |f(x_i)|$. Apply this modified scheme to redo Problem 17.

Section 5.6

*22. Generalize Newton's method to quadratic rather than linear approximating polynomials. Thus if $f(x)$ is the function whose root we seek, and x_0 is our initial approximation, find the roots y_1 and y_2 of the quadratic Taylor's polynomial expansion of $f(x)$ about x_0. Let x_1 be that root y_1 or y_2 which minimizes $|f(x)|$. Encode this algorithm and try it out.

*23. (Continuation of Problem 22.) Prove that the convergence rate for the method sketched above is cubic under appropriate differentiability assumptions.

*24. Consider the simple scheme: Select x_1 and x_2 arbitrarily and recursively let

$$x_{k+1} = x_k - \frac{f(x_k)}{k} \operatorname{sign} \left[\frac{f(x_k) - f(x_{k-1})}{x_k - x_{k-1}} \right].$$

It will find a root, if one exists, of any monotonic function having a derivative $f'(x)$ bounded away from 0 for all x. Modify NEIT to implement this method and redo Problem 20. [*Note:* This rule lies at the heart of a well-known statistical scheme, "stochastic approximation," which finds roots from inaccurate measurements of $f(x_k)$.]

Section 5.7

25. The complex number $z_1 = 1 + i$ is a root of the polynomial equation $p(z) = z^4 - 2z^3 + 3z^2 - 2z + 2$. By synthetic division, find the cubic polynomial $q(z)$ such that $p(z) = (z - z_1)q(z)$.

26. The quadratic polynomial $Q(z) = z^2 + 1$ is a factor of the polynomial

$$p(z) = z^4 - 2z^3 + 3z^2 - 2z + 2.$$

By synthetic division, find the polynomial $q(z)$ such that

$$p(z) = Q(z)q(z).$$

27. Use NTPOL to find all the roots of

$$p(z) = z^6 + z^5 + z^4 - z^3 + z^2 + z + 1.$$

Verify that these points are roots by evaluating $p(z)$ at these values.

28. Find a polynomial for which NTPOL fails.

*Section 5.8

*29. Formulate the scheme of the multivariate Newton's method for the nonlinear system of equations

$$\cos (x) - y = 0$$
$$x - \sin (y) = 0.$$

Find an initial value of the solution by the graphical method. [**HINT:** Draw the graphs of the functions $y = \cos (x)$, $x = \sin (y)$ and approximate their intersection.]

*30. Find roots of the following nonlinear system of equations, to four significant decimals. We leave the selection of initial points to the reader.

(a) $x^2 + y^2 = 1$
     $x^3 - y = 0.$

(b) $\sin (x) - y = 0$
     $e^y - x = 0.$

(c) $x^2 - y^2 = 1$
     $y - e^x = 0.$

CHAPTER 6

Function Approximation and Data Fitting

6.1
PRELIMINARIES

The motivation and foundations for function approximation and data fitting stem from considerations introduced in Chapter 2. Specifically, one desires a computer-amenable function for approximating a given function $f(x)$, or for fitting a curve to given data points. As in Chapter 2, our focus will be on use of polynomials as the computer-amenable approximating functions. In the closing section, however, Fourier methods using sinusoids will be introduced. Chapter 2 restricted interest to interpolation; the functions we constructed were required to pass through the data. In "approximation" we drop the interpolation restriction.

Let us now distinguish between the interpolation problems studied in Chapter 2 and the class of tasks for which the methods of the present chapter are intended. First, interpolation is appropriate only if the data are believed to consist of accurate values from a well-behaved function. For example, if the graph of the data points was found to resemble Figure 6.1, then interpolation is sensible. Such a set of data, which almost calls out to be connected by a smooth curve, might represent radiation intensity from a radioactive source, measured at equally spaced times. Moreover, the interpolation polynomial for these data ought to give a very good estimate of the radiation intensity that would have been found if an observation had been taken at time point t not among the data values.

Still, as we saw in the Runge example (see Example 2.5), even when the data are from a fairly smooth and regular curve, polynomial interpolation methods are not always adequate. Although spline techniques can be expected to handle some of these situations, splines are nevertheless sometimes bothersome to use, in comparison to polynomials, especially when evaluation speed and minimal programming and memory requirements are of prime consideration. By way of evidence, splines are seldom used for intrinsic computer library functions.

Some data sets arise from phenomena having random components or other unpredictable mechanisms. In Figure 6.2 we have plotted test scores of the mathematics section of the Scholastic Aptitude Test as reported in U.S. Bureau of

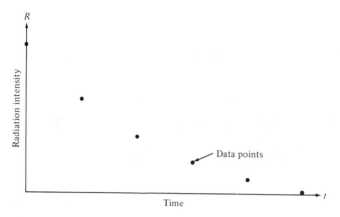

FIGURE 6–1   Data Well-Suited to Interpolation

Census (1984, p. 158). Although these points do seem to contain trend information, interpolation of these points by the Lagrange polynomial (via subroutine LAGR, Table 2.5) resulted in the meaningless plot of Figure 6.3.

The approximation methods of this chapter are suited to finding a polynomial of relatively low degree (thus assuring ''smoothness'') which ''fits'' these data in the sense that the curve passes as close as possible to the data points, subject to the constraint that the polynomial degree not exceed a specified number m.

Outside the realm of interpolatory methods such as those discussed in Chapter 2, the three most popular techniques for function approximation are

1. Least-squares polynomials.
2. Uniform approximation polynomials.
3. Fourier methods.

These are the subjects of Sections 6.2, 6.3, and 6.4, respectively, of this chapter. These methods can be related in terms of measures of ''fit.'' Let (x_1, f_1), (x_2, f_2), . . ., (x_n, f_n) denote the data pairs to be approximated, and let m be the desired maximal degree of the polynomial approximation. Then the *least-squares polynomial* for the data is any polynomial of degree m or less that minimizes the

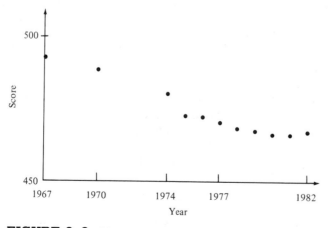

FIGURE 6–2   Plot of SAT Scores

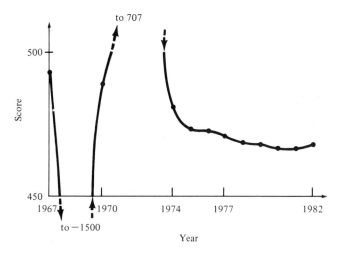

FIGURE 6–3 Polynomial Interpolation of SAT Test Scores

criterion

$$Q(f, p) = \sum_{i=1}^{n} [f_i - p(x_i)]^2 \qquad (6.1)$$

over all polynomials $p(x)$ of degree m or less. Thus the least-squares polynomial minimizes the sum of the squares of the approximation errors at the data points. In the terminology above, the *uniform approximation* criterion has us choose the polynomial, among all polynomials $p(x)$ of degree not exceeding m, that minimizes the quantity

$$U(f, p) = \max_{l \le i \le n} |f_i - p(x_i)|. \qquad (6.2)$$

Hence the uniform approximation minimizes the maximum among the errors occurring at the data points. Figure 6.4 compares these two criteria.

6.2
LEAST-SQUARES APPROXIMATION

6.2.1 Construction of Least-Squares Polynomials

Let $x_1, \ldots, x_n$ be distinct domain points and let $f_1, \ldots, f_n$ be the corresponding values. Assume that $m \le n - 1$ is a nonnegative integer. Suppose that we wish to approximate these data by a polynomial $p(x)$ of degree not exceeding m. If $m = n - 1$, the unique interpolating polynomial provides perfect "fit" to the data. But if $m < n - 1$, typically no interpolation polynomial of degree m exists. In such cases we seek a polynomial of degree m that is "closest," in some sense, to the given data set. In the method of least squares, the discrepancy of the data points $(x_1, f_1), (x_2, f_2), \ldots, (x_n, f_n)$ and approximating polynomial $p(x)$ is

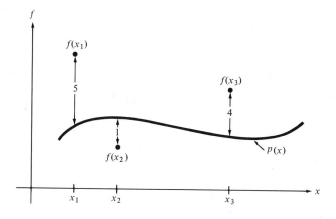

Least squares criterion:
$$Q(f, p) = [f(x_1) - p(x_1)]^2 + [f(x_2) - p(x_2)]^2 + [f(x_3) - p(x_3)]^2$$
$$= 5^2 + 1^2 + 4^2 = 42$$

Uniform criterion:
$$U(f, p) = \max \{|f(x_1) - p(x_1)|, |f(x_2) - p(x_2)|, |f(x_3) - p(x_3)|\}$$
$$= \max \{5, 1, 4\} = 5$$

FIGURE 6–4 Comparison of Least-Squares and Uniform Criteria

measured by the sum of the squared residuals:

$$Q(f, p) = \sum_{i=1}^{n} [f_i - p(x_i)]^2.$$

Under the least-squares principle, this quantity $Q(f, p)$ has to be minimized, the minimization being over the coefficients $a_0, a_1, \ldots, a_m$ of the polynomial $p(x) = a_0 + a_1 x + \cdots + a_m x^m$. The polynomial $p(x)$ minimizing $Q(f, p)$ over all polynomials p of degree m or less is called the *least-squares polynomial* approximation of degree m. Toward seeing the principle behind the general method, let us study the simplest case.

Assume first that $m = 0$. That is, the least-squares constant polynomial is to be determined. The only parameter to be chosen is a_0. In this case

$$Q(f, p) = \sum_{i=1}^{n} (f_i - a_0)^2 = \sum_{i=1}^{n} (f_i^2 - 2f_i a_0 + a_0^2)$$

$$= n a_0^2 - 2a_0 \left(\sum_{i=1}^{n} f_i \right) + \sum_{i=1}^{n} f_i^2,$$

which is a quadratic function of a_0. Since the coefficient of a_0^2 is positive, there is a minimum point of this parabola that can be obtained by differentiation with respect to a_0. Thus the optimum point is the root of the equation

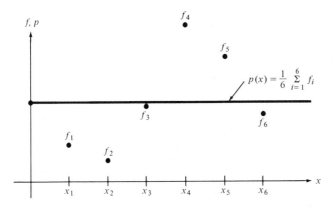

FIGURE 6–5 Approximation by the Least-Squares Constant Polynomial

$$\frac{d}{da_0} Q(f, p) = 2na_0 - 2 \sum_{i=1}^{n} f_i = 0.$$

That is,

$$a_0 = \frac{1}{n} \sum_{i=1}^{n} f_i. \tag{6.3}$$

The zero-degree least-squares polynomial (illustrated in Figure 6.5) is, then, the constant polynomial which for every x equals the average of the function values $f_1, f_2, \ldots, f_n$.

We turn to the general case, in which the degree is any given nonnegative integer $m < n - 1$. To minimize the quantity $Q(f, p)$, view the objective function $Q(f, p)$ as an explicit function of the unknown coefficients $a_0, a_1, \ldots, a_m$. Since

$$p(x) = a_0 + a_1 x + a_2 x^2 + \cdots + a_m x^m = \sum_{k=0}^{m} a_k x^k,$$

we have

$$Q(f, p) = \sum_{i=1}^{n} \left(f_i - \sum_{k=0}^{m} a_k x_i^k \right)^2. \tag{6.4}$$

By taking the partial derivatives of this function with respect to the unknown coefficients a_j ($j = 0, 1, \ldots, m$), we get the necessary conditions (Kaplan, 1952, Sec. 2-15) for minimizing $Q(f, p)$:

$$\frac{\partial}{\partial a_j} Q(f, p) = -2 \sum_{i=1}^{n} \left(f_i - \sum_{k=0}^{m} a_k x_i^k \right) x_i^j = 0, \qquad (0 \le j \le m). \tag{6.5}$$

After some obvious algebra, the above becomes

$$\sum_{k=0}^{m}\left(\sum_{i=1}^{n}x_i^{k+j}\right)a_k = \sum_{i=1}^{n}f_i x_i^j, \qquad (0 \le j \le m). \tag{6.6}$$

which are called *normal equations*. It is known (Szidarovszky and Yakowitz, 1978) that the normal equations have a unique solution $a_0, a_1, \ldots, a_m$, which indeed are the coefficients of the least-squares polynomial.

Observe that (6.6) comprise a system of linear algebraic equations. If it is sufficiently well-conditioned, this system can be efficiently solved by the Gaussian elimination method, discussed in Chapter 4.

Table 6.1 presents a listing of the subroutine LSQM for least-squares polynomial approximation. The subroutine compiles the normal equations from the data and calls on GAUS1 (Table 4.5) to provide the solution, which is available as an output array C. The numbers in this array are related to the least-squares polynomial coefficients by $a_k = C(k + 1)$. The structure of subroutine LSQM is as follows. DO loops with labels 1 and 2 compute the right-hand side of the first normal equation ($j = 0$). The DO loop with label 3 computes the right-hand sides of all other normal equations. DO loop with label 4 computes the sums $s_j = \sum_{i=1}^{n} x_i^j$ and DO loop with label 5 accumulates these sums as the coefficients of the normal equations. The solution of the normal equations is then obtained by Gaussian elimination (through subroutine GAUS1), and the DO loop with label 6 produces the output array C from the solution. Example 6.5 in Section 6.2.2 includes a calling program (Table 6.2) that uses LSQM.

TABLE 6.1   Subroutine LSQM for the Least-Squares Method

```
      SUBROUTINE LSQM(N,M,X,F,C)
C
C     ****************************************************************
C     *  FUNCTION: THIS SUBROUTINE COMPUTES THE COEFFICIENTS OF     *
C     *            THE LEAST-SQUARES POLYNOMIAL                      *
C     *            P(X)=C(1)+C(2)*X+...+C(M+1)*X**M                  *
C     *            FOR A SET OF FUNCTIONAL VALUES F AND DOMAIN       *
C     *            POINTS X                                          *
C     *  USAGE:                                                      *
C     *      CALL SEQUENCE: CALL LSQM(N,M,X,F,C)                     *
C     *      EXTERNAL FUNCTIONS/SUBROUTINES:                         *
C     *                     SUBROUTINE GAUS1                         *
C     *  PARAMETERS:                                                 *
C     *      INPUT:                                                  *
C     *          N=NUMBER OF POINTS (NOT GREATER THAN 1000)          *
C     *          M=POLYNOMIAL DEGREE (LESS THAN 30)                  *
C     *          X=N BY 1 ARRAY OF DOMAIN POINTS                     *
C     *          F=N BY 1 ARRAY OF FUNCTIONAL VALUES                 *
C     *      OUTPUT:                                                 *
C     *          C=M+1 BY 1 ARRAY OF COEFFICIENTS FOR THE            *
C     *            LEAST-SQUARES POLYNOMIAL                          *
C     ****************************************************************
C
      DIMENSION X(N),F(N),C(M+1),U(1000),V(1000),A(30,31),S(100)
C     *** INITIALIZATION ***
      DO 1 I=1,N
         U(I)=1.0
         V(I)=F(I)
    1 CONTINUE
```

TABLE 6.1   (Continued)

```
C       *** COMPUTE COEFFICIENTS OF THE NORMAL EQUATION ***
        A(1,M+2)=0.0
        DO 2 I=1,N
          A(1,M+2)=A(1,M+2)+V(I)
      2 CONTINUE
        S(1)=N
        DO 3 I=2,M+1
          S(I)=0.0
          A(I,M+2)=0.0
          DO 3 J=1,N
            U(J)=U(J)*X(J)
            V(J)=V(J)*X(J)
            S(I)=S(I)+U(J)
            A(I,M+2)=A(I,M+2)+V(J)
      3 CONTINUE
        DO 4 I=M+2,2*(M+1)
          S(I)=0.0
          DO 4 J=1,N
            U(J)=U(J)*X(J)
            S(I)=S(I)+U(J)
      4 CONTINUE
        DO 5 I=1,M+1
          DO 5 J=1,M+1
            A(I,J)=S(I+J-1)
      5 CONTINUE
C       *** PERFORM GAUSSIAN ELIMINATION ***
        N1=M+1
        M1=1
        ND=30
C       *** USER MAY WANT TO RESET EPS ***
        EPS=0.000001
        CALL GAUS1(N1,M1,ND,A,EPS)
C       *** THE SOLUTION IS PLACED IN THE ARRAY C ***
        DO 6 I=1,M+1
          C(I)=A(I,M+2)
      6 CONTINUE
        RETURN
        END
```

─────── **EXAMPLE 6.1** ───────────────────

Let $n = 5$ and suppose that the data points are specified by

$$x_1 = -2, \quad x_2 = -1, \quad x_3 = 0, \quad x_4 = 1, \quad x_5 = 2$$
$$f_1 = 0, \quad f_2 = 1, \quad f_3 = 2, \quad f_4 = 1, \quad f_5 = 0.$$

A graph of these points (Figure 6.6) suggests that a quadratic ($m = 2$) approximation should be attempted. Toward constructing the normal equations (6.6), we first compute the coefficients:

$$\sum_{i=1}^{5} x_i^0 = 5, \quad \sum_{i=1}^{5} x_i^1 = 0, \quad \sum_{i=1}^{5} x_i^2 = 10,$$

$$\sum_{i=1}^{5} x_i^3 = 0, \quad \sum_{i=1}^{5} x_i^4 = 34, \quad \sum_{i=1}^{5} x_i^0 f_i = 4, \quad \sum_{i=1}^{5} x_i^1 f_i = 0, \quad \sum_{i=1}^{5} x_i^2 f_i = 2,$$

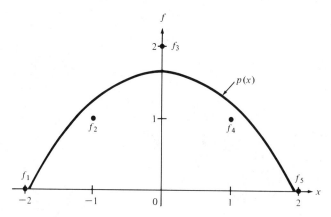

FIGURE 6–6 Data Points and Quadratic Least-Squares Approximation for Example 6.1

which implies that the system of normal equations is as follows:

$$\text{(for } j = 0)\qquad 5a_0 + 0a_1 + 10a_2 = 4$$
$$\text{(for } j = 1)\qquad 0a_0 + 10a_1 + 0a_2 = 0$$
$$\text{(for } j = 2)\qquad 10a_0 + 0a_1 + 34a_2 = 2.$$

The solution to these equations is

$$a_0 = \tfrac{58}{35}, \qquad a_1 = 0, \qquad a_2 = -\tfrac{3}{7},$$

and the corresponding least-squares polynomial is

$$p(x) = -\tfrac{3}{7}x^2 + \tfrac{58}{35}.$$

The curve shown in Figure 6.6 is actually this polynomial.

To introduce the matrix representation of the least-squares method, let us adopt the following definitions:

$$\mathbf{X} = \begin{bmatrix} 1 & x_1 & \cdots & x_1^m \\ 1 & x_2 & \cdots & x_2^m \\ \cdots\cdots\cdots\cdots \\ 1 & x_n & \cdots & x_n^m \end{bmatrix}, \qquad \mathbf{a} = \begin{bmatrix} a_0 \\ a_1 \\ \vdots \\ a_m \end{bmatrix}, \qquad \mathbf{f} = \begin{bmatrix} f_1 \\ f_2 \\ \vdots \\ f_n \end{bmatrix}. \qquad (6.7)$$

With this notation, the normal equations (6.6) can be expressed compactly as

$$(\mathbf{X}^T\mathbf{X})\mathbf{a} = \mathbf{X}^T\mathbf{f}, \qquad (6.8)$$

where superscript T denotes "transpose" as defined in item 7 of Appendix A. By using property 18 of Appendix A, from (6.8) we obtain

$$\boxed{\mathbf{a} = (\mathbf{X}^T\mathbf{X})^{-1}\mathbf{X}^T\mathbf{f}.} \tag{6.9}$$

From discussions in Chapter 4 we recognize, however, that solution of the linear equation (6.8) is more efficient than matrix inversion and a matrix-vector multiplication as suggested by (6.9).

From the form of (6.8) we may explain why equations (6.6) are called "normal equations." Observe first that (6.8) can be rewritten as

$$\mathbf{x}^{(j)}(\mathbf{X}\mathbf{a} - \mathbf{f}) = 0 \qquad (0 \le j \le m) \tag{6.10}$$

where the rows of matrix $\mathbf{X}^T$ have been denoted by

$$\mathbf{x}^{(j)} = (x_1^j, x_2^j, \ldots, x_n^j) \qquad (j = 0, 1, \ldots, m),$$

and the coordinates of vector $\mathbf{r} = \mathbf{X}\mathbf{a} - \mathbf{f}$ equal the residual error values

$$r_i = (1, x_i, x_i^2, \ldots, x_i^m) \begin{bmatrix} a_0 \\ a_1 \\ a_2 \\ \vdots \\ a_m \end{bmatrix} - f_i = \sum_{j=0}^{m} a_j x_i^j - f_i = p(x_i) - f_i.$$

With this notation (6.10) implies that the row vectors $\mathbf{x}^{(j)}$ are perpendicular, in other words, *normal,* to the residual error vector $\mathbf{r}$. That is,

$$\mathbf{x}^{(j)}\mathbf{r} = 0, \, 0 \le j \le m.$$

EXAMPLE 6.2

In the case of Example 6.1,

$$\mathbf{X} = \begin{bmatrix} 1 & -2 & 4 \\ 1 & -1 & 1 \\ 1 & 0 & 0 \\ 1 & 1 & 1 \\ 1 & 2 & 4 \end{bmatrix}, \qquad \mathbf{a} = \begin{bmatrix} a_0 \\ a_1 \\ a_2 \end{bmatrix}, \qquad \mathbf{f} = \begin{bmatrix} 0 \\ 1 \\ 2 \\ 1 \\ 0 \end{bmatrix}.$$

Then

$$\mathbf{X}^T\mathbf{X} = \begin{bmatrix} 1 & 1 & 1 & 1 & 1 \\ -2 & -1 & 0 & 1 & 2 \\ 4 & 1 & 0 & 1 & 4 \end{bmatrix} \begin{bmatrix} 1 & -2 & 4 \\ 1 & -1 & 1 \\ 1 & 0 & 0 \\ 1 & 1 & 1 \\ 1 & 2 & 4 \end{bmatrix} = \begin{bmatrix} 5 & 0 & 10 \\ 0 & 10 & 0 \\ 10 & 0 & 34 \end{bmatrix}$$

and

$$\mathbf{X}^T\mathbf{f} = \begin{bmatrix} 1 & 1 & 1 & 1 & 1 \\ -2 & -1 & 0 & 1 & 2 \\ 4 & 1 & 0 & 1 & 4 \end{bmatrix} \begin{bmatrix} 0 \\ 1 \\ 2 \\ 1 \\ 0 \end{bmatrix} = \begin{bmatrix} 4 \\ 0 \\ 2 \end{bmatrix}.$$

Consequently, (6.8) has the form

$$\begin{bmatrix} 5 & 0 & 10 \\ 0 & 10 & 0 \\ 10 & 0 & 34 \end{bmatrix} \begin{bmatrix} a_0 \\ a_1 \\ a_2 \end{bmatrix} = \begin{bmatrix} 4 \\ 0 \\ 2 \end{bmatrix},$$

which coincides with the normal equations obtained in Example 6.1. One easily calculates from the least-squares polynomial constructed in that example that the residual

$$\mathbf{r} = \begin{pmatrix} 0.057 \\ -0.2285 \\ 0.343 \\ -0.2285 \\ 0.057 \end{pmatrix}$$

and confirms that this vector indeed is normal to the rows $\mathbf{x}^{(j)}$ of $\mathbf{X}$.

The linear approximating polynomials obtained by the use of the least-squares method are sometimes called *linear regression functions*. In this particular case, $m = 1$, and the normal equations have the special form

$$a_0 n + a_1 \sum_{i=1}^{n} x_i = \sum_{i=1}^{n} f_i$$

$$a_0 \sum_{i=1}^{n} x_i + a_1 \sum_{i=1}^{n} x_i^2 = \sum_{i=1}^{n} x_i f_i.$$

One may verify that the solution coefficients a_0 and a_1 of the linear regression function $p(x) = a_0 + a_1 x$ are given by the equations

$$a_1 = \frac{\overline{xf} - (\bar{x})(\bar{f})}{\overline{x^2} - (\bar{x})^2} \qquad (6.11)$$

$$a_0 = \bar{f} - a_1 \bar{x}, \qquad (6.12)$$

where

$$\bar{x} = \frac{1}{n} \sum_{i=1}^{n} x_i, \qquad \overline{x^2} = \frac{1}{n} \sum_{i=1}^{n} x_i^2, \qquad \bar{f} = \frac{1}{n} \sum_{i=1}^{n} f_i, \qquad \overline{xf} = \frac{1}{n} \sum_{i=1}^{n} x_i f_i.$$

$$(6.13)$$

—— **EXAMPLE 6.3** ——

The foregoing procedure for linear regression functions is exemplified. Here, we take $n = 4$,

$$x_1 = 0, \quad x_2 = 1, \quad x_3 = 2, \quad x_4 = 3,$$

$$f_1 = 0 \quad f_2 = 2, \quad f_3 = 1, \quad f_4 = 2.$$

Then the average values defined in (6.13) are

$$\bar{x} = \tfrac{1}{4}(0 + 1 + 2 + 3) = \tfrac{6}{4} = 1.5$$

$$\overline{x^2} = \tfrac{1}{4}(0 + 1 + 4 + 9) = \tfrac{14}{4} = 3.5$$

$$\bar{f} = \tfrac{1}{4}(0 + 2 + 1 + 2) = \tfrac{5}{4} = 1.25$$

$$\overline{xf} = \tfrac{1}{4}(0 \cdot 0 + 1 \cdot 2 + 2 \cdot 1 + 3 \cdot 2) = \tfrac{10}{4} = 2.5.$$

Thus (6.11) and (6.12) imply that

$$a_1 = \frac{2.5 - 1.5 \cdot 1.25}{3.5 - 1.5^2} = 0.5$$

$$a_0 = 1.25 - 0.5 \cdot 1.5 = 0.5.$$

Consequently, the linear least-squares polynomial is

$$p(x) = 0.5 + 0.5x.$$

Figure 6.7 shows the data points and the approximating linear polynomial.

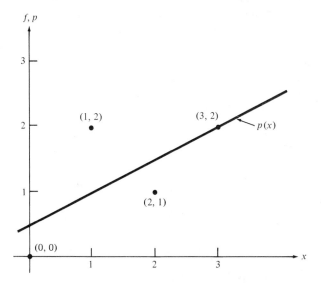

FIGURE 6–7 Data and Approximating Polynomial
of Example 6.3

With respect to least-squares approximation, an important procedural issue is the selection of m, the approximation polynomial degree. Unfortunately, only in certain very specialized statistical settings is there a definitive theory for this issue. In some instances, visual inspection of a graph of the data can be useful. For example, the points in Figure 6.6 suggested a quadratic approximation. With respect to the data set illustrated in Figure 6.2, one would anticipate that a constant ($m = 0$) is not as sensible as a line ($m = 1$) with negative slope. Ralston and Rabinowitz (1978, Sec. 6.3-2) offer some statistically motivated selection procedures based on increasing m until the squared residual error

$$Q(f, p) = \sum_{i=1}^{n} [f_i - p_m(x_i)]^2$$

of approximation by $p_m(x)$, the mth-degree least-squares polynomial, does not decrease "sufficiently" with further increase in m.

6.2.2 Practical Considerations and Case Studies

In this section we put previous methodology into action. Low-degree least-squares polynomials are found from various data sets. The least-squares approach, and data smoothing in general, raises some procedural concerns, some of which are encountered even in the simple studies here.

EXAMPLE 6.4

This study illustrates the least-squares polynomial application to curve fitting of the scattered data shown in Figure 6.8. Figure 6.9 reproduces these data points, which serve as input for the subroutine LSQM (Table 6.1), and the resulting least-squares polynomial approximations $p_0(x)$, $p_1(x)$, and $p_8(x)$ of degrees 0, 1, and 8.

Visually, the constant polynomial $p_0(x)$ does not seem well centered with respect to the data. This illustrates *underfitting*. On the other hand, the degree 8 approximation is so irregular as to give rise to the suspicion that it is trying to fit the random irregularities of the data. Such a situation is referred to as *overfitting*.

The data were, in fact, generated by adding scaled random numbers to the exponentially decaying curve shown in Figure 6.10.

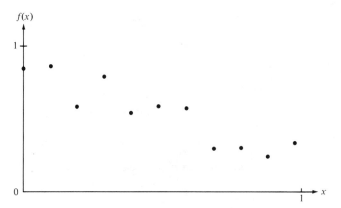

FIGURE 6–8 A Data Set

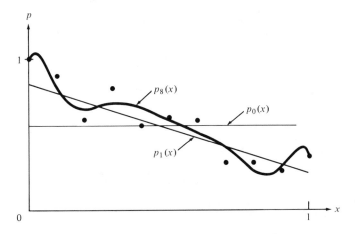

FIGURE 6–9   Least-Squares Fitting of Scattered Data

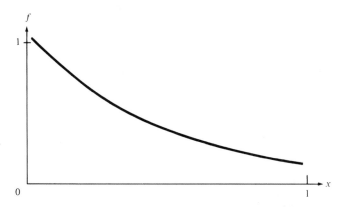

FIGURE 6–10   Underlying Function for Example 6.4

EXAMPLE 6.5

In this case study the least-squares method has been applied to the problem (studied also in Examples 2.2 and 2.3) of finding an approximating polynomial for sin (x), over the interval $I = [0, 2]$. For this example 50 data points are used, and domain points $x_i (1 \le i \le 50)$ are evenly spaced along the interval I. The calling program and the output for this experiment are presented in Tables 6.2 and 6.3. The values of the least-squares polynomial are compared with those of sin (x) at 10 points evenly spaced between 0 and 2. The subscript m of $p_m(x)$ in Table 6.3 indicates the degree m of the least-squares approximating polynomial. In Figure 6.11 the graphs of these approximating polynomials are plotted against the "target" function sin (x). The reader may wish to compare this figure with Figures 2.2 and 2.4, which approximate the same target function by Taylor's and interpolation polynomials, respectively. In comparing these figures, we see that the linear approximation $p_1(x)$ by least squares seems much better positioned than the linear Taylor's and Lagrange approximations. The approximation errors listed in Table 6.3 are similarly, almost without exception, less than their Taylor's and Lagrange counterparts listed in Tables 2.3 and 2.8.

TABLE 6.2   Program for Example 6.5

```
C       PROGRAM LSQUARE
C
C       ****************************************************************
C       THIS PROGRAM FINDS LEAST-SQUARES POLYNOMIALS FOR SIN(X) ON
C       THE INTERVAL [0,2]. THE DATA CONSIST OF VALUES SIN(2.*J/50.)
C       FOR J=1,...,50
C       CALLS:    LSQM,GAUS1
C       OUTPUT:
C               X=VALUE OF X
C               TRUE=VALUE OF SIN(X)
C               P=POLYNOMIAL APPROXIMATION AT X
C               DIFF=DIFFERENCE (TRUE-P) FOR EACH X
C       ****************************************************************
C
        DIMENSION A(5,6),XX(50),F(50),C(6)
        N=50
        Y=0.
        H=2./50.
C
C       *** CONSTRUCT DATA XX AND SIN(XX) FOR THE APPROXIMATION   ***
C
        DO 10 J=1,N
          Y=Y+H
          XX(J)=Y
          F(J)=SIN(XX(J))
     10 CONTINUE
C
C       *** M IS THE DEGREE OF THE LEAST-SQUARES POLYNOMIAL       ***
C
        DO 40 M=1,3
          WRITE(10,1)
     1    FORMAT(////)
C
C       *** SUBROUTINE LSQM WILL CALCULATE THE M-TH DEGREE LEAST  ***
C       *** SQUARES POLYNOMIAL COEFFICIENTS                       ***
C
          CALL LSQM(N,M,XX,F,C)
C
C       *** NEXT COMPUTE THE VALUE OF THE POLYNOMIAL AT 10 EQUALLY***
C       *** SPACED POINTS AND COMPARE THESE VALUES WITH SIN(X) AT ***
C       *** THESE POINTS                                          ***
C
          X=0.
          DO 30 I=1,10
            X=X+.2
            P=0.
            DO 20 K=1,M+1
              K1=K-1
              T=C(K)*X**K1
              P=P+T
     20     CONTINUE
            TRUE=SIN(X)
            DIFF=TRUE-P
            WRITE(10,*) X,TRUE,P,DIFF
     30   CONTINUE
     40 CONTINUE
        STOP
        END
```

TABLE 6.3   Application of Least-Squares Polynomials

x	sin (x)	$p_1(x)$	sin (x) $- p_1(x)$	$p_2(x)$	sin (x) $- p_2(x)$	$p_3(x)$	sin (x) $- p_3(x)$
0.2	0.198669	0.3292865	-0.1306172	0.1947901	3.8792044E-03	0.1997269	-1.0576099E-03
0.4	0.389418	0.4238683	-3.4449995E-02	0.4035662	-1.4147878E-02	0.3925681	-3.1497478E-03
0.6	0.564642	0.5184502	4.6192288E-02	0.5806217	-1.5979230E-02	0.5659643	-1.3218522E-03
0.8	0.717356	0.6130320	0.1043240	0.7259566	-8.6005330E-03	0.7159558	1.4002919E-03
1.0	0.841471	0.7076139	0.1338571	0.8395711	1.8998981E-03	0.8385826	2.8883815E-03
1.2	0.932039	0.8021957	0.1298434	0.9214648	1.0574281E-02	0.9298849	2.1541715E-03
1.4	0.985450	0.8967776	8.8672161E-02	0.9716381	1.3811588E-02	0.9859029	-4.5317411E-04
1.6	0.999574	0.9913595	8.2141161E-03	0.9900907	9.4828606E-03	1.002676	-3.1028986E-03
1.8	0.973848	1.085941	-0.1120937	0.9768227	-2.9751658E-03	0.9762460	-2.3984313E-03
2.0	0.909297	1.180523	-0.2712259	0.9318342	-2.2536874E-02	0.9026516	6.6457391E-03

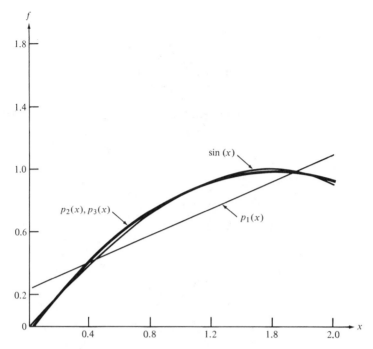

FIGURE 6–11 Least-Square Approximation Graph

ENGINEERING EXAMPLE

Mathematics Scholastic Aptitude Test scores are thought to be good indicators of the proportion of entering freshmen nationwide who are qualified for serious programs in computer science, engineering, mathematics, physics, and other quantitative disciplines. With this in mind, we have used least-squares polynomials, as constructed by subroutine LSQM (Table 6.1), to examine the SAT score data of Figure 6.2 in the hope of discovering some sort of trend and thereby a suggestion as to what may be expected in the future.

Unfortunately, the graphs (Figure 6.12) of polynomials of varying degree m give different suggestions. The linear ($m = 1$) least-squares polynomial shows a depressing monotonic decline, but this polynomial has a very limited structure and is incapable of detecting changes in "trend." Polynomials of degrees $m = 2$ through 4 seem to suggest that a leveling off has occurred and a rise in average SAT scores may be anticipated. The cases $m = 3$ and 4 are optimistic in this regard.

Of course, the difficulty now is selecting which of the graphs to believe. Here strict scientific methodology is not available, and personal judgment must be exercised; all the graphs fit the data fairly closely. The authors' inclination is to accept the middle graphs and say only that the decline in scores seems to be leveling off. There is some statistical thinking (e.g., Savage, 1954) which bears on how to draw conclusions from statistical evidence. On a less profound but more amusing level, the reader may be interested in *How to Lie with Statistics* by Huff (1954). (We mention parenthetically that in our initial calculations, the fourth-degree polynomial did not come close to the data. We quickly recognized that

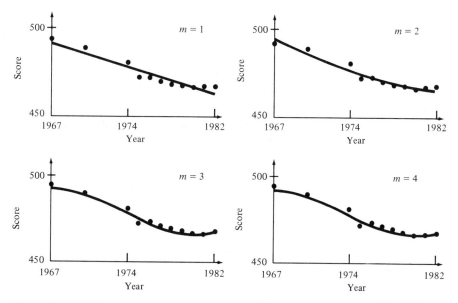

FIGURE 6–12 Least-Squares Polynomial Fitting of Mathematics SAT Scores

this was due to numerical instabilities which are discussed in Section 6.2.4. When the calculations were redone in double precision on a DEC 10, we obtained the graph shown.) ∎

*6.2.3 Linear and Nonlinear Least-Squares Problems

The problem of finding the polynomial $p(x) = a_0 + a_1x + \cdots + a_mx^m$ that minimizes the squared error criterion $Q(f, p) = \sum_{i=1}^{n} [f_i - p(x_i)]^2$ is said to be a *linear* least-squares problem because $p(x)$ is, in the terminology of algebra, a linear function of the unknown coefficients $a_0, a_1, \ldots, a_m$.

It is sometimes useful to observe that if other "basis" functions $\varphi_0(x)$, $\varphi_1(x)$, $\ldots, \varphi_m(x)$ are substituted for $1, x, \ldots, x^m$ in the least-squares approach, the methods and results of the preceding section apply with very little modification. Assume that a least-squares approximation of the form

$$p(x) = a_0\varphi_0(x) + a_1\varphi_1(x) + \cdots + a_m\varphi_m(x)$$

is desired. Then the matrix $\mathbf{X}$ of (6.7) and (6.9) is replaced by the matrix

$$\mathbf{\Phi} = \begin{bmatrix} \varphi_0(x_1) & \varphi_1(x_1) & \cdots & \varphi_m(x_1) \\ \varphi_0(x_2) & \varphi_1(x_2) & \cdots & \varphi_m(x_2) \\ \vdots & \vdots & & \vdots \\ \varphi_0(x_n) & \varphi_1(x_n) & \cdots & \varphi_m(x_n) \end{bmatrix}. \tag{6.14}$$

The normal equations for the problem of finding the coefficients $a_0, a_1, \ldots, a_m$ to minimize

$$Q(f, \varphi) = \sum_{i=1}^{n} \left[f_i - \sum_{j=0}^{m} a_j \varphi_j(x_i) \right]^2$$

lead [in the notation of (6.9) and (6.14)] to the normal equations (in matrix form)

$$(\mathbf{\Phi}^T \mathbf{\Phi}) \mathbf{a} = \mathbf{\Phi}^T \mathbf{f}. \tag{6.15}$$

The vector $\mathbf{a}$ of least-squares coefficients can be represented explicitly as

$$\boxed{\mathbf{a} = (\mathbf{\Phi}^T \mathbf{\Phi})^{-1} \mathbf{\Phi}^T \mathbf{f}.} \tag{6.16}$$

Here $\mathbf{a} = (a_0, a_1, \ldots, a_m)^T$ and $\mathbf{f} = (f_1, \ldots, f_n)^T$ again.

Linear least-squares problems using basis functions other than polynomials arise in approximation of periodic and diurnal data such as traffic intensity or height of tides, where sinusoidal $\varphi_j(x)$ are sensible, or in approximating radioactivity after a nuclear event, in which for physical reasons it would be natural to use negative exponential functions $\varphi_j(x) = \exp(-c_j x)$ as "basis" functions. In Section 6.4 we will find that even when we desire a polynomial least-squares approximation, for reasons of numerical stability, it is useful to consider a polynomial basis $\varphi_0(x)$, $\varphi_1(x), \ldots, \varphi_m(x)$ other than $1, x, \ldots, x^m$.

The vector $\mathbf{a}$ can be viewed as a parameter that determines an approximating function according to the rule

$$\varphi_{\mathbf{a}}(x) = \sum_{j=0}^{n} a_j \varphi_j(x).$$

A *nonlinear* least-squares problem arises when the approximating function $\varphi_{\mathbf{a}}(x)$ does not depend linearly on $\mathbf{a}$. For example, consider the parameter pair $\mathbf{a} = (a_1, a_2)$ and the approximating function

$$\varphi_{(a_1, a_2)}(x) = a_1 \exp(a_2 x), \qquad -\infty < a_1, a_2 < \infty. \tag{6.17}$$

Such a family of approximating functions is appropriate if for physical reasons the values $f_j = f(x_j)$ are known to be points on a solution curve of the linear differential equation

$$\frac{d}{dx} f(x) = \alpha f(x) \tag{6.18}$$

with coefficient α and the initial condition unknown. In this case, a nonlinear least-squares problem occurs if we wish to "identify" the initial condition $a_1 = f(0)$ and the differential equation parameter $a_2 = \alpha$ by finding the parameter vector $\mathbf{a} = (a_1, a_2)$ giving the least-squares fit to the data $(x_1, f_1), (x_2, f_2), \ldots,$

(x_n, f_n). That is, we solve for unknowns a_1 and a_2 which minimize

$$Q(f, \mathbf{a}) = \sum_{i=1}^{n} [f_i - a_1 \exp (a_2 x_i)]^2. \tag{6.19}$$

[In the identification theory literature, it is typically presumed that the observation values f_i are not exactly equal to the solution values of (6.18) but are corrupted by measurement error.] The resulting nonlinear normal equations for (6.19), obtained by taking first partial derivatives with respect to the parameters a_1 and a_2, respectively, are

$$\sum_{i=1}^{n} \{[f_i - a_1 \exp (a_2 x_i)] \exp (a_2 x_i)\} = 0$$
$$\sum_{i=1}^{n} \{[f_i - a_1 \exp (a_2 x_i)]a_1 x_i \exp (a_2 x_i)\} = 0. \tag{6.20}$$

The multivariable nonlinear equation technique given in Section 5.8 may be employed to find the root of (6.20).

EXAMPLE 6.6 ─────────────────────────

Here we illustrate implementation of the nonlinear least-squares identification technique discussed above. The procedure is to obtain data points (x_i, f_i) of the function (6.17) with known parameters a_1 and a_2, and then see how well the least-squares technique can identify these known parameters from the data. The parameters were selected to be

$$a_1 = 1 = f(0) \qquad \text{and} \qquad a_2 = \alpha = -0.25,$$

respectively. Thus $f(x) = \exp (-0.25x)$. The data consisted of the points (x_i, f_i), with $x_i = 0.2(i - 1)$ and, of course, $f_i = f(x_i)$, $1 \le i \le 20$. The data points are plotted in Figure 6.13. The multivariable Newton's method (subroutine MVNE in Section 5.8) was used for solving the normal equations (6.20), and the listing of our implementation is given in Table 6.4. The array DAT contains the

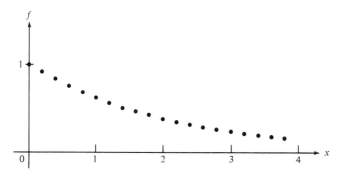

FIGURE 6–13   Noiseless Data for Nonlinear Least-Squares Identification

values f_i, and the function subprogram F is used in the sense of the MVNE comment lines. Otherwise, we have followed the notation in the discussion above. The initial parameter estimates were taken as $a_1 = 1.5$ and $a_2 = -0.4$, and the successive MVNE iterates are displayed in Table 6.5. The parameters were recovered perfectly.

TABLE 6.4   Listing of Nonlinear Least-Squares Systems Identification Program

```
C       PROGRAM NONLINLS
C
C       *****************************************************************
C       THIS PROGRAM GENERATES A SET OF 20 DATA POINTS USING THE
C       FUNCTIONAL VALUES EXP(-.25X) FOR X=0,.2,.4,...,3.8
C       THESE DATA ARE USED IN THE FUNCTIONS F AND DF TO DETERMINE THE
C       NORMAL EQUATIONS FOR THE NON-LINEAR LEAST-SQUARES PROBLEM
C       OF THE FORM A0(1)*EXP(A0(2)*X)
C       CALLS:   MVNE,GAUS1
C       OUTPUT(PRINTED FROM SUBROUTINE MVNE):
C               A(0)=SUCCESSIVE NEWTON APPROXIMATIONS OF THE FIRST
C                   LEAST-SQUARES COEFFICIENT ABOVE
C               A(1)=SUCCESSIVE NEWTON APPROXIMATIONS OF THE SECOND
C                   LEAST-SQUARES COEFFICIENT ABOVE
C       *****************************************************************
C
        COMMON DAT(20),X(20),NT
        DIMENSION A0(2),A(2)
C
C       *** SPECIFY INPUT PARAMETERS FOR MVNE                    ***
C
        N=2
        DATA (A0(I),I=1,2)/1.5,-0.4/
        EPS=1.E-04
C
C       *** GENERATE THE DATA TO BE FIT                          ***
C
        DO 10 I=1,20
          X(I)=.2*(I-1)
          DAT(I)=EXP(-.25*X(I))
    10  CONTINUE
        NT=20
C
C       *** SUBROUTINE MVNE WILL PERFORM NEWTONS METHOD USING    ***
C       *** THE FUNCTION SUBROUTINES F AND DF BELOW              ***
C
        CALL MVNE(N,A0,A,EPS)
        STOP
        END


        FUNCTION F(K,A)
C
C       *** THIS FUNCTION IS CALLED BY SUBROUTINE MVNE TO        ***
C       *** CALCULATE FUNCTION VALUES NEEDED FOR NEWTON'S METHOD ***
```

TABLE 6.4   (Continued)

```
C
      DIMENSION A(2)
      COMMON DAT(20),X(20),NT
      S=0.
      IF(K.EQ.1) THEN
          DO 10 I=1,NT
              S=S+(DAT(I)-A(1)*EXP(A(2)*X(I)))*EXP(A(2)*X(I))
   10     CONTINUE
      ELSE
          DO 20 I=1,NT
              S=S+(DAT(I)-A(1)*EXP(A(2)*X(I)))*A(1)*X(I)*EXP(A(2)*X(I))
   20     CONTINUE
      END IF
      F=S
      RETURN
      END


      FUNCTION DF(K1,K2,A)
C
C     *** THIS FUNCTION IS CALLED BY SUBROUTINE MVNE TO          ***
C     *** CALCULATE DERIVATIVE VALUES NEEDED FOR NEWTONS METHOD  ***
C
      DIMENSION A(2)
      COMMON DAT(20),X(20),NT
      S=0.
      IF(K1.EQ.1) THEN
          IF(K2.EQ.1) THEN
              DO 10 I=1,NT
                  S=S-EXP(2*A(2)*X(I))
   10         CONTINUE
          ELSE
              DO 20 I=1,NT
                  Q=(DAT(I)-A(1)*EXP(A(2)*X(I)))*X(I)*EXP(A(2)*X(I))
                  Q=Q-X(I)*A(1)*EXP(2*A(2)*X(I))
                  S=S+Q
   20         CONTINUE
          END IF
      ELSE IF(K1.EQ.2) THEN
          IF(K2.EQ.1) THEN
              DO 30 I=1,NT
                  Q=(DAT(I)-A(1)*EXP(A(2)*X(I)))*X(I)*EXP(A(2)*X(I))
                  Q=Q-X(I)*A(1)*EXP(2*A(2)*X(I))
                  S=S+Q
   30         CONTINUE
          ELSE
              DO 40 I=1,NT
                  Q=(X(I)**2)*
     1              (EXP(A(2)*X(I))*(DAT(I)-A(1)*EXP(A(2)*X(I))))*A(1)
                  Q=Q-EXP(2*A(2)*X(I))*(A(1)*X(I))**2
                  S=S+Q
   40         CONTINUE
          END IF
      END IF
      DF=S
      RETURN
      END
```

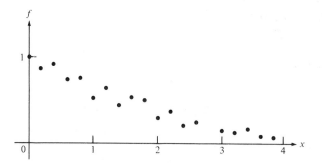

FIGURE 6–14 Noisy Data for Nonlinear Least-Squares Identification

TABLE 6.5 Nonlinear Least-Squares Identification with Noiseless Measurement

a_1	a_2
1.500000	-0.4000000
0.8643785	-0.2130418
0.9845823	-0.2429858
0.9998832	-0.2499678
1.000000	-0.2500000

In a second computation, we "simulated" measurement error by using the computer random number generator to corrupt the function values f_i. The noisy data are displayed in Figure 6.14. The least-squares estimates provided by our program on this data set are given in Table 6.6.

TABLE 6.6 Nonlinear Least-Squares Identification with Measurement Error

a_1	a_2
1.500000	-0.4000000
0.9087929	-0.2279210
1.011362	-0.2633236
1.016372	-0.2651989
1.016384	-0.2652051

6.2.4 Ill-Conditioning in Least-Squares Methods

Let us return our attention to the least-squares polynomial approximation problem determined by (6.1). Practitioners find that when the degree of the approximating polynomial is moderate (say 8) to large, solution of the normal equations (6.8)

for the polynomial coefficients often leads to seriously erroneous answers. The heart of the problem is that the matrix $S = X^T X$, with X defined by (6.7), tends to be ill-conditioned in the sense discussed in Section 4.4. As a consequence, the computed solution of the linear equation (6.8) tends to differ substantially from its exact solution.

To demonstrate that ill-conditioning can arise in very natural least-squares settings, consider the case in which the x_i values are evenly spaced along the unit interval. Thus let

$$x_i = \frac{i}{n}, \qquad 1 \le i \le n.$$

Regardless of X, from (6.7), we see that if $S = X^T X$, and s_{kj} is the (k, j) coordinate of S and x_{kj} the (k, j) coordinate of X, then

$$s_{kj} = \sum_{i=1}^{n} x_{ik} x_{ij} = \sum_{i=1}^{n} (x_i^{k-1})(x_i^{j-1}) = \sum_{i=1}^{n} x_i^{k+j-2}. \qquad (6.21)$$

For $x_i = i/n$, we readily calculate that

$$s_{kj} = n \sum_{i=1}^{n} \left(\frac{i}{n}\right)^{k+j-2} \frac{1}{n} \approx n \int_0^1 t^{k+j-2} \, dt = \frac{n}{k+j-1}. \qquad (6.22)$$

We have used the fact that

$$\sum_{i=1}^{n} \left(\frac{i}{n}\right)^{k+j-2} \frac{1}{n} \qquad (6.23)$$

is a Riemann sum approximation for the integral

$$\int_0^1 t^{k+j-2} \, dt = \frac{1}{k+j-1}.$$

From (6.22) one readily concludes that $S \approx nH$, where H is the notoriously unstable nth-order Hilbert segment matrix the (k, j) coordinate of which equals $1/(k + j - 1)$. The ill-conditioning of H, for moderate n, was established in Problem 9 of Chapter 4.

―――― **EXAMPLE 6.7** ――――――――――――――――――――――――――――――

Here we exemplify normal equation ill-conditioning by making a simple modification of the computation in Example 6.5 of least-squares polynomial approximation of the sine function. Specifically, the only change we made was to set the degree m of the least-squares polynomial to 10 in the program listed in Table 6.2, and to modify the program and subroutines for double precision to increase the accuracy. For $m = 10$ we received an overflow message. We then experimented with smaller degree m. For $m = 9$ a successful run was completed, and the errors at the 10 output points were on the order of 10^{-8}. The conclusion is that on the computer we used for this experiment (a DEC 10), by the conventional equation (6.8), it is impossible to get an accuracy of 12 significant decimals in the least-squares approximation of sin (x), $0 \leq x \leq 2$, using 50 equally spaced data points on this interval. If m is less than 10, then the degree is not high enough to obtain such accuracy, and if m is 10 or larger, the accuracy deteriorates as a result of $\mathbf{S} = \mathbf{X}^T\mathbf{X}$ being ill-conditioned. To some extent, the limits of accuracy depend on the computer used, but we would be surprised if any computer is able to achieve a 15-significant-decimal-place accuracy.

A standard way of alleviating the least-squares ill-conditioning problem is to make use of a different polynomial basis $\varphi_0(x)$, $\varphi_1(x)$, . . ., $\varphi_m(x)$. As stated in Section 6.2.3, for any set of basis polynomials,

$$\varphi_0(x), \ \varphi_1(x), \ . \ . \ ., \ \varphi_m(x),$$

we may define the matrix $\mathbf{\Phi}$ as in (6.14) with (i, j) coordinate

$$\varphi_{ij} = \varphi_j(x_i), \ 1 \leq i \leq n, \qquad 0 \leq j \leq m,$$

and solve the normal equations $(\mathbf{\Phi}^T\mathbf{\Phi})\mathbf{a} = \mathbf{\Phi}^T\mathbf{f}$ to obtain the coefficient vector $\mathbf{a} = (a_0, a_1, \ . \ . \ ., a_m)^T$. Then $p(x) = \sum_{j=0}^{m} a_j\varphi_j(x)$ is the least-squares polynomial. The value of this insight is that one can find basis functions for which $\mathbf{S} = \mathbf{\Phi}^T\mathbf{\Phi}$ is well-conditioned. In Section 6.4.1 we will see how "orthogonal" polynomials can serve to achieve well-conditioned least-squares equations. In that section we return to the sine function approximation problem that stymied us in the preceding example. Through the use of orthogonal polynomials, we will be able to achieve a least-squares polynomial approximation of the sine function having an accuracy commensurate with the double-precision computer word length (about 17 significant decimal places) of the DEC 10.

*6.2.5. Links Between the Least-Squares Theory and Statistics

In this section we assume that the reader has completed introductory course work in probability and statistics.

The idea behind polynomial approximation is that the coefficients $a_0, a_1, \ . \ . \ .,$ a_m are to be chosen so that the polynomial $p(x)$ "fits" the data (x_1, f_1), (x_2, f_2), . . ., (x_n, f_n) in the sense that

$$p(x_i) = a_0 + a_1x_i + \cdot \cdot \cdot + a_mx_i^m \approx f_i, \qquad 1 \leq i \leq n.$$

In statistical theory, we give more structure to the "fitting" problem by writing the system of equations

$$f_i = a_0 + a_1x_i + \cdot \cdot \cdot + a_mx_i^m + e_i, \qquad 1 \leq i \leq n,$$

where the e_i's are regarded as random variables representing measurement error. In fact, analogous to this polynomial case, we enlarge the scope of possible "models" and representations by the more general formulation

$$f_i = a_0x_{i0} + a_1x_{i1} + \cdot \cdot \cdot + a_mx_{im} + e_i, \qquad 1 \leq i \leq n, \qquad (6.24)$$

where $x_{i0}, x_{i1}, \ . \ . \ ., x_{im}$, and f_i, $1 \leq i \leq n$, are the given data. In the polynomial case, $x_{ij} = x_i^j$, and in the particular case of (6.14), $x_{ij} = \varphi_j(x_i)$. By introducing the matrix notation

$$
\mathbf{X} = \begin{bmatrix} x_{10} & x_{11} & \cdots & x_{1m} \\ x_{20} & x_{21} & \cdots & x_{2m} \\ \vdots & \vdots & & \vdots \\ x_{n0} & x_{n1} & \cdots & x_{nm} \end{bmatrix}, \quad \mathbf{a} = \begin{bmatrix} a_0 \\ a_1 \\ \vdots \\ a_m \end{bmatrix}, \quad \mathbf{e} = \begin{bmatrix} e_1 \\ e_2 \\ \vdots \\ e_n \end{bmatrix}
$$

equations (6.24) can be rewritten

$$
\mathbf{f} = \mathbf{X}\mathbf{a} + \mathbf{e}. \tag{6.25}
$$

In this equation, the vector $\mathbf{f} = (f_1, \ldots, f_n)^T$ and the matrix $\mathbf{X}$ are to be regarded as observed data, and the parameter vector $\mathbf{a} = (a_0, a_1, \ldots, a_m)^T$ is an unknown to be determined. In the terminology of statistics, (6.25) is the *general linear model* if the e_j's are presumed independent normal random variables with zero mean and a common variance σ^2. From developments in Section 6.2.1, we may readily confirm that the vector $\hat{\mathbf{a}}$ which minimizes the least-squares criterion

$$
Q(\mathbf{f}, \mathbf{a}) = \sum_{i=1}^{n} \left(f_i - \sum_{j=0}^{m} a_j x_{ij} \right)^2 \tag{6.26}
$$

is determined by the normal equation

$$
\boxed{\hat{\mathbf{a}} = (\mathbf{X}^T\mathbf{X})^{-1}\mathbf{X}^T\mathbf{f}.} \tag{6.27}
$$

The components of vector $\hat{\mathbf{a}}$ are viewed as estimates of the unknown parameters a_j $(0 \le j \le m)$ in (6.25). The statistical properties of these estimates will be examined next.

Let E denote expectation with respect to the variables e_i and $\mathbf{0}$ the column vector of zeros. Then one calculates that

$$
\begin{aligned}
E[\hat{\mathbf{a}}] &= E[(\mathbf{X}^T\mathbf{X})^{-1}\mathbf{X}^T\mathbf{f}] = (\mathbf{X}^T\mathbf{X})^{-1}\mathbf{X}^T E[\mathbf{f}] \\
&= (\mathbf{X}^T\mathbf{X})^{-1}\mathbf{X}^T E[\mathbf{X}\mathbf{a} + \mathbf{e}] = (\mathbf{X}^T\mathbf{X})^{-1}\mathbf{X}^T\mathbf{X}\mathbf{a} + (\mathbf{X}^T\mathbf{X})^{-1}\mathbf{X}^T\mathbf{0} = \mathbf{a}.
\end{aligned} \tag{6.28}
$$

Thus the expected value of the estimator equals the value to be estimated. In statistical terms we say that estimator $\hat{\mathbf{a}}$ is *unbiased*. A famous result of statistics (the Gauss–Markov theorem) states that $\hat{\mathbf{a}}$, as determined by (6.27), is also the minimum variance estimator for $\mathbf{a}$, among all unbiased linear functions of the vector $\mathbf{f}$. That is, if $\mathbf{M}$ is any $(m + 1) \times n$-order matrix and the estimator as given in the linear form $\mathbf{M}\mathbf{f}$ is such that

$$
E[\mathbf{M}\mathbf{f}] = \mathbf{a}, \tag{6.29}
$$

the expected sum of the squared differences of the components of the unknown parameter $\mathbf{a}$ and the estimator is minimal for $\mathbf{M} = (\mathbf{X}^T\mathbf{X})^{-1}\mathbf{X}^T$. That is, for arbitrary matrix $\mathbf{M}$ satisfying (6.29),

$$E[(\mathbf{a} - \mathbf{Mf})^T(\mathbf{a} - \mathbf{Mf})] \geq E[(\mathbf{a} - \hat{\mathbf{a}})^T(\mathbf{a} - \hat{\mathbf{a}})].$$

We refer the reader to Bickel and Doksum (1977, Chap. 7) for a proof and for supplementary information. The Gauss–Markov theorem is true regardless of the law governing the independent variables e_i, provided only that their means are all zero and that they have a common variance σ^2. If, in addition, the e_i's are normally distributed, it is known that $\hat{\mathbf{a}}$ is also the maximum-likelihood estimator of the parameter.

From (6.27) and the definition (6.25) of "general linear model," we may verify that

$$E[(\mathbf{a} - \hat{\mathbf{a}})(\mathbf{a} - \hat{\mathbf{a}})^T] = \sigma^2(\mathbf{X}^T\mathbf{X})^{-1},$$

and by summing up the diagonal elements on both sides, we find that the expected sum of the squared error is

$$E\left[\sum_{j=0}^{m} (a_j - \hat{a}_j)^2\right] = \sigma^2 \text{ trace } [(\mathbf{X}^T\mathbf{X})^{-1}],$$

where the trace of a square matrix is defined to be the sum of the diagonal elements.

Instability, discussed in the preceding section in connection with least-squares polynomials, plagues computation of the minimum-variance estimator as well. The orthogonal polynomial remedy to be presented in Section 6.4 is not applicable to the more general linear model (6.24). A methodology based on "singular-value decompositions" is the state-of-the-art technique. Forsythe et al. (1977) devote Chapter 9 to this topic. Lawson and Hanson (1974) contains advanced topics on these and related matters.

*6.3
UNIFORM APPROXIMATION

Again let us denote our data by (x_1, f_1), (x_2, f_2), . . ., (x_n, f_n) and presume that we wish to "fit" these data by a polynomial $p(x)$ of degree not exceeding m, where m is a given nonnegative integer. To avoid the triviality of the interpolation polynomial fitting the data perfectly, assume that $m < n - 1$. The settings of uniform and least-squares approximation differ only in that in the latter case, the criterion of fit was measured by $Q(f, p) = \sum_{i=1}^{n} [f_i - p(x_i)]^2$, whereas in this section the criterion for "fit" is the uniform distance between $\{f_i\}$ and $\{p(x_i)\}$. That is,

$$U(f, p) = \max_{1 \leq i \leq n} |f_i - p(x_i)|. \qquad (6.30)$$

In words, $U(f, p)$ is the maximum discrepancy between the data and the approximating polynomial, the maximum being taken over all the data points. The mth-degree polynomial $p(x)$ that minimizes $U(f, p)$ is called the *best approximating polynomial*. Let us continue to denote the approximating polynomial as $p(x) = a_0 + a_1x + \cdots + a_mx^m$. Then the quest for the best approximating polynomial can be rephrased as the task of finding the coefficients $a_0, a_1, \ldots, a_m$ that minimize

$$U(f, p) = \max_{1 \leq i \leq n} |f_i - a_0 - a_1x_i - \cdots - a_mx_i^m|.$$

Our attention now focuses on the numerical construction of the minimizing coefficients $a_0, a_1, \ldots, a_m$. In the case of the least-squares method, the objective function $Q(f, p)$ is differentiable and therefore the standard methods of calculus can be used to obtain the minimum. In the uniform case, however, the function $U(f, p)$ to be minimized is not differentiable. To verify this observation, recall from calculus the fact that the absolute value function does not have a derivative at zero. Hence, an alternative approach to minimization must be devised.

Before describing the general construction of the best approximating polynomial, by way of orientation, the elementary case of constant approximation ($p(x) = a_0$) is examined. In this case

$$U(f, p) = \max_{1 \leq i \leq n} |f_i - a_0|.$$

Since $U(f, p)$ is the maximum of the discrepancies $|f_i - a_0|$, each of these terms is less than or equal to $U(f, p)$. That is,

$$|f_i - a_0| \leq U(f, p), \qquad (1 \leq i \leq n).$$

We rearrange the previous inequality as

$$-U(f, p) \leq f_i - a_0 \leq U(f, p), \qquad (1 \leq i \leq n),$$

or, upon adding a_0 to all terms,

$$\left. \begin{aligned} a_0 - U(f, p) &\leq f_i \\ a_0 + U(f, p) &\geq f_i \end{aligned} \right\} \qquad (1 \leq i \leq n).$$

Since in (6.30), the quantity $U(f, p)$ should be minimized, we have derived the following constrained optimization problem with two unknowns, a_0 and $U = U(f, p)$:

Minimize U
subject to

$$\left. \begin{aligned} a_0 - U &\leq f_i \\ a_0 + U &\geq f_i \end{aligned} \right\} (1 \leq i \leq n). \tag{6.31}$$

This problem has the following special properties. The term to be minimized and the inequalities are linear functions of the unknowns; that is, they are sums of scalar multipliers of the unknowns. In our case the function to be minimized is

$$0 \cdot a_0 + 1 \cdot U$$

and the left-hand sides of the inequalities (6.31) are similarly linear in the two unknowns a_0 and U.

Minimization problems of this form are called *linear programming* problems. In the special case (6.31), one can verify (see Problem 10) that the optimal solution is

$$a_0 = \frac{\min_i (f_i) + \max_i (f_i)}{2}, \qquad U(f, p) = \frac{\max_i (f_i) - \min_i (f_i)}{2}.$$

This solution is plotted in Figure 6.15. Notice that the uniform criterion has led us to a constant approximation polynomial different from the one engendered by the least-squares criterion, which was [equation (6.3)] the function average

$$a_0 = \frac{1}{n} \sum_{i=1}^{n} f_i.$$

We proceed now to the general case. Whereas the expressions will be lengthy, the rationale is exactly that of the constant polynomial case: We simply rephrase the approximation problem as a linear programming problem. In order to minimize $U(f, p)$, observe first that (6.30) implies the system of inequalities

$$\left| f_i - a_0 - a_1 x_i - \cdots - a_m x_i^m \right| \leq U(f, p), \qquad (1 \leq i \leq n),$$

where equality holds for at least one value of i. The absolute values on the left-hand sides can be eliminated by rewriting the inequalities as

$$-U(f, p) \leq f_i - a_0 - a_1 x_i - \cdots - a_m x_i^m \leq U(f, p), \qquad 1 \leq i \leq n.$$

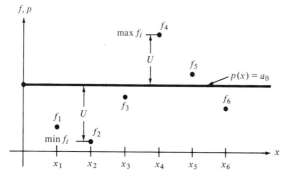

FIGURE 6–15   Best Approximating Constant Polynomial

This, in turn, can be restated as

$$a_0 + a_1 x_i + \cdots + a_m x_i^m - U(f, p) \leq f_i$$
$$a_0 + a_1 x_i + \cdots + a_m x_i^m + U(f, p) \geq f_i \qquad (1 \leq i \leq n). \qquad (6.32)$$

We seek coefficients a_0, a_1, . . ., a_m that make $U(f, p)$ as small as possible. In view of (6.32), the problem of finding the best approximating polynomial has been rephrased as the following linear programming problem:

Minimize U
subject to

$$a_0 + a_1 x_i + \cdots + a_m x_i^m - U \leq f_i$$
$$a_0 + a_1 x_i + \cdots + a_m x_i^m + U \geq f_i \qquad (1 \leq i \leq n), \qquad (6.33)$$

where the variables are a_0, a_1, . . ., a_m and U. The solution values a_0, a_1, . . ., a_m give the coefficients of the best approximating polynomial $p(x)$, and U is the uniform distance $U(f, p)$ for that polynomial.

In our problem (6.33) we have n constraints of both $\leq$ and $\geq$ type, but the unknown polynomial coefficients a_0, a_1, . . ., a_m may be negative. The standard techniques for solving linear programming problems (see Luenberger, 1973, Chap. 2) require that all the unknowns be nonnegative. We can overcome this difficulty by introducing two new nonnegative variables a_i^+ and a_i^- for each a_i, and rewriting each original unknown as the difference of the corresponding new variables:

$$a_i = a_i^+ - a_i^- . \qquad (6.34)$$

Then, by replacing the variables a_i in problem (6.33) by these differences, we obtain the following problem:

Minimize U
subject to

$$a_0^+ - a_0^- + a_1^+ x_i - a_1^- x_i + \cdots$$
$$+ a_m^+ x_i^m - a_m^- x_i^m - U \leq f_i$$
$$a_0^+ - a_0^- + a_1^+ x_i - a_1^- x_i + \cdots \qquad (1 \leq i \leq m) \qquad (6.35)$$
$$+ a_m^+ x_i^m - a_m^- x_i^m + U \geq f_i$$

where all variables are presumed nonnegative.

Subroutine BAP, Table 6.7, first performs the foregoing conversion of the uniform approximation data (x_1, f_1), (x_2, f_2), . . ., (x_n, f_n) to the corresponding linear programming problem and then calls the linear programming subroutine LINPRO (Appendix C) to solve for the best approximating polynomial coefficients.

**TABLE 6.7   Subroutine BAP for the Best
Approximating Polynomials**

```
      SUBROUTINE BAP(N,M,X,F,A,E)
C
C     ****************************************************************
C     *  FUNCTION: THIS SUBROUTINE COMPUTES THE COEFFICIENTS         *
C     *            OF THE BEST APPROXIMATING POLYNOMIAL FOR A        *
C     *            SET OF PAIRS (X,F)                                *
C     *  USAGE:                                                      *
C     *        CALL SEQUENCE: CALL BAP(N,M,X,F,A,E)                  *
C     *        EXTERNAL FUNCTIONS/SUBROUTINES:                       *
C     *                SUBROUTINE LINPRO(SEE APPENDIX C)             *
C     *  PARAMETERS:                                                 *
C     *     INPUT:                                                   *
C     *         N=NUMBER OF DATA POINTS(MAXIMUM 25)                  *
C     *         M=DEGREE OF THE APPROXIMATING POLYNOMIAL             *
C     *            (MAXIMUM 24)                                      *
C     *         X=N BY 1 ARRAY OF INDEPENDENT DATA VALUES            *
C     *         F=N BY 1 ARRAY OF DEPENDENT FUNCTIONAL VALUES        *
C     *     OUTPUT:                                                  *
C     *         A=M+1 BY 1 ARRAY OF APPROXIMATING POLYNOMIAL         *
C     *            COEFFICIENTS                                      *
C     *         E=VALUE OF THE MAXIMUM POLYNOMIAL DEVIATION          *
C     ****************************************************************
C
      DIMENSION X(N),F(N),A(M+1),A1(151)
      COMMON AA(52,151),B(50),C(50)
      KONST=2*M+3
      DO 1 K=1,N
        U=1.0
        V=-1.0
        DO 2 I=1,M+1
           J=2*I
           AA(K,J-1)=U
           AA(K,J)=V
           AA(K+N,J-1)=U
           AA(K+N,J)=V
           U=U*X(K)
           V=V*X(K)
    2   CONTINUE
        AA(K,KONST)=-1.0
        AA(K+N,KONST)=1.0
        B(K)=F(K)
        B(K+N)=F(K)
    1 CONTINUE
      DO 3 I=1,M+1
        C(I)=0.0
        C(I+M+1)=0.0
    3 CONTINUE
      NN=KONST
      C(NN)=-1.0
      M1=N
      M2=0
      M3=N
      EPS=.10E-05
      NNN=NN
C     *** PERFORM LINEAR PROGRAM REDUCTION ***
      CALL LINPRO(NN,M1,M2,M3,EPS,A1,KOD)
      E=A1(NNN)
C     *** COMPUTE APPROXIMATING POLYNOMIAL COEFFICIENTS ***
      DO 4 I=1,M+1
        A(I)=A1(2*I-1)-A1(2*I)
    4 CONTINUE
      RETURN
      END
```

Following the next example, which describes the details of a hand calculation of a simple uniform approximation problem, we present and discuss a program that finds best approximating polynomials through calls to subroutines BAP and LINPRO.

────── **EXAMPLE 6.8** ──

Here the best approximating polynomial of degree 2 will be determined for the data of Example 6.1 by the linear programming method. Since

$$x_1 = -2, \quad x_2 = -1, \quad x_3 = 0, \quad x_4 = 1, \quad x_5 = 2$$
$$f_1 = 0, \quad f_2 = 1, \quad f_3 = 2, \quad f_4 = 1, \quad f_5 = 0,$$

the associated linear programming problem has the form

Minimize U
subject to

$$
\begin{array}{ll}
a_0 - 2a_1 + 4a_2 - U \leq 0 \\
a_0 - 2a_1 + 4a_2 + U \geq 0 \\
a_0 - a_1 + a_2 - U \leq 1 \\
a_0 - a_1 + a_2 + U \geq 1 \\
a_0 \qquad\qquad - U \leq 2 \\
a_0 \qquad\qquad + U \geq 2 & (6.36)\\
a_0 + a_1 + a_2 - U \leq 1 \\
a_0 + a_1 + a_2 + U \geq 1 \\
a_0 + 2a_1 + 4a_2 - U \leq 0 \\
a_0 + 2a_1 + 4a_2 + U \geq 0.
\end{array}
$$

Finally, to convert these relations to the standard linear programming problem, we have to introduce two unknowns for each variable a_0, a_1, and a_2. Let

$$a_0 = a_0^+ - a_0^-$$
$$a_1 = a_1^+ - a_1^-$$
$$a_2 = a_2^+ - a_2^-,$$

where we now assume that the variables a_i^+, a_i^- ($i = 0, 1, 2$) are all nonnegative. The last variable U is necessarily nonnegative because of its definition. By substituting these variables into problem (6.36), we get

Minimize U
subject to

$$a_0^+ - a_0^- - 2a_1^+ + 2a_1^- + 4a_2^+ - 4a_2^- - U \le 0$$

$$a_0^+ - a_0^- - 2a_1^+ + 2a_1^- + 4a_2^+ - 4a_2^- + U \ge 0$$

$$a_0^+ - a_0^- - a_1^+ + a_1^- + a_2^+ - a_2^- - U \le 1$$

$$a_0^+ - a_0^- - a_1^+ + a_1^- + a_2^+ - a_2^- + U \ge 1$$

$$a_0^+ - a_0^- \qquad\qquad\qquad\qquad - U \le 2$$

$$a_0^+ - a_0^- \qquad\qquad\qquad\qquad + U \ge 2$$

$$a_0^+ - a_0^- + a_1^+ - a_1^- + a_2^+ - a_2^- - U \le 1$$

$$a_0^+ - a_0^- + a_1^+ - a_1^- + a_2^+ - a_2^- + U \ge 1$$

$$a_0^+ - a_0^- + 2a_1^+ - 2a_1^- + 4a_2^+ - 4a_2^- - U \le 0$$

$$a_0^+ - a_0^- + 2a_1^+ - 2a_1^- + 4a_2^+ - 4a_2^- + U \ge 0$$

$$a_i^+, a_i^- \ge 0.$$

It can be shown that the variables

$$a_0^+ = 1.75, \qquad a_0^- = 0$$

$$a_1^+ = 0, \qquad a_1^- = 0$$

$$a_2^+ = 0, \qquad a_2^- = 0.5$$

$$U = 0.25$$

give the optimal solution, which implies that

$$a_0 = a_0^+ - a_0^- = 1.75$$

$$a_1 = a_1^+ - a_1^- = 0$$

$$a_2 = a_2^+ - a_2^- = -0.5.$$

Thus the best approximating polynomial of degree 2 is

$$p(x) = 1.75 - 0.5x^2$$

and the largest discrepancy between the data f_i and the corresponding polynomial values equals 0.25. The data and polynomial $p(x)$ are shown in Figure 6.16. It is interesting to note that in the case of this example, $|f_i - p(x_i)| = U$ for all i. Also note that as in the case of constant approximation, the uniform criterion, on the same data, has yielded a quadratic approximation polynomial differing from the least-squares quadratic polynomial (Example 6.1).

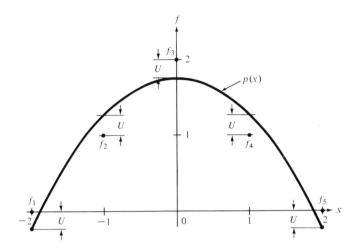

FIGURE 6–16   Illustration of Data and Their Best Second-Degree Approximating Polynomial

━━━━━━━ **EXAMPLE 6.9** ━━━━━━━

In the computer study we now describe, we have found best approximating polynomials of degrees 1, 2, and 3, for sin (x), $0 \le x \le 2$. This approximation problem had previously been attacked by Taylor's, interpolation, and least-squares polynomials in Examples 2.2, 2.4, and 6.5, respectively. As data, we took sin (x) values at 22 evenly spaced points over the interval [0, 2]. The calling program is given in Table 6.8, and its output, comparing the sine function and its best approximating polynomial, is displayed in Table 6.9. Functions sin (x) and the best approximating polynomials $p_1(x)$, $p_2(x)$, $p_3(x)$ are illustrated in Figure 6.17. The computations were done in double precision.

TABLE 6.8   Program for Example 6.9

```
C       PROGRAM LINEAR
C
C       ******************************************************************
C       THIS PROGRAM COMPUTES THE BEST APPROXIMATING POLYNOMIAL FOR 22
C       POINTS XX EQUALLY SPACED OVER THE INTERVAL [0,2] WITH
C       FUNCTIONAL VALUES COMPUTED AS SIN(XX). IT COMPARES THIS TO
C       THE FUNCTIONAL VALUES COMPUTED BY THE FUNCTION SIN
C       CALLS:   BAP,LINPRO (BOTH MODIFIED FOR DOUBLE PRECISION)
C       OUTPUT:
C                X = THE VALUE OF X AT WHICH THE FUNCTIONS ARE COMPARED
C                TRUE = THE VALUE OF THE SINE FUNCTION AT X
C                P = THE POLYNOMIAL APPROXIMATION AT X
C                DIFF = THE DIFFERENCE (TRUE-P) AT X
C       ******************************************************************
C
```

TABLE 6.8   (Continued)

```
        IMPLICIT DOUBLE PRECISION(A-H,O-Z)
        DIMENSION A(21),XX(25),F(25)
        Y=0.D0
        N=22
        H=2.D0/22.D0
C
C     *** CALULATE THE POINTS XX AND SIN(XX) TO BE USED IN        ***
C     *** THE POLYNOMIAL APPROXIMATION                           ***
C
        DO 10 J=1,N
          Y=Y+H
          XX(J)=Y
          F(J)=SIN(XX(J))
    10 CONTINUE
C
C     *** IN THIS LOOP WE COMPUTE THE APPROXIMATING POLYNOMIAL FOR ***
C     *** DEGREES M=1,2,3                                         ***
C
        DO 100 M=1,3
          WRITE(10,1)
     1    FORMAT(////)
C
C     *** SUBROUTINE BAP COMPUTES THE COEFFICIENTS FOR THE        ***
C     *** APPROXIMATING POLYNOMIAL                               ***
C
          CALL BAP(N,M,XX,F,A,E)
          X=0.D0
          DO 30 I=1,10
            X=X+.2D0
            P=0.D0
            DO 20 K=1,M+1
              K1=K-1
              P=P+A(K)*X**K1
    20      CONTINUE
            TRUE=SIN(X)
            DIFF=TRUE-P
            WRITE(10,2)X,TRUE,P,DIFF
     2      FORMAT(5X,F6.2,2(5X,F11.5),5X,E13.6)
    30    CONTINUE
   100 CONTINUE
        STOP
        END
```

TABLE 6.9   Application of Best Approximating Polynomials

x	$\sin(x)$	$p_1(x)$	$\sin(x) - p_1(x)$	$p_2(x)$	$\sin(x) - p_2(x)$	$p_3(x)$	$\sin(x) - p_3(x)$
0.20	0.19867	0.32132	-0.122646E+00	0.19910	-0.433899E-03	0.19972	-0.105503E-02
0.40	0.38942	0.40706	-0.176458E-01	0.40584	-0.164220E-01	0.39302	-0.359852E-02
0.60	0.56464	0.49281	0.718293E-01	0.58111	-0.164702E-01	0.56633	-0.168842E-02
0.80	0.71736	0.57856	0.138794E+00	0.72492	-0.756420E-02	0.71592	0.143259E-02
1.00	0.84147	0.66431	0.177160E+00	0.83726	0.420787E-02	0.83805	0.341926E-02
1.20	0.93204	0.75006	0.181979E+00	0.91814	0.138979E-01	0.92897	0.306648E-02
1.40	0.98545	0.83581	0.149640E+00	0.96755	0.178952E-01	0.98494	0.506553E-03
1.60	0.99957	0.92156	0.780153E-01	0.98550	0.140706E-01	1.00222	-0.264688E-02
1.80	0.97385	1.00731	-0.334597E-01	0.97199	0.186082E-02	0.97706	-0.321395E-02
2.00	0.90930	1.09306	-0.183759E+00	0.92701	-0.177084E-01	0.90572	0.357394E-02

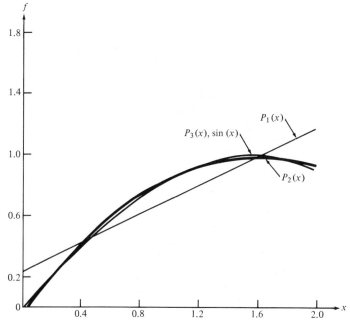

FIGURE 6–17   Best Approximating Polynomial Graphs for sin(x)

---

*6.4
ORTHOGONAL FUNCTION AND FOURIER APPROACHES TO LEAST-SQUARES APPROXIMATIONS

*6.4.1   Application of Orthogonal Polynomials

Orthogonal function methods are a powerful safeguard against the least-squares instability phenomenon discussed in Section 6.2.4. Additionally, they have served well in a myriad of other pursuits in pure and applied mathematics, especially in differential equation theory, stochastic processes, communication and control, and probability theory. Our focus in this first subsection will be on stable least-squares procedures.

Orthogonal functions are defined in terms of "inner products." Let $X = \{x_1, \ldots, x_n\}$ be first a finite set of points, and let functions $f(x)$ and $g(x)$ be defined on X. Then the (discrete) *inner product* of functions $f(x)$ and $g(x)$ is defined to be

$$(f, g) = \sum_{k=1}^{n} f(x_k)g(x_k). \tag{6.37}$$

Alternatively, if X is any (finite or infinite) interval on the real line, then the (continuous) *inner product* is

$$(f, g) = \int_X f(x)g(x) \, dx. \tag{6.38}$$

Functions $f(x)$ and $g(x)$ are said to be *orthogonal* with respect to a given inner product if $(f, g) = 0$. Any finite or infinite sequence of functions $\varphi_0(x)$, $\varphi_1(x)$, . . . is called an *orthogonal system* if and only if for all i and j,

$$(\varphi_i, \varphi_j) = 0 \qquad (i \neq j) \qquad \text{and} \qquad (\varphi_i, \varphi_i) \neq 0. \qquad (6.39)$$

Let us now specialize to an orthogonal function system, $\varphi_0(x)$, $\varphi_1(x)$, . . ., $\varphi_m(x)$ of polynomials with subscript i indicating the degree of $\varphi_i(x)$, and the leading coefficient of $\varphi_i(x)$ (i.e., the coefficient of x^i) being 1. For any particular inner product, such a system is unique and is known (Ralston and Rabinowitz, 1978, Sec. 6.4) to satisfy the *three-term recurrence relation*

$$\varphi_i(x) = (x - b_i)\varphi_{i-1}(x) - c_i\varphi_{i-2}(x), \qquad (6.40)$$

where

$$b_i = \frac{(x\varphi_{i-1}, \varphi_{i-1})}{(\varphi_{i-1}, \varphi_{i-1})} \qquad (i = 1, 2, . . .) \qquad (6.41)$$

and

$$c_i = \frac{(x\varphi_{i-1}, \varphi_{i-2})}{(\varphi_{i-2}, \varphi_{i-2})} \qquad (i = 2, 3, . . .; c_1 = 0). \qquad (6.42)$$

Construction proceeds by defining the initial terms to be $\varphi_0 = 1$ and $\varphi_{-1} = 0$. The recursion (6.40)–(6.42) is highly useful for computing the entire orthogonal polynomial system and for the implementation of orthogonal polynomial approximation.

Let us now demonstrate that orthogonal polynomials can effectively alleviate the instability of least-squares polynomial approximation observed in Section 6.2.4. Let $\varphi_0(x)$, $\varphi_1(x)$, . . ., $\varphi_m(x)$ be the system of orthogonal polynomials for the discrete inner product associated with the given set $\{x_1, . . ., x_n\}$ of points. Then the least-squares polynomial approximation for the data (x_1, f_1), (x_2, f_2), . . ., (x_n, f_n) bears the representation,

$$p_m(x) = \sum_{j=0}^{m} a_j\varphi_j(x), \qquad (6.43)$$

where, in view of (6.16),

$$\mathbf{a} = (\mathbf{\Phi}^T\mathbf{\Phi})^{-1}\mathbf{\Phi}^T\mathbf{f} \qquad (6.44)$$

and according to (6.14),

$$\mathbf{\Phi} = \begin{bmatrix} \varphi_0(x_1) & \varphi_1(x_1) & \cdots & \varphi_m(x_1) \\ \varphi_0(x_2) & \varphi_1(x_2) & \cdots & \varphi_m(x_2) \\ \vdots & \vdots & & \vdots \\ \varphi_0(x_n) & \varphi_1(x_n) & \cdots & \varphi_m(x_n) \end{bmatrix}, \quad \mathbf{f} = \begin{bmatrix} f_1 \\ f_2 \\ \vdots \\ f_n \end{bmatrix}, \quad \mathbf{a} = \begin{bmatrix} a_0 \\ a_1 \\ \vdots \\ a_m \end{bmatrix}. \qquad (6.45)$$

The main property of orthogonal polynomials that overcomes the ill-conditioning is that $\mathbf{S} = (\mathbf{\Phi}^T\mathbf{\Phi})$ here is a diagonal matrix, for the (i, j) coordinate s_{ij} of matrix $\mathbf{S}$ is given by

$$s_{ij} = \sum_{k=1}^{n} \varphi_i(x_k)\varphi_j(x_k) = (\varphi_i, \varphi_j) = 0, \qquad \text{for } i \neq j. \qquad (6.46)$$

The inversion of a diagonal matrix is trivial; it is the diagonal matrix the diagonal coordinates of which are the reciprocals of the corresponding diagonal elements in the original matrix. That is,

$$(\mathbf{\Phi}^T\mathbf{\Phi})^{-1} = \begin{bmatrix} \dfrac{1}{(\varphi_0, \varphi_0)} & 0 & \cdots & 0 \\ 0 & \dfrac{1}{(\varphi_1, \varphi_1)} & \cdots & 0 \\ \vdots & 0 & & \vdots \\ & & \vdots & \\ 0 & 0 & \cdots & \dfrac{1}{(\varphi_m, \varphi_m)} \end{bmatrix} \qquad (6.47)$$

and from (6.47) we easily compute the coefficients a_j in (6.44) to be

$$a_j = \frac{(f, \varphi_j)}{(\varphi_j, \varphi_j)} \qquad (j = 0, 1, \ldots, m). \qquad (6.48)$$

In summary, through the orthogonal polynomial approach we have avoided the numerical solution of simultaneous linear equations altogether, and obtained the coefficients a_j in closed form.

———— EXAMPLE 6.10 ————————————————————————

The least-squares problem that was the subject of Example 6.1 will now be solved by orthogonal polynomials. In this case $x_1 = -2$, $x_2 = -1$, $x_3 = 0$, $x_4 = 1$, and $x_5 = 2$. We start the procedure by initializing $\varphi_{-1} = 0$, $\varphi_0 = 1$. Then $\varphi_1(x)$ can be obtained from (6.40), where

$$b_1 = \frac{(x\varphi_0, \varphi_0)}{(\varphi_0, \varphi_0)}, \qquad c_1 = 0.$$

To obtain the inner products in the numerator and denominator, note that the values of function φ_0 are all 1's at the points x_i and the values of $x\varphi_0(x) = x$ are

$$-2, -1, 0, 1, 2.$$

Thus

$$(x\varphi_0, \varphi_0) = (x, 1) = (-2) \cdot 1 + (-1) \cdot 1 + 0 \cdot 1 + 1 \cdot 1 + 2 \cdot 1 = 0$$

$$(\varphi_0, \varphi_0) = (1, 1) = 1 \cdot 1 + 1 \cdot 1 + 1 \cdot 1 + 1 \cdot 1 + 1 \cdot 1 = 5.$$

Consequently, $b_1 = 0$ and

$$\varphi_1(x) = (x - 0)\varphi_0(x) - 0 \cdot 0 = x.$$

One similarly gets

$$b_2 = \frac{(x\varphi_1, \varphi_1)}{(\varphi_1, \varphi_1)} = 0, \qquad c_2 = \frac{(x\varphi_1, \varphi_0)}{(\varphi_0, \varphi_0)} = 2,$$

since

$$(x\varphi_1, \varphi_1) = (x^2, x) = 4 \cdot (-2) + 1 \cdot (-1) + 0 \cdot 0 + 1 \cdot 1 + 4 \cdot 2 = 0,$$

$$(\varphi_1, \varphi_1) = (x, x) = (-2) \cdot (-2) + (-1) \cdot (-1) + 0 \cdot 0 + 1 \cdot 1 + 2 \cdot 2$$
$$= 10,$$

$$(x\varphi_1, \varphi_0) = (x^2, 1) = 4 \cdot 1 + 1 \cdot 1 + 0 \cdot 1 + 1 \cdot 1 + 4 \cdot 1 = 10.$$

Thus

$$\varphi_2(x) = (x - b_2)\varphi_1(x) - c_2\varphi_0(x) = [(x - 0) \cdot x] - 2 \cdot 1 = x^2 - 2.$$

The quadratic least-squares polynomial can then be determined by using (6.48) and (6.43) with $m = 2$. In our case the values of $f(x)$ at the data points x_i are

$$0, 1, 2, 1, 0,$$

respectively, which imply that

$$(f, \varphi_0) = 0 \cdot 1 + 1 \cdot 1 + 2 \cdot 1 + 1 \cdot 1 + 0 \cdot 1 = 4$$

$$(f, \varphi_1) = 0 \cdot (-2) + 1 \cdot (-1) + 2 \cdot 0 + 1 \cdot 1 + 0 \cdot 2 = 0$$

$$(f, \varphi_2) = 0 \cdot 2 + 1 \cdot (-1) + 2 \cdot (-2) + 1 \cdot (-1) + 0 \cdot 2 = -6$$

$$(\varphi_2, \varphi_2) = 2 \cdot 2 + (-1) \cdot (-1) + (-2) \cdot (-2) + (-1) \cdot (-1)$$
$$+ 2 \cdot 2 = 14,$$

since

$$\varphi_2(x_1) = x_1^2 - 2 = (-2)^2 - 2 = 2, \qquad \varphi_2(x_2) = (-1)^2 - 2 = -1,$$

$$\varphi_2(x_3) = 0^2 - 2 = -2, \qquad \varphi_2(x_4) = 1^2 - 2 = -1,$$

$$\varphi_2(x_5) = 2^2 - 2 = 2.$$

Then, from (6.48), we conclude that

$$a_0 = \frac{(f, \varphi_0)}{(\varphi_0, \varphi_0)} = \frac{4}{5}, \qquad a_1 = \frac{(f, \varphi_1)}{(\varphi_1, \varphi_1)} = 0, \qquad a_2 = \frac{(f, \varphi_2)}{(\varphi_2, \varphi_2)} = -\frac{3}{7},$$

that is,

$$p_2(x) = a_0\varphi_0(x) + a_1\varphi_1(x) + a_2\varphi_2(x)$$
$$= \tfrac{4}{5} \cdot 1 + 0 \cdot x - \tfrac{3}{7} \cdot (x^2 - 2) = -\tfrac{3}{7}x^2 + \tfrac{58}{35},$$

which coincides with the result obtained in Example 6.1.

The gist of the situation is that although the correct least-squares solution must be the same in both cases, the orthogonal polynomial method is less subject to roundoff error. This advantage can be appreciated only when more demanding problems are tackled. ∎

Subroutine ORTH (Table 6.10) finds the least-squares polynomial approximation $p_m(x)$ for any given data set (x_1, f_1), (x_2, f_2), . . ., (x_n, f_n) and any specified degree m. In this code we have made use of a property (Problem 15) that if $\varphi_0(x)$, $\varphi_1(x)$, . . . is an orthogonal system of polynomials, and $p_0(x)$, $p_1(x)$, . . . the least-squares approximation of degree indicated by the subscript, then

$$p_{j+1}(x) = a_{j+1}\varphi_{j+1}(x) + p_j(x). \tag{6.49}$$

TABLE 6.10   Subroutine ORTH for Orthogonal Least-Squares Approximation

```
      SUBROUTINE ORTH(N,M,NT,X,F,T,Y)
C
C   ****************************************************************
C   *   FUNCTION: THIS SUBROUTINE COMPUTES THE LEAST- SQUARES     *
C   *             POLYNOMIAL VALUES Y ON A SET T OF DOMAIN        *
C   *             POINTS FOR THE DATA PAIRS (X,F) BY THE USE OF   *
C   *             AN ORTHOGONAL POLYNOMIAL BASIS                  *
C   *   USAGE:                                                    *
C   *        CALL SEQUENCE: CALL ORTH(N,M,NT,X,F,T,Y)            *
C   *   PARAMETERS:                                               *
C   *        INPUT:                                               *
C   *             N=NUMBER OF DATA POINTS                         *
C   *             M=DEGREE OF THE LEAST SQUARES POLYNOMIAL        *
C   *             NT=NUMBER OF LEAST SQUARES POLYNOMIAL           *
C   *                EVALUATIONS                                  *
C   *             X=N BY 1 ARRAY OF INDEPENDENT DATA VALUES       *
C   *             F=N BY 1 ARRAY OF DEPENDENT FUNCTIONAL VALUES   *
C   *             T=NT BY 1 ARRAY OF POINTS FOR LEAST-SQUARES     *
C   *                EVALUATION                                   *
C   *        OUTPUT:                                              *
C   *             Y=NT BY 1 ARRAY OF VALUES OF LEAST-SQUARES      *
C   *                POLYNOMIALS                                  *
C   ****************************************************************
C
      DIMENSION X(N),F(N),P(100),PO(100),Y(NT),PT(100),PTO(100),T(NT)
C     *** INITIALIZATION ***
      S1=0.0
      COE=0.0
      BE=0.
      DO 1 I=1,N
        P(I)=1.0
        PO(I)=0.
        S1=S1+X(I)
        COE=COE+F(I)
    1 CONTINUE
      COE=COE/N
      DO 2 I=1,NT
        PT(I)=1.0
        PTO(I)=0.
        Y(I)=COE
    2 CONTINUE
      AL=S1/FLOAT(N)
```

TABLE 6.10   (Continued)

```
C      *** START ITERATIONS ***
       DO 3 K=1,M
         S1=0.0
         S2=0.0
         S3=0.0
         S4=0.0
         S5=0.0
C      *** SET S1=(XPK,PK),S2=(PK,PK),S3=(XPK,PK-1) ***
C      ***     SET S4=(FKPK,PK),S5=(PK-1,PK-1)     ***
         DO 4 I=1,N
           W=(X(I)-AL)*P(I)-BE*PO(I)
           S1=S1+X(I)*W**2
           S2=S2+W**2
           S3=S3+X(I)*W*P(I)
           S4=S4+F(I)*W
           S5=S5+P(I)**2
           PO(I)=P(I)
           P(I)=W
     4   CONTINUE
C      *** UPDATE LEAST-SQUARES ESTIMATE Y=P(T) ON VALUES T ***
         COE=S4/S2
         DO 5 I=1,NT
           W=(T(I)-AL)*PT(I)-BE*PTO(I)
           Y(I)=COE*W+Y(I)
           PTO(I)=PT(I)
           PT(I)=W
     5   CONTINUE
         AL=S1/S2
         BE=S3/S5
     3 CONTINUE
       RETURN
       END
```

Through this recursion, the least-squares approximates at specified points in the ORTH array T are computed as the successive orthogonal polynomials are constructed. If for some reason the user wants to get the coefficients a_j in (6.43), they can be obtained by printing out values of the variable COE every time it is updated. The structure of subroutine ORTH is as follows. DO loops with labels 1 and 2 do the initialization. The DO loop with label 3 computes the least-squares polynomials $p_k(x)$ ($k = 1, 2, \ldots, m$), and for each value of k, the DO loop with label 4 computes the inner products needed in using (6.41) and (6.42), and then DO loop with label 5 evaluates these relations.

──────── **EXAMPLE 6.11** ────────

We reconsider Example 6.7, in which we found it impossible to achieve 12-significant-decimal-place accuracy in approximating the sine function on the interval [0, 2], using the conventional least-squares representation (6.9). By means of the calling program in Table 6.11, we applied subroutine ORTH in double precision, using the same 50 data and 10 readout points as in the earlier example. For degrees $m = 0$ through 3, the results were identical to those in Table 6.3. As we increased m, the accuracy increased, the error being about 10^{-11} for $m = 10$, and decreasing to about 10^{-16} for $m = 14$. This is the approximate limit of the double-precision word length of the computer used, and for higher-degree m, the error did not decrease further, but neither did it increase. Seventeen-significant-figure accuracy was observed for $m = 25$, the highest degree that we attempted. The printout for $m = 15$ is displayed in Table 6.12.

TABLE 6.11 Computes Orthogonal Least-Squares Approximation of sin (x)

```
C       PROGRAM ORTHPOLY
C
C       *******************************************************************
C       THIS PROGRAM FINDS HIGH DEGREE ORTHOGONAL POLYNOMIAL
C       APPROXIMATIONS FOR SIN(X) ON THE INTERVAL [0,2].
C       THE DATA CONSIST OF VALUES SIN(2*J/50), J=1,...,50
C       CALLS:    ORTH (MODIFIED FOR DOUBLE PRECISION)
C       OUTPUT:
C               X(I)=VALUE OF X AT CURRENT ITERATION
C               TRUE=VALUE OF SIN(X)
C               P(I)=POLYNOMIAL APPROXIMATION AT X
C               DIFF=DIFFERENCE (TRUE-P) FOR EACH X
C       *******************************************************************
C
        IMPLICIT DOUBLE PRECISION (A-H,O-Z)
        DIMENSION X(10),XX(50),F(50),P(10)
        N=50
        NT=10
C
C       *** CONSTRUCT DATA ARRAYS XX AND SIN(XX) FOR USE IN THE      ***
C       *** APPROXIMATION. THE POLYNOMIAL IS TO BE EVALUATED AT      ***
C       *** POINTS IN THE ARRAY X                                    ***
C
        DO 10 J=1,N
            XX(J)=.04D0*J
            F(J)=DSIN(XX(J))
     10 CONTINUE
        DO 20 I=1,10
            X(I)=I*.2D0
     20 CONTINUE
C
C       *** M IS THE DEGREE OF THE POLYNOMIAL REQUESTED              ***
C
        DO 40 M=1,25
            WRITE(10,1)
      1     FORMAT(/////)
C
C       *** SUBROUTINE ORTH WILL CALCULATE THE M-TH DEGREE LEAST     ***
C       *** SQUARES POLYNOMIAL USING ORTHOGONAL POLYNOMIALS          ***
C       *** RETURNING POLYNOMIAL VALUES P FOR EACH OF THE NT         ***
C       *** POINTS SPECIFIED IN X                                    ***
C
            CALL ORTH(N,M,NT,XX,F,X,P)
C
C       *** NEXT COMPARE THE VALUE OF THE POLYNOMIAL AT 10 EQUALLY   ***
C       *** SPACED POINTS WITH THE VALUES OF SIN(X) AT THESE         ***
C       *** POINTS                                                   ***
C
            DO 30 I=1,10
                TRUE=DSIN(X(I))
                DIFF=TRUE-P(I)
                WRITE(10,25) X(I),TRUE,P(I),DIFF
     25         FORMAT(5X,F8.2,5X,D15.5,3X,D15.5,3X,D15.5)
     30     CONTINUE
     40 CONTINUE
        STOP
        END
```

TABLE 6.12 Errors of Fifteenth-Degree Orthogonal Least-Squares Polynomial Approximation

x	Error, $\sin(x) - p_{15}(x)$
0.2	-0.69389D-17
0.4	0.34694D-15
0.6	-0.30531D-15
0.8	0.99920D-15
1.0	-0.15127D-14
1.2	-0.37470D-15
1.4	0.41633D-15
1.6	-0.11102D-15
1.8	-0.55511D-16
2.0	-0.55511D-16

∎

*6.4.2 Fourier Analysis

The orthogonal polynomial approach to the least-squares approximation problem is a particular example of *Fourier analysis*. The general framework of Fourier analysis is now sketched.

First, define generalized versions of the inner products (6.37) and (6.38) as follows:

$$(f, g) = \sum_{k=1}^{n} f(x_k)g(x_k)w(x_k) \tag{6.50}$$

and

$$(f, g) = \int_X f(x)g(x)w(x) \, dx, \tag{6.51}$$

where $w(x) > 0$. Consider a finite or infinite system $\varphi_0(x)$, $\varphi_1(x)$, . . . of real orthogonal functions [i.e., a system satisfying (6.39)] and a function $f(x)$ for which (f, f) is finite. Then the series

$$f_m(x) = \sum_{j=0}^{m} \frac{(f, \varphi_j)}{(\varphi_j, \varphi_j)} \varphi_j(x), \tag{6.52}$$

with m either finite or infinite, is called the (generalized) *Fourier series* of $f(x)$. The ratios $\beta_j = (f, \varphi_j)/(\varphi_j, \varphi_j)$ are termed *Fourier coefficients*. In this terminology, then, (6.43) and (6.48) imply that the orthogonal polynomial representation of the least-squares polynomial is a special Fourier series with the discrete inner product and $w(x) = 1$. More generally, in the discrete inner product case, the Fourier series minimizes the sum

$$\sum_{i=1}^{n} \left[f(x_i) - \sum_{j=0}^{m} a_j \varphi_j(x_i) \right]^2 w(x_i) \tag{6.53}$$

over all possible choices of coefficients a_j (Apostol, 1957, Sec. 15.5). An analogous statement holds with respect to minimizing the integral

$$\int_X \left[f(x) - \sum_{j=0}^{m} a_j \varphi_j(x) \right]^2 w(x) \, dx$$

in the case of the "continuous" inner products (6.51).

The best-known Fourier series is that determined by the inner product (6.51) over the finite interval $(-\pi, \pi)$, the orthogonal series being $\frac{1}{2}$, sin (x), cos (x), sin $(2x)$, cos $(2x)$, . . ., sin (mx), cos (mx), ... The expression "trigonometric series" or, for finite series, "trigonometric polynomials" refers to this setting, which has played a prominent role in pure and applied mathematics, engineering, and physics since the nineteenth century. Such series also lead to major computational methods for linear systems, time series, signal processing, and in their extension to two-dimensional domains, for digital image processing.

─────── **EXAMPLE 6.12** ───

Let us construct a trigonometric Fourier series of the "square-wave" function

$$f(x) = \begin{cases} -1 & \text{if } -\pi < x < 0 \\ 1 & \text{if } 0 < x < \pi. \end{cases}$$

For this orthogonal system, the Fourier representation (6.52) takes on the specific form

$$\frac{a_0}{2} + \sum_{k=1}^{\infty} [a_k \cos (kx) + b_k \sin (kx)],$$

where

$$a_k = \frac{1}{\pi} \int_{-\pi}^{\pi} f(x) \cos (kx) \, dx, \qquad (k = 0, 1, 2, \ldots)$$

and

$$b_k = \frac{1}{\pi} \int_{-\pi}^{\pi} f(x) \sin (kx) \, dx, \qquad (k = 1, 2, 3, \ldots).$$

[Here we used that (sin (mx), sin (mx)) = (cos (mx), cos (mx)) = π, if $m \neq 0$.] This special Fourier series is often called the *continuous Fourier transformation*, in contrast to the discrete Fourier transformation discussed in the next section.

For our particular square-wave function, the Fourier coefficients are

$$a_k = \frac{1}{\pi} \left\{ -\int_{-\pi}^{0} \cos(kx)\, dx + \int_{0}^{\pi} \cos(kx)\, dx \right\} = 0$$

and

$$b_k = \frac{1}{\pi} \left\{ -\int_{-\pi}^{0} \sin(kx)\, dx + \int_{0}^{\pi} \sin(kx)\, dx \right\} = \frac{2}{\pi k}[1 - (-1)^k]$$

$$= \begin{cases} 0 & \text{if } k \text{ even} \\ \dfrac{4}{\pi k} & \text{if } k \text{ odd.} \end{cases}$$

In summary, for odd m, the mth-order trigonometric polynomial of $f(x)$ is

$$f_m(x) = \frac{4}{\pi} \sin(x) + \frac{4}{3\pi} \sin(3x) + \cdots + \frac{4}{m\pi} \sin(mx).$$

It is known (Apostol, 1957) that $f_m(x) \to f(x)$ for $m \to \infty$, except at x values that are multiples of π. In Figure 6.18 we compare $f_1(x)$, $f_3(x)$, $f_5(x)$, and $f_7(x)$ with $f(x)$. A curiosity is that since $f(x) = 1$, $0 < x < \pi$,

$$\frac{\pi}{4} = \sin(x) + \frac{\sin(3x)}{3} + \frac{\sin(5x)}{5} + \cdots, \qquad (0 < x < \pi).$$

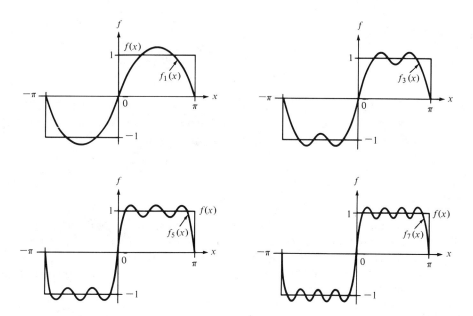

FIGURE 6–18 Fourier Series Approximation of the Square-Wave Function

*6.4.3   Discrete and Fast Fourier Transforms

There is a certain Fourier series representation that has attracted attention in recent years because it is relatively inexpensive to compute. Assume that we wish to approximate a function $f(x)$ over the unit interval $[0, 1]$. The representation now discussed is commonly known as the *discrete Fourier transform*. It has as a basis complex-valued orthogonal functions of the form

$$\varphi_k(x) = \exp(i2\pi kx) = \cos(2\pi kx) + i\sin(2\pi kx), \tag{6.54}$$

$$(k = 0, 1, -1, 2, -2, \ldots).$$

Here, as before, i is $\sqrt{-1}$. These orthogonal functions are closely allied to the trigonometric polynomials in the preceding section. Specifically, for any integer k,

$$\cos(2\pi kx) = \frac{1}{2}[\varphi_k(x) + \varphi_{-k}(x)]$$

and $\hspace{10cm}$ (6.55)

$$\sin(2\pi kx) = \frac{1}{2i}[\varphi_k(x) - \varphi_{-k}(x)].$$

One could thereby avoid complex functions by using the trigonometric functions $\sin(2\pi kx)$ and $\cos(2\pi kx)$ in this section, but this would lead to more complicated notation.

The inner product by which orthogonality and the Fourier coefficients are defined in the discrete Fourier transform is a generalization of (6.50) to the complex case. In this section, N being a given positive integer, the points x_j are determined by

$$x_j = \frac{j}{N}, \qquad 0 \le j \le N - 1. \tag{6.56}$$

That is, the x_j's comprise a uniformly spaced grid on the unit interval. With respect to N and these points x_j, the discrete Fourier transform of two complex-valued functions $f(x)$ and $g(x)$ has the discrete inner product

$$(f, g) = \sum_{j=0}^{N-1} f(x_j)\bar{g}(x_j), \tag{6.57}$$

where the overbar denotes the complex conjugate of $g(x_j)$.

It is important to note that the $\varphi_k(x)$'s in (6.54) are orthogonal with respect to the inner product (6.57). Toward showing this assertion, first recall that for any complex number r other than $1 + i \cdot 0$, the sum of the initial part of a geometric series satisfies the relation

$$\sum_{k=0}^{N-1} r^k = \frac{1 - r^N}{1 - r}. \tag{6.58}$$

From this we see that for $0 \leq k, m \leq N - 1, k \neq m$,

$$(\varphi_k, \varphi_m) = \sum_{j=0}^{N-1} \exp{(i2\pi k x_j)} \exp{(-i2\pi m x_j)} = \sum_{j=0}^{N-1} \exp{(i2\pi(k - m)/N)^j}.$$

If we select $r = \exp{(i2\pi(k - m)/N)}$, then by using (6.58) we get that

$$(\varphi_k, \varphi_m) = \frac{1 - \exp{(i2\pi(k - m))}}{1 - \exp{(i2\pi(k - m)/N)}}. \tag{6.59}$$

But from (6.54), the numerator is readily seen to be 0. Also, it is a simple matter to confirm that for every k, $(\varphi_k, \varphi_k) = N$.

We now turn attention to discrete Fourier transforms as approximators and interpolators for a function $f(x)$ defined on the unit interval. Let S be a finite set of integers. In view of (6.52) and the orthogonality of the φ_k's just established, the trigonometric polynomial

$$\boxed{p(x) = \sum_{k \in S} \beta_k \exp{\left(\frac{i2\pi k x}{N}\right)}} \tag{6.60}$$

is a Fourier series for $f(x)$, with respect to φ_k, $k \in S$, if

$$\beta_k = \frac{(f, \varphi_k)}{(\varphi_k, \varphi_k)} = \frac{(f, \varphi_k)}{N}, \tag{6.61}$$

or, what is the same thing,

$$\boxed{\beta_k = \frac{1}{N} \sum_{j=0}^{N-1} f\left(\frac{j}{N}\right) \exp{\left(\frac{-i2\pi k j}{N}\right)}, \qquad k \in S.} \tag{6.62}$$

Incidentally, the nomenclature "trigonometric polynomial" is more transparent here than in the preceding section. For if $S = \{0, 1, \ldots, N - 1\}$, then in view of (6.60),

$$p(x) = \sum_{k=0}^{N-1} \beta_k \left[\exp{\left(\frac{i2\pi x}{N}\right)}\right]^k = \tilde{p}\left(\exp{\left(\frac{i2\pi x}{N}\right)}\right). \tag{6.63}$$

where $\tilde{p}(z) = \sum_{k=0}^{N-1} \beta_k z^k$, and in (6.63) the variable

$$z = \exp{\left(\frac{i2\pi x}{N}\right)} = \cos{\left(\frac{2\pi x}{N}\right)} + i \sin{\left(\frac{2\pi x}{N}\right)}.$$

In summary, the discrete Fourier transformation of the data $f(j/N)$, $0 \leq j \leq N - 1$, is the trigonometric polynomial $p(x)$ defined by (6.60), where the coefficients are calculated according to (6.62).

Some elementary observations about the discrete Fourier transform include:

1. If S has at least N integers, $p(x)$ interpolates $f(x)$ at the points $x_j = j/N$, $0 \leq j \leq N - 1$. [Thus there is never any reason for using more basis functions $\varphi_j(x)$ than data points.]

2. If S has fewer than N numbers, $p(x)$ gives that approximation of $f(x)$ which minimizes the least-squares criterion,

$$Q(f, p) = \sum_{k=0}^{N-1} |f(k/N) - p(k/N)|^2 = (f - p, f - p)$$

over all possible complex linear combinations of $\varphi_j(x)$, $j \in S$.

3. If $f(x)$ is real and even [i.e., $f(x) = f(-x)$, all x], then $\beta_k = \beta_{-k}$ for all $k \geq 0$. If $f(x)$ is real and odd [i.e., $f(x) = -f(-x)$], then $\beta_k = -\beta_{-k}$.

4. The discrete Fourier transform can be regarded as a Riemann sum approximation of the continuous Fourier transformation discussed in Example 6.12. The point is that the discrete Fourier transform is directly computer imple-

TABLE 6.13   Subroutine SFT for Discrete Fourier Transform

```
      SUBROUTINE SFT(N,A,B,F,X,P)
C
C     ****************************************************************
C     *   FUNCTION: THIS SUBROUTINE COMPUTES AND EVALUATES          *
C     *             DISCRETE FOURIER TRANSFORMS                     *
C     *   USAGE:                                                    *
C     *       CALL SEQUENCE: CALL SFT(N,A,B,F,X,P)                  *
C     *   PARAMETERS:                                               *
C     *       INPUT:                                                *
C     *           N=NUMBER OF DATA POINTS                           *
C     *           A=LEFT ENDPOINT OF THE INTERVAL OF INTERPOLATION  *
C     *           B=RIGHT ENDPOINT OF THE INTERVAL OF INTERPOLATION *
C     *           F=N ARRAY OF FUNCTIONAL VALUES                    *
C     *           X=THE REAL ARGUEMENTAT WHICH THE TRIGNOMETRIC     *
C     *             POLYNOMIAL IS TO BE EVALUATED                   *
C     *       OUTPUT:                                               *
C     *           P=THE TRIGNOMETRIC POLYNOMIAL VALUE AS A FUNCTION *
C     *             OF X                                            *
C     ****************************************************************
C
      IMPLICIT COMPLEX(A-H,O-Z)
      REAL A,B,T,PI,X,FLAG
      DIMENSION F(N),BETA(100)
C     *** INITIALIZATION ***
      PI=4.*ATAN(1.)
      N2=N/2
      EYE=(0.,1.0)
      CONS=2.*PI*EYE/(B-A)
C     *** COMPUTE FOURIER COEFFICIENT BETA***
      DO 10 J=1,N
        BETA(J)=(0.,0.)
        CONS2=(B-A)/FLOAT(N)
        DO 10 K=1,N
          X1=A+(K-1)*CONS2
          BETA(J)=BETA(J)+F(K)*CEXP(-CONS*(-N2+J-1)*X1)
   10 CONTINUE
C     *** EVALUATE THE TRIGNOMETRIC POLYNOMIAL AT X***
      P=(0.,0.)
      DO 20 J=1,N
        P=P+BETA(J)*CEXP(CONS*X*(-N2+J-1))
   20 CONTINUE
      P=P/FLOAT(N)
      RETURN
      END
```

mentable, whereas we still face methodological problems and truncation error in performing the integration required by the continuous Fourier transformation. As the number N increases, the discrete Fourier transformation should in theory converge to the continuous Fourier transformation if $f(x)$ is continuous and roundoff effects are negligible.

In Table 6.13 we offer subroutine SFT, which constructs the discrete Fourier transform associated with a selected odd number N and an interval $[A, B]$ of approximation. The subroutine returns the value $p(x)$ as the calling parameter P, where x is the argument specified in the calling parameter X. The acronym SFT stands for "slow Fourier transform," in contrast to a faster discrete Fourier transformation algorithm to be described after the next example.

EXAMPLE 6.13

Here we find the discrete Fourier transform of the square-wave function. In Example 6.12 we have already found and plotted its continuous Fourier transformation. In the present example, we took N, the number of points to be interpolated, to be 11. The calling program for our calculation is listed in Table 6.14.

TABLE 6.14 Program for Discrete Fourier Approximation Example

```
C       PROGRAM DISFOUR
C
C       *********************************************************************
C       THIS PROGRAM COMPUTES THE DISCRETE FOURIER TRANSFORM OF A SQUARE
C       WAVE FUNCTION DEFINED ON THE INTERVAL [0,10]
C       CALLS:    SFT
C       OUTPUT:
C                 X=VALUE OF X AT CURRENT ITERATION
C                 P=VALUE OF POLYNOMIAL APPROXIMATION AT X
C       *********************************************************************
C
        COMPLEX F(11),P
        N=11
C
C       *** FIRST COMPUTE THE FUNCTIONAL VALUES TO DEFINE THE        ***
C       *** SQUARE WAVE FUNCTION                                     ***
C
        F(1)=0.
        DO 10 J=2,6
           F(J)=1.0
           F(5+J)=-1.0
     10 CONTINUE
C
C       *** NEXT CALCULATE THE POLYNOMIAL APPROXIMATION P(X) FOR     ***
C       *** VALUES OF X RANGING FROM 0 TO 2                          ***
C
        A=0.
        B=10.
        X=0.
        DO 20 J=1,100
C
C       *** SUBROUTINE SFT WILL COMPUTE THE DISCRETE FOURIER         ***
C       *** TRANSFORM                                               ***
C
           CALL SFT(N,A,B,F,X,P)
           X=X+10./101.
           WRITE(10,30) X,REAL(P)
     30    FORMAT(5X,F20.12,5X,F20.12)
     20 CONTINUE
        STOP
        END
```

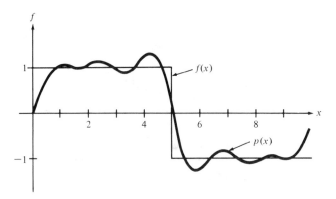

FIGURE 6-19 Discrete 11-Point Fourier Approximation to the Square Wave

Figure 6.19 is a plot of the output values. This should be compared with the plot for $m = 5$ in Figure 6.18. These two graphs depend on exactly the same basis functions; the only difference is that the former uses the continuous inner product (6.51), and the latter, the discrete approximation (6.57) of that inner product. In our actual computation, we put a flag in the subroutine SFT so that the coefficients BETA(J) were computed only once. ■

For some while it was thought that determination of the coefficients of the N-point discrete Fourier transform required in the order of N^2 operations: N multiplications and $(N - 1)$ additions for each value of β_j, and repeated for each of the j values. It was therefore considered a major breakthrough when in the mid-1960s, Cooley, Tukey, Sande, and others found a trick by which operations grow as $N \log (N)$, as N increases. The schemes that compute the discrete Fourier transformation with this efficiency are commonly referred to as *fast Fourier transforms*. Since for large N, $\log (N) \ll N$, fast Fourier transforms have much computational appeal. The reader should bear firmly in mind, however, that the fast Fourier transform is a fast way for getting the discrete Fourier transformation; it is not a different transformation and in fact, a fast Fourier transform should end up giving exactly the same answers as our subroutine SFT; it just gets them quicker.

We describe in detail the Cooley–Tukey fast Fourier transformation. Other methods have similar motivation and depend in some manner on factorization of N. Here we assume that N is a power of 2; that is, for some integer n, $N = 2^n$. Our interpolation index set will be $S = \{0, 1, \ldots, N - 1\}$, and the interpolation points are $x_j = j/N$, $0 \leq j \leq N - 1$.

By formulas given next, one can calculate β_j, $-N/2 \leq j < 0$, in terms of β_j, $j > N/2$, and thereby use the more conventional index set $\{-(N/2 - 1), \ldots, N/2\}$. For observe that in view of (6.54), $\exp (i2\pi k) = 1$, and

$$1 \cdot \exp \left[\frac{i2\pi k(-j)}{N} \right] = \exp \left(\frac{i2\pi kN}{N} \right) \exp \left[\frac{i2\pi k(-j)}{N} \right] = \exp \left[\frac{i2\pi k(N - j)}{N} \right].$$

Consequently,

$$
\begin{aligned}
\beta_{-j} &= \sum_{k=0}^{N-1} f\left(\frac{k}{N}\right) \exp\left[\frac{i2\pi k(-j)}{N}\right] \\
&= \sum_{k=0}^{N-1} f\left(\frac{k}{N}\right) \exp\left[\frac{i2\pi k(N-j)}{N}\right] = \beta_{N-j}, \qquad 0 < j < N.
\end{aligned}
\tag{6.64}
$$

The Cooley–Tukey form of the fast Fourier transform hinges on noticing that if somehow we have constructed polynomials $q(x)$ and $r(x)$ which for some even positive integer M satisfy

$$
\begin{aligned}
q\left(\frac{2j}{M}\right) &= f\left(\frac{2j}{M}\right) \\
r\left(\frac{2j}{M}\right) &= f\left(\frac{2j+1}{M}\right), \qquad 0 \le j \le M/2,
\end{aligned}
\tag{6.65}
$$

then evidently

$$
p(x) = \frac{1 + \exp(i\pi x M)}{2} q(x) + \frac{1 - \exp(i\pi x M)}{2} r\left(x - \frac{1}{M}\right) \tag{6.66}
$$

interpolates $f(j/M)$, $0 \le j \le M - 1$. This is because $\exp(i\pi(j/M)M)$ is $+1$ or -1 according to whether j is even or odd. Armed with this formula, we begin by constructing N constant polynomials,

$$
p_j^{(0)}(x) = f\left(\frac{j}{N}\right), \qquad (0 \le j < N) \tag{6.67}
$$

and then apply (6.66) step by step, with increasing index m. Define n so that $2^n = N$, and let

$$
R = 2^{n-m} \qquad \text{and} \qquad M = 2^m. \tag{6.68}
$$

Assume inductively that for some nonnegative integer m, we have R polynomials $p_j^{(m)}(x)$, which for each j, $0 \le j \le R$, satisfy

$$
p_j^{(m)}\left(\frac{k}{M}\right) = f\left(\frac{j}{N} + \frac{k}{M}\right), \qquad 0 \le k < M. \tag{6.69}
$$

Clearly, this relation holds for the constant polynomials $p_j^{(0)}(x)$ in (6.67), since $m = 0$ and $M = 1$, and $R = 2^n = N$, and $k = 0$. For m, R, M, and $p_j^{(m)}(x)$, $0 \le j < R$ as in (6.68) and (6.69), define a successor set of polynomials by

$$
\begin{aligned}
p_j^{(m+1)}(x) = {}& \frac{1 + \exp(i2\pi x M)}{2} p_j^{(m)}(x) \\
& + \frac{1 - \exp(i2\pi x M)}{2} p_{j+(R/2)}^{(m)}\left(x - \frac{1}{2M}\right).
\end{aligned}
\tag{6.70}
$$

These new polynomials satisfy relations (6.69) with m replaced by $m + 1$, and therefore with M replaced by $2M$. For if k is even, then set $k' = k/2$ and from (6.69) and (6.70), calculate that

$$
p_j^{(m+1)}\left(\frac{2k'}{2M}\right) = p_j^{(m)}\left(\frac{2k'}{2M}\right)
$$

$$
= p_j^{(m)}\left(\frac{k'}{M}\right) = f\left(\frac{j}{N} + \frac{k'}{M}\right) = f\left(\frac{j}{N} + \frac{2k'}{2M}\right).
$$

Alternatively, if k is odd, we define $k' = (k - 1)/2$. Then

$$
p_j^{(m+1)}\left(\frac{2k' + 1}{2M}\right) = p_{j+(R/2)}^{(m)}\left(\frac{2k'}{2M}\right)
$$

$$
= p_{j+(R/2)}^{(m)}\left(\frac{k'}{M}\right) = f\left(\frac{j+(R/2)}{N} + \frac{k'}{M}\right)
$$

$$
= f\left(\frac{j}{N} + \frac{R}{2N} + \frac{2k'}{2M}\right) = f\left(\frac{j}{N} + \frac{2k' + 1}{2M}\right),
$$

where we have used $R/N = 1/M$. Thus (6.69) holds for $m + 1$ and by induction for all m, $0 \le m \le n$. The recursion terminates with $p_0^{(n)}(x)$, which, in view of (6.69) and using the fact that here, $M = N$, satisfies the complete interpolation requirement

$$
p_0^{(n)}\left(\frac{k}{N}\right) = f\left(\frac{k}{N}\right), \qquad 0 \le k < N.
$$

The construct can be described entirely in terms of the Fourier coefficients $\beta_{jk}^{(m)}$ of $p_j^{(m)}(x)$. From (6.70) we can conclude that

$$
\boxed{
\begin{aligned}
2\beta_{jk}^{(m+1)} &= \beta_{jk}^{(m)} + \beta_{j+(R/2),k}^{(m)} \exp\left(-i\pi\left(\frac{j}{2M}\right)\right) \\[2mm]
2\beta_{j,M+k}^{(m+1)} &= \beta_{jk}^{(m)} - \beta_{j+(R/2),k}^{(m)} \exp\left(-i\pi\left(\frac{j}{2M}\right)\right), \qquad (0 \le k < M)
\end{aligned}
}
\tag{6.71}
$$

The process begins by observing that the only coefficient of each polynomial $p_j^{(0)}(x)$ is

$$
\boxed{\beta_{j,0}^{(0)} = f\left(\frac{j}{N}\right).}
$$

Let us assess the computational effort of the fast Fourier transform. At each stage m there are $R = 2^{n-m}$ polynomials $p_j^{(m)}(x)$, each having 2^m Fourier coefficients $\beta_{jk}^{(m)}$. For each such coefficient one addition and one multiplication are required in (6.71). Therefore, at stage m, there are $3 \cdot 2^n$ operations, and there are n stages. Thus the grand total is $3n \cdot 2^n = 3nN$ operations. But $n = \log_2(N)$, so this is how we conclude that with the fast Fourier transform, the processing effort grows proportionally to $N \log (N)$.

EXAMPLE 6.14

Suppose that $N = 4 = 2^2$, so that $n = 2$. We wish to interpolate a function $f(x)$ such that $f(j/4) = j$, $0 \le j \le 3$. The construction begins by defining $m = 0$, and $\beta_{j,0}^{(0)} = j$, $0 \le j < R = 2^{2-0} = 4$. Now apply (6.71),

$$\beta_{0,0}^{(1)} = \tfrac{1}{2}(\beta_{0,0}^{(0)} + \beta_{2,0}^{(0)}) = 1$$

$$\beta_{0,1}^{(1)} = \tfrac{1}{2}(\beta_{0,0}^{(0)} - \beta_{2,0}^{(0)}) = -1$$

$$\beta_{1,0}^{(1)} = \tfrac{1}{2}(\beta_{1,0}^{(0)} + \beta_{3,0}^{(0)}) = 2$$

$$\beta_{1,1}^{(1)} = \tfrac{1}{2}(\beta_{1,0}^{(0)} - \beta_{3,0}^{(0)}) = -1.$$

Finally, for the $m = 2$ terms, after updating, $R = 1$, $M = 2$, and

$$\beta_{0,0}^{(2)} = \frac{1}{2}(\beta_{0,0}^{(1)} + \beta_{1,0}^{(1)}) = \frac{3}{2}$$

$$\beta_{0,1}^{(2)} = \frac{1}{2}\left(\beta_{0,1}^{(1)} + \beta_{1,1}^{(1)} \exp\left(-\frac{2\pi i}{4}\right)\right) = \frac{1}{2}\left[-1 - \exp\left(-\frac{2\pi i}{4}\right)\right]$$

$$\beta_{0,2}^{(2)} = \frac{1}{2}(\beta_{0,0}^{(1)} - \beta_{1,0}^{(1)}) = -\frac{1}{2}$$

and

$$\beta_{0,3}^{(2)} = \frac{1}{2}\left[\beta_{0,1}^{(1)} - \beta_{1,1}^{(1)} \exp\left(-\frac{2\pi i}{4}\right)\right] = \frac{1}{2}\left[-1 + \exp\left(-\frac{2\pi i}{4}\right)\right].$$

As one may verify, trigonometric polynomial with these coefficients does interpolate $f(x)$ at 0, $\tfrac{1}{4}$, $\tfrac{1}{2}$, and $\tfrac{3}{4}$.

■

We refer the reader to Bloomfield (1976, pp. 75–76) for FORTRAN code for a fast Fourier transform. Also this reference has much supplementary information and contains studies of specific data sets. Stoer and Bulirsch (1980, Sec. 2.3) describe several other such transforms. The IMSL routine FFTCC also computes fast Fourier transforms.

Unless the reader is wishing to interpolate more than 100 points, or is designing an algorithm that will be used repeatedly, he or she will probably find the SFT subroutine (Table 6.13) satisfactory for discrete Fourier series interpolation.

6.5
SUPPLEMENTARY NOTES AND DISCUSSIONS

In this chapter we resumed a path initiated in Chapter 2, namely, the quest for computer-amenable representations of functions that are given only at a finite set of data. In that chapter there were some loose ends and here we attended to some of these. In particular, if there are a great many data points, polynomial interpolation is not suitable since under the best of circumstances, it tends to be unstable for high degree (say 20). Spline interpolation does not suffer this defect, but even after the spline function coefficients have been computed for a given data set, spline evaluation requires more programming, memory and running time than polynomials. Therefore, splines are not suitable for use as library functions, for example. Moreover, interpolation is unsatisfactory when the data are irregular, indicating a random or highly oscillatory character in the data.

The methods of least squares and uniform approximation, which are the central topics of this chapter, do overcome these difficulties to a large extent. However, they too can exhibit drawbacks and numerical instabilities. We saw that in its most simplistic form, least-squares approximation tends to result in ill-conditioned linear equation problems, and consequent numerical instability, when the degree of the approximating polynomial is moderate to large. We drew connections between the least-squares approximation problem and the general linear model in statistics. (Statisticians, too, are aware of instabilities associated with the least-squares approach, this instability being referred to by them as "collinearity.") The orthogonal polynomial approach for overcoming this instability was described and implemented in our routine ORTH. This approach is taken by the IMSL least-squares approximation routine RLFOTH.

Uniform approximation displays many of the advantages of the least-squares approach, but its implementation requires a program for solving linear programming problems. Such programs are available in many computer centers. A linear programming code is given in Appendix C.

The orthogonal polynomial scheme for achieving stability in least-squares approximation was viewed as a particular example of the Fourier approach. Fourier methods play an important role in communication theory and time-series analysis, and two-dimensional versions are central to many of the techniques in the fast-growing discipline of digital image analysis and enhancement. By presentation of Fourier trigonometric polynomials and a version of fast Fourier transforms, we hope to provide the reader with a foundation for understanding recent computational developments in these areas.

PROBLEMS

Section 6.2

1. For the following four points, find the constant and linear least-squares polynomials.

x_i	-2	-1	0	1
f_i	6	4	3	3

Show your calculations.

2. Consider the data domain points $x_i = (i - 1)h$ $(i = 1, 2, \ldots, N)$, and the functional values $f_i = \sin(x_i)$, where N is a given integer and $h = \pi/(N - 1)$. Compute the constant, linear and quadratic, and fifth-degree least-squares polynomials based on these data and check the accuracy at the test points $t_k = k\pi/100$ $(k = 0, 1, \ldots, 100)$. Repeat the computations for $N = 7, 10, 20, 30$.

3. Perform the following least-squares fits (with respect to parameters a and b). The data pairs designate (x_i, f_i).
(a) $y = a + bx$, points $(0, 1)$, $(1, 2)$, $(2, 4)$.
(b) $y = a \sin(x) + b \cos(x)$, points $(0, 0)$, $(\pi/4, 1)$, $(\pi/2, 0)$.
(c) $y = a + be^x$, points $(0, 1)$, $(1, 2)$, $(2, 2)$.
(d) $y = ae^x + be^{-x}$, points $(0, 0)$, $(1, 1)$, $(2, 1)$, $(3, 0)$.
[**HINT FOR PARTS (b), (c), AND (d):** Form normal equations just as in the polynomial case, but in terms of values $\sin(x_i)$, and so on. You should still end up with a linear equation.]

4. By increasing the degree m, attempt to approximate $f(x) = \sin(x)$, $\exp(x)$, and $\sqrt{x}$ as accurately as possible. Use data points (x_i, f_i), with $x_i = i/100$, $1 \leq i \leq 100$, and $f_i = f(x_i)$. Use subroutine LSQM. Find the maximum error (against library functions) at points $t_j = (j + 0.5)/100$, $0 \leq j \leq 99$.

5. Compare the Hilbert segment matrix of order n and the matrix S given by (6.21) for $n = 5, 10, 15, 20, 50$. Print out the differences, $h_{ij} - 1/n \, s_{ij}$, of the matrix coefficients, of $H - 1/n \, S$.

*6. Consider the data abscissa points $x_i = -1 + i/n$ $(i = 0, 1, \ldots, 2n)$ and the corresponding functional values $f_i = (x_i + |x_i|)/2$. Find the least-squares polynomial of degree zero. Determine its limit if $n \to \infty$.

7. Find the quadratic least-squares polynomial of the function $f(x) = x^3$ with respect to inner product (6.38). Take $X = [0, 1]$. [**HINT:** Minimize the function

$$Q(a, b, c) = \int_0^1 [x^3 - a - bx - cx^2]^2 \, dx.$$

Section 6.3

*8. Solve Problem 1 by uniform approximation. Construct the corresponding linear programming problem. Show your work.

9. Find the best (with respect to degree) best approximating polynomials for the tasks in Problem 4. But take $x_i = i/15$, $1 \leq i \leq 15$.

*10. Prove that the optimal solution of the problem in (6.31) is

$$a_0 = \frac{\max_i (f_i) + \min_i (f_i)}{2}$$

and

$$U(f, p) = \frac{\max_i (f_i) - \min_i (f_i)}{2}.$$

[**HINT:** Demonstrate and then use the fact that the constraints are equivalent to the inequalities

$$a_0 - U(f, p) \leq \min_i (f_i)$$

$$a_0 + U(f, p) \geq \max_i (f_i).]$$

*11.   Solve Problem 6 by uniform approximation. Compare the result by that obtained for the least-squares method. Explain the difference.

*12.   Solve Problem 2 for $N = 7, 10, 20$ by uniform approximation and compare the results by that of the least-squares method. (**HINT:** Use double precision.)

Section 6.4

*13.   Find the linear and quadratic orthogonal polynomials $\varphi_1(x)$ and $\varphi_2(x)$ if
        (a) $n = 3$, and $x_j = j$, $1 \leq j \leq 3$, $w(x) = 1$.
        (b) $T = [1, 0]$, the unit interval, and $w(x) = 1$.

*14.   Redo Problem 4 using subroutine ORTH.

*15.   Verify (6.49). [**HINT:** Combine relations (6.43) and (6.48).]

*16.   Verify that the functions 1, cos (x), cos $(2x)$, . . . , sin (x), sin $(2x)$, . . . , are orthogonal for the inner product (6.38) with $T = [-\pi, \pi]$.

*17.   Construct the first two members, φ_0, φ_1, of the orthogonal polynomial system based on the points $x_1 = -2$, $x_2 = -1$, $x_3 = 0$, and $x_4 = 1$. [**HINT:** Use relations (6.40)–(6.42).]

*18.   Construct the least-squares polynomial given in Example 6.1 by using orthogonal polynomials. (**HINT:** Use the result of Problem 17.)

*19.   Find the trigonometric series for the saw-tooth function

$$f(x) = x - [x],$$

        where $[x]$ means the "integral part of x." For this, take as the inner product (6.38), with X the unit interval. Use 1, $\sin(2\pi kx)$, $\cos(2\pi kx)$, $k = 1, 2, . . .$, as the orthogonal function sequence.

*20.   Find the discrete Fourier transformation for

$$f(x) = \exp(x), \qquad 0 \leq x \leq 1.$$

        Take various values of N, and find the approximation error at

$$y_i = \frac{i}{100}, \qquad 1 \leq i \leq 100.$$

CHAPTER 7

The Solution of Ordinary Differential Equations

7.1
INTRODUCTION

Many models of physics, engineering, economics, and essentially every other quantitative science fall within the category of differential equations. That is, the evolutionary behavior of certain quantities satisfies some equation that depends not only on the quantities themselves, but also on their derivatives. For example, the law governing the velocity $v(t)$ of an unforced sliding mass subject only to viscous frictional effects and Newton's second law of motion is postulated to be

$$m \frac{d}{dt} v(t) = -Dv(t), \tag{7.1}$$

m being the mass of the object and D the damping coefficient. By substituting one may verify that the function

$$v(t) = C \exp\left(-\frac{D}{m} t\right) \tag{7.2}$$

satisfies (7.1), since

$$m \frac{d}{dt} v(t) = -mC \frac{D}{m} \exp\left(-\frac{D}{m} t\right) = -Dv(t).$$

If the velocity at time $t = 0$ is known to be v_0, we must have that $C = v_0$ in order for (7.2) to be numerically consistent with this initial information. In this context, v_0 is an initial condition, and the parameter C in the solution (7.2) of the differential equation (7.1) allows us a means of incorporating this initial condition.

If all derivatives are taken with respect to the same variable, the differential equation is said to be *ordinary*. Thus (7.1) is ordinary, the independent variable being t. Partial differential equations, the other category, are not treated in this book.

In this chapter we offer numerical methods for solving ordinary differential equations. We study two central problems in the computational theory for ordinary differential equations: initial- and boundary-value problems.

Initial-value problems are differential equation problems in which information about the solution is given at a single time point. This information determines the solution parameters. The damping problem above is an example of an initial-value problem inasmuch as the "initial value" $v(0) = v_0$ is sufficient to determine the only solution parameter, C, in (7.2).

Boundary-value problems, by contrast, provide solution information at two or more time points. For example, if instead of specifying above that $v(0) = v_0$ we had known that $v(0) + v(1) = 1.0$, then we would have had an example of a boundary-value problem. In this case we could have solved for C by noting that

$$v(0) + v(1) = C + C \exp\left(-\frac{D}{m}\right) = 1.0.$$

The major thrust of our studies is directed at initial-value problems. In Section 7.6 we examine ways to recast boundary-value problems as initial-value problems.

7.2
THE MOST ELEMENTARY METHODS
AND THEIR LIMITATIONS

The *initial value problem* of an ordinary first-order differential equation has the form

$$y'(x) = f(x, y(x)), \qquad y(x_0) = y_0, \tag{7.3}$$

where x is a scalar variable, $y(x)$ and $f(x, y)$, are real-valued functions, and y_0 is a given real number. The reason this equation is called *first* order is that only the unknown function and its *first* derivative are included in the equation. In (7.3) and in what follows, $y'(x)$ represents $dy(x)/dx$. The second derivative is denoted by $y''(x)$, and in general the kth derivative, by $y^{(k)}(x)$. The solution of equation (7.3) will be determined on a finite interval $[x_0, b]$ starting with the initial point x_0.

It is the tradition of differential equation literature to freely write f in place of $f(x, y)$, y for $y(x)$, $y^{(k)}$ for $y^{(k)}(x)$, and so on. We will also use this simplifying notation.

7.2.1   The Taylor's Series Method

Taylor's series, introduced in Chapter 2, lead to methods for solving (7.3). The motivation for Taylor's series as a tool for solving ordinary differential equations is that the derivatives of the solution function are sometimes easily found from the differential equation itself. We begin by applying the Taylor's series procedure in a particularly simple setting.

——— **EXAMPLE 7.1** ———

Consider the differential equation

$$y' = x^2 + y^2, \qquad y(0) = 0.$$

By substituting $x = 0$ we get

$$y'(0) = 0^2 + y(0)^2 = 0^2 + 0^2 = 0.$$

By differentiating both sides of the differential equation in accordance with the chain rule (Appendix B, item 36), we find the relation

$$y'' = 2x + 2yy'.$$

So

$$y''(0) = 2 \cdot 0 + 2 \cdot y(0) \cdot y'(0) = 0.$$

After further differentiation, we find that

$$y^{(3)} = 2 + 2y'y' + 2yy'',$$

which implies that

$$y^{(3)}(0) = 2 + 2 \cdot 0 \cdot 0 + 2 \cdot 0 \cdot 0 = 2.$$

Continuing this process, we obtain the relations

$$\begin{aligned}
y^{(4)} &= 2y''y' + 2y'y'' + 2y'y'' + 2yy^{(3)} \\
&= 6y'y'' + 2yy^{(3)} \\
y^{(5)} &= 6y''y'' + 6y'y^{(3)} + 2y'y^{(3)} + 2yy^{(4)} \\
&= 6(y'')^2 + 8y'y^{(3)} + 2yy^{(4)},
\end{aligned}$$

which imply

$$\begin{aligned}
y^{(4)}(0) &= 6 \cdot 0 \cdot 0 + 2 \cdot 0 \cdot 2 = 0 \\
y^{(5)}(0) &= 6 \cdot 0^2 + 8 \cdot 0 \cdot 2 + 2 \cdot 0 \cdot 0 = 0.
\end{aligned}$$

With these derivatives in hand, in view of (2.6) with "y" replacing "f," we see that the fifth-degree Taylor's polynomial approximation of the solution function is given by

$$y(x) \approx \frac{2x^3}{3!} = \frac{x^3}{3}.$$

At $x = 0.5$ and $x = 1.0$, this approximation gives, respectively, 0.04166 and 0.3333, whereas to the accuracy shown, the correct answers are $y(0.5) = 0.04179$ and $y(1.0) = 0.3502$.

■

The methodology illustrated in the example above can be extended. In the following discussions we assume that all derivatives mentioned do exist. The derivatives of the solution can be determined recursively by the use of the chain rule of elementary calculus (Appendix B, item 36). In the representations below, subscripts indicate that partial derivatives have been taken with respect to the displayed variables. For instance,

$$f_x(x, y) = \frac{\partial}{\partial x} f(x, y), \qquad f_y(x, y) = \frac{\partial}{\partial y} f(x, y), \qquad f_{xx}(x, y) = \frac{\partial^2}{\partial x^2} f(x, y),$$

and so on. Then from (7.3), $y'(x) = f(x, y(x))$ and by successive differentiation, we recursively construct the relations

$$
\begin{aligned}
y'(x) &= f(x, y(x)) \\
y''(x) &= f_x(x, y(x)) + f_y(x, y(x))y'(x) \\
y^{(3)}(x) &= f_{xx}(x, y(x)) + 2f_{xy}(x, y(x))y'(x) + f_{yy}(x, y(x))y'(x)^2 \\
&\qquad\qquad\qquad\qquad\qquad + f_y(x, y(x))y''(x)
\end{aligned}
\tag{7.4}
$$

$$\vdots \qquad\qquad \vdots$$

Using the expressions above, evaluated at $x = x_0$ and $y = y_0$, the approximation of $y(x)$ can be computed from the Taylor's polynomial

$$y(x) \approx y(x_0) + \frac{y'(x_0)}{1!} (x - x_0) + \cdots + \frac{y^{(m)}(x_0)}{m!} (x - x_0)^m. \tag{7.5}$$

The error of this approximation is $O((x - x_0)^{m+1})$, in the notation of Section 3.1.

If $|x - x_0|$ is large, the error of the approximation (7.5) can also become large, and consequently the accuracy of this method in an interval $[x_0, b]$ of moderate length is often found to be unsatisfactory. Accuracy can be improved through a modification in which Taylor's expansions are successively computed along a grid $x_0, x_1, x_2, \ldots$ of increasing x values. By this device, $|x - x_j|$ need never be large. Let step size $h = (b - x_0)/N$, where $N > 1$ is a given positive integer, and $[x_0, b]$ be the solution interval. Define the grid points

$$x_j = x_0 + jh \qquad (j = 0, 1, 2, \ldots, N).$$

Then between grid points x_j and x_{j+1} the solution can be approximated by

$$y(x) \approx y(x_j) + \frac{y'(x_j)}{1!} (x - x_j) + \cdots + \frac{y^{(m)}(x_j)}{m!} (x - x_j)^m.$$

On the basis of this approximation the following algorithm is offered for solving the initial-value problem by "marching" along the x-axis from x_0: For x between

x_0 and x_1, apply this Taylor polynomial with $j = 0$ to approximate $y(x)$, where the derivatives are computed by using (7.4). For $x = x_1$, in particular, we calculate

$$y_1 = y_0 + \frac{y_0'}{1!} h + \cdots + \frac{y_0^{(m)}}{m!} h^m,$$

where $y_0', \ldots, y_0^{(m)}$ denote the derivative values $y'(x_0), \ldots, y^{(m)}(x_0)$. If x is between x_1 and x_2, the solution function is approximated by the Taylor polynomial with $j = 1$. For $j = 1$, use (7.4) with $y = y_1$, as computed above, and $x = x_1$. Then between x_1 and x_2 approximate the solution $y(x)$ by

$$y(x) \approx y_1 + \frac{y_1'}{1!} (x - x_1) + \cdots + \frac{y_1^{(m)}}{m!} (x - x_1)^m.$$

In particular, at $x = x_2$, the approximation of $y(x_2)$ is given by

$$y_2 = y_1 + \frac{y_1'}{1!} h + \cdots + \frac{y_1^{(m)}}{m!} h^m.$$

If we continue in this fashion, the solution can be extended to the further subintervals $[x_j, x_{j+1}]$, $j = 2, \ldots, N - 1$, by the relation

$$y(x) \approx y_j + \frac{y_j'}{1!} (x - x_j) + \cdots + \frac{y_j^{(m)}}{m!} (x - x_j)^m.$$

Then the approximation of $y(x_{j+1})$ is

$$y_{j+1} = y_j + \frac{y_j'}{1!} h + \cdots + \frac{y_j^{(m)}}{m!} h^m, \tag{7.6}$$

where the derivatives $y_j^{(k)}$ are determined by (7.4) evaluated at $x = x_j$, $y = y_j$. This formula can be used recursively to construct piecewise polynomial approximations of the solution. We thereby create separate Taylor's polynomial for each subinterval $[x_j, x_{j+1}]$. In applications one often seeks a table of grid points $x_0, x_1, x_2, \ldots$ and corresponding solution values $y_0, y_1, y_2, \ldots$. These values can be computed recursively by using (7.6), with y_j obtained in the previous step, and the approximating derivative values $y_j', \ldots, y_j^{(m)}$ are computed according to (7.4), at $x = x_j$ and $y = y_j$.

EXAMPLE 7.2

The equation

$$y' = x^2 + y^2, \qquad y(0) = 0$$

of Example 7.1 was solved by the recursive method (7.6), in the interval $[0, 1]$. We chose $N = 10$, so $x_j = (0.1)j$ ($j = 0, 1, \ldots, 10$). Table 7.1 compares the

TABLE 7.1   Computations of Examples 7.1 and 7.2

x_j	$y(x_j) = \frac{1}{3}x_j^3$ [Method (7.5)]	y_j Obtained by Recursion (7.6)	Exact Value of $y(x_j)$
0.0	0.0	0.0	0.0
0.1	0.000333	0.000333	0.000333
0.2	0.002667	0.002667	0.002667
0.3	0.009000	0.009001	0.009003
0.4	0.213333	0.021346	0.021359
0.5	0.041666	0.041734	0.041791
0.6	0.072000	0.072268	0.072448
0.7	0.114333	0.115182	0.115660
0.8	0.170667	0.172959	0.174080
0.9	0.243000	0.248501	0.250907
1.0	0.333333	0.345392	0.350232

values of the solution obtained by applying the Taylor's series methods (7.4) and (7.6) with $m = 3$ to the above differential equation.

The improvement in accuracy with increasing x obtained by the recursions is obvious in this case.

Since the higher-order derivatives of $y(x)$ can seldom be determined conveniently [because in most cases the recursive equations (7.4) become very complicated], the Taylor's series approach is not presently competitive with the techniques to follow, but it does provide insight into their construction.

7.2.2   Euler's Method

Consider the Taylor series method (7.6) with $m = 1$. This gives the recursive relation

$$y_{j+1} = y_j + hy_j',$$

where the approximating derivative value y_j' is obtained from the first equation of (7.4), with $y(x_j)$ replaced by y_j. That is, this recursion can be rewritten as

$$\boxed{y_{j+1} = y_j + hf(x_j, y_j).} \tag{7.7}$$

Equation (7.7) recursively provides estimates $y_1, y_2, y_3, \ldots$ of $y(x_1), y(x_2), y(x_3), \ldots$, starting from the initial condition $y_0 = y(x_0)$. This particular successive Taylor's series formula is called *Euler's method*. Observe that no differentiation is required by (7.7).

A geometric interpretation of Euler's method is shown in Figure 7.1, in which one step at $j = 0$ is illustrated. Consider the linear function

$$z(x) = y_0 + y'(x_0)(x - x_0) = y_0 + f(x_0, y_0)(x - x_0),$$

which is the tangent line at $x = x_0$ of that solution of the differential equation which passes through the point (x_0, y_0). Then y_1 is the value of $z(x)$ at $x = x_1$.

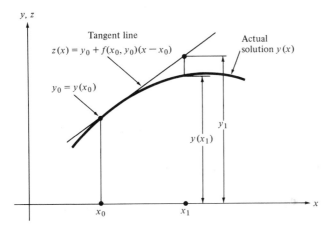

FIGURE 7–1 Euler's Method

For larger values of j, the approximation y_{j+1} is the value at x_{j+1} of the tangent line of that solution curve which passes through the preceding point (x_j, y_j).

Subroutine EULER, given in Table 7.2, allows us to implement Euler's method by simply specifying the domain $[x_0, b]$ on which the solution is desired, the initial value y_0, the number N of Euler's steps, and the function $f(x, y)$. This function may be specified as a function subprogram. The initial values x_0 and y_0 are trans-

TABLE 7.2   Subroutine for Euler's Method

```
          SUBROUTINE EULER(X,Y,N,B)
C
C     *************************************************************
C     *  FUNCTION: THIS SUBROUTINE COMPUTES THE SOLUTION OF THE   *
C     *            FIRST ORDER DIFFERENTIAL EQUATION Y'=F(X,Y)    *
C     *            OVER THE INTERVAL [X(1),B] USING EULER'S METHOD*
C     *  USAGE:                                                   *
C     *      CALL SEQUENCE: CALL EULER(X,Y,N,B)                   *
C     *      EXTERNAL FUNCTIONS/SUBROUTINES: FUNCTION F(U,V)      *
C     *  PARAMETERS:                                              *
C     *      INPUT:                                               *
C     *          N=NUMBER OF STEPS                                *
C     *          X(1)=INITIAL IDEPENDENT VARIABLE VALUE           *
C     *          B=SOLUTION INTERVAL ENDPOINT (LAST X VALUE)      *
C     *          Y(1)=INITIAL DEPENDENT VARIABLE VALUE            *
C     *      OUTPUT:                                              *
C     *          X=N+1 BY 1 ARRAY OF INDEPENDENT VARIABLE VALUES  *
C     *          Y=N+1 BY 1 ARRAY OF DEPENDENT VARIABLE SOLUTION  *
C     *          VALUES                                           *
C     *************************************************************
C
          DIMENSION X(N+1),Y(N+1)
C     *** INITIALIZATION ***
          H=(B-X(1))/N
C     *** RECURSIVELY COMPUTE THE SOLUTION VALUES ***
          DO 1 I=2,N+1
              X(I)=X(I-1)+H
              U=X(I-1)
              V=Y(I-1)
C     *** EULER'S STEP ***
              Y(I)=Y(I-1)+H*F(U,V)
       1 CONTINUE
          RETURN
          END
```

mitted to the subroutine as the numbers $X(1)$ and $Y(1)$ in the arrays X and Y. The calling parameter N and the step size in (7.7) are, of course, related by

$$h = \frac{b - x_0}{N}.$$

Whereas Euler's method is attractive for its simplicity and ease of implementation, when accuracy counts, other methods employing greater sophistication are vastly more efficient. Theoretical and computational evidence supporting this claim is offered in the sections to follow.

EXAMPLE 7.3

We have applied Euler's method to the differential equation

$$y' = x^2 + y^2, \qquad y(0) = 0,$$

of the previous examples.

TABLE 7.3　Illustration of Euler's Method

```
C       PROGRAM EMETH
C
C       ******************************************************************
C       THIS PROGRAM SOLVES THE DIFFERENTIAL EQUATION Y'=X**2+Y**2
C       WITH THE INITIAL CONDITION Y(0.)=0. THE SOLUTION IS OBTAINED
C       AT 11 POINTS OVER THE INTERVAL [0,1] USING EULERS METHOD
C       CALLS:   EULER
C       OUTPUT:
C                X(I)=VALUE OF X FOR I=1,11
C                Y(I)=APPROXIMATED VALUE OF Y AT X(I)
C       ******************************************************************
C
        DIMENSION X(11),Y(11)
C
C       *** INITIALIZE                                              ***
C
        X(1)=0.
        Y(1)=0.
        B=1.
        N=10
        N1=N+1
C
C       *** SUBROUTINE EULER WILL APPROXIMATE THE SOLUTION          ***
C       *** RETURNING N1 VALUES IN ARRAYS X AND Y                   ***
C
        CALL EULER(X,Y,N,B)
        WRITE(10,1)(X(I),Y(I),I=1,N1)
      1 FORMAT(5X,F4.1,5X,F10.6)
        STOP
        END
C
C       *** FUNCTION F SPECIFIES THE DIFFERENTIAL EQUATION. IT IS   ***
C       *** CALLED BY SUBROUTINE EULER                              ***
C
        FUNCTION F(X,Y)
        F=X**2+Y**2
        RETURN
        END
```

TABLE 7.4 Output of Euler's Computation

x_j	y_j	Exact Value, $y(x_j)$
0.0	0.000000	0.000000
0.1	0.000000	0 000333
0.2	0.001000	0.002667
0.3	0.005000	0.009003
0.4	0.014003	0.021359
0.5	0.030022	0.041791
0.6	0.055112	0.072448
0.7	0.091416	0.115660
0.8	0.141252	0.174080
0.9	0.207247	0.250907
1.0	0.292542	0.350232

The initial condition implies that $y_0 = 0$. Then from the recursion (7.7) with $j = 0$ we conclude that

$$y_1 = y_0 + hf(x_0, y_0) = y_0 + 0.1(x_0^2 + y_0^2) = 0,$$

and with $j = 1$ we get the value

$$y_2 = y_1 + hf(x_1, y_1) = y_1 + 0.1(x_1^2 + y_1^2) = 0 + 0.1(0.01 + 0) = 0.001.$$

Continuing this calculation yields the values reported in Table 7.4. A calling program for these computations is given in Table 7.3. The function $f(x, y)$ is specified by means of a function subprogram, and the step size h is taken to be 0.1. That is, $N = 10$ is selected. We remark that when we took $h = 10^{-4}$ (i.e., $N = 10,000$), the computed Euler estimate (in double precision) of $y(1)$ was 0.3501692. This answer is correct to only four significant decimal places. ∎

*7.2.3 Foundations for More Accurate Methods

The linear Taylor's polynomial (7.6) is the basis for Euler's method. The main advantages of Euler's method over the higher-order Taylor's methods of Section 7.2.1 are: (1) no differentiation is required, and (2) its computer implementation is simple. A weakness of Euler's method is that larger truncation error can be anticipated. Because of this weakness, Euler's method is seldom used for serious computational tasks.

It is possible to construct approximation polynomials that coincide, up to a higher order, with the Taylor's polynomial, but which nevertheless do not require derivative information and the complicated formulas (7.4). The basic trick is that functional values at different values of x can be made to "substitute" for derivative information. The two prominent categories of numerical differential equation methods, the Runge–Kutta and the Adams rules, are based on different exploitations of this idea. We illustrate these exploitations now by deriving some simple formulas that are far more accurate than Euler's method. Then in Sections 7.3 and 7.4, the general principles are abstracted.

Consider the Taylor's series method (7.6) with $m = 2$:

$$y_{j+1} = y_j + \frac{y_j'}{1!} h + \frac{y_j''}{2!} h^2. \tag{7.8}$$

On the right-hand side, y_j' can be approximated by

$$y_j' \approx y'(x_j) = f(x_j, y(x_j)) \approx f(x_j, y_j).$$

Let h^* be a small step (not necessarily equal to h), and approximate y_j'' by the numerical differentiation formula (3.2) of Chapter 3 to obtain

$$
\begin{aligned}
y_j'' \approx y''(x_j) &\approx \frac{y'(x_j + h^*) - y'(x_j)}{h^*} \\
&= \frac{f(x_j + h^*, y(x_j + h^*)) - f(x_j, y(x_j))}{h^*} \\
&\approx \frac{f(x_j + h^*, y(x_j) + h^*y'(x_j)) - f(x_j, y(x_j))}{h^*} \\
&\approx \frac{f(x_j + h^*, y_j + h^*f(x_j, y_j)) - f(x_j, y_j)}{h^*}.
\end{aligned}
\tag{7.9}
$$

If we choose $h^* = \lambda h$ (λ being a constant), then by substituting this approximation into (7.8) we get the following relation:

$$y_{j+1} = y_j + h\left[\left(1 - \frac{1}{2\lambda}\right)f(x_j, y_j) + \frac{1}{2\lambda} f(x_j + \lambda h, y_j + \lambda h f(x_j, y_j))\right]. \tag{7.10}$$

In order to simplify this lengthy equation, we introduce the following notation:

$$
\boxed{
\begin{aligned}
k_1 &= f(x_j, y_j) \\
k_2 &= f(x_j + \lambda h, y_j + \lambda h k_1) \\
\alpha_1 &= 1 - \frac{1}{2\lambda}, \qquad \alpha_2 = \frac{1}{2\lambda}.
\end{aligned}
}
\tag{7.11}
$$

Then (7.10) can be rewritten,

$$\boxed{y_{j+1} = y_j + h(\alpha_1 k_1 + \alpha_2 k_2).} \tag{7.12}$$

This formulation is at the heart of Runge–Kutta-type methods. In the section to follow we will see that Heun's method is the special case of (7.11)–(7.12) in which $\lambda = \frac{2}{3}$. When applied to the test problem of Example 7.3, Heun's method calculates $y(1) \approx 0.349640$, whereas the true value is $y(1) = 0.350232$. Euler's method with $h = 0.1$, in Example 7.3, calculated $y(1) \approx 0.292542$. To be fair to Euler's method, it requires only 10 calls to the function $f(x, y)$, whereas Heun's method requires 20. At $h = 0.05$, Euler's method also makes 20 calls, and gives the approximation $y(1) \approx 0.3202117$, which still has about 50 times the error of Heun's method.

Consider again the Taylor's series representation (7.8), and approximate the derivatives y_j' and y_j'' as follows:

$$y_j' \approx y'(x_j) = f(x_j, y(x_j)) \approx f(x_j, y_j),$$

and by using the Chapter 3 derivative approximation (3.2), with h replaced by $-h$, we write

$$y_j'' = \frac{y'(x_j) - y'(x_j - h)}{h} = \frac{y'(x_j) - y'(x_{j-1})}{h}$$

$$= \frac{f(x_j, y(x_j)) - f(x_{j-1}, y(x_{j-1}))}{h}$$

$$\approx \frac{f(x_j, y_j) - f(x_{j-1}, y_{j-1})}{h}.$$

Substituting these estimates into (7.8) gives the recursion

$$y_{j+1} = y_j + h[\tfrac{3}{2} f(x_j, y_j) - \tfrac{1}{2} f(x_{j-1}, y_{j-1})]. \tag{7.13}$$

By introducing the notation

$$f_{j-1} = f(x_{j-1}, y_{j-1}), \qquad f_j = f(x_j, y_j)$$

$$\beta_2 = -\tfrac{1}{2}, \qquad \beta_1 = \tfrac{3}{2},$$

recursion (7.13) can be rewritten as

$$\boxed{y_{j+1} = y_j + h(\beta_1 f_j + \beta_2 f_{j-1}).} \tag{7.14}$$

This idea generalizes to give the Adams methods, the subject of Section 7.4. In fact, (7.14) is the simplest Adams–Bashforth formula. This formula, with step size $h = 0.1$, computes $y(1) = 0.33873$ for the problem in Example 7.3. This is also much closer than the Euler's method approximation.

7.3
RUNGE–KUTTA METHODS

7.3.1 Runge–Kutta Formulas

Recall from Section 7.2.1 that the recursive Taylor's series approach, while seemingly sufficiently accurate, was hampered by the inconvenience of requiring higher-order derivatives of the function $f(x, y)$. The Runge–Kutta method succeeds in approximating the Taylor's polynomial, without taking derivatives. The idea behind the technique was outlined in Section 7.2.3: Various "tentative" steps are taken from the current location $x = x_j$ and solution estimate $y = y_j$, and the function $f(x, y)$ is evaluated at these nearby locations. These estimates are combined in such a fashion that the sum must agree with the Taylor's expansion of the solution, up to a certain power in step size h.

The (explicit) *Runge–Kutta* method for numerical solution of differential equations has us compute the approximating solution values y_1, y_2, . . . on a grid $x_1 = x_0 + h$, $x_2 = x_0 + 2h$, . . . starting from the initial condition $y_0 = y(x_0)$. The successor estimate y_{j+1} is computed recursively from y_j ($j = 0, 1, . . .$) by the formula

$$y_{j+1} = y_j + h\left(\sum_{i=1}^{t} \alpha_i k_i\right), \tag{7.15}$$

where the terms k_i are computed recursively according to

$$k_1 = f(x_j, y_j), \tag{7.16}$$

and for $i = 2, . . ., t$,

$$k_i = f\left(x_j + h\mu_i, y_j + h\left(\sum_{m=1}^{i-1} \lambda_{im} k_m\right)\right). \tag{7.17}$$

In (7.15), (7.16), and (7.17), α_i, μ_i and λ_{im}, $1 \le m \le i - 1$, $1 \le i \le t$, are parameters to be chosen to make the method as accurate as possible. The integer t in (7.15) determines the number of *stages* of the rule. Further motivation of the Runge–Kutta method and detailed analysis of the selection of the parameters for certain cases will be related in the section to follow. For now, let us simply note that the Runge–Kutta formula (7.15) allows us to obtain a Taylor's polynomial approximation of the solution $y(x)$ about x_j, and evaluated at x_{j+1}, without having to find derivatives. At this point, since (7.15) and (7.17) are a bit complicated, we offer some well-known examples of Runge–Kutta formulas and investigate their performance on our standard test problem.

───── **EXAMPLE 7.4** ─────

For one stage rules ($t = 1$), in view of (7.15) and (7.17), the Runge–Kutta method necessarily takes the form

$$y_{j+1} = y_j + h\alpha_1 f(x_j, y_j).$$

It will be seen in the section to follow that for small h, the error is minimized by taking $\alpha_1 = 1$. Therefore, the one-stage Runge–Kutta formula coincides with Euler's method, the performance of which was examined in Example 7.3. ∎

———— EXAMPLE 7.5 ————

The *corrected Euler's formula* is a two-stage ($t = 2$) rule. If we define

$$
\begin{aligned}
k_1 &= f(x_j, y_j) \\
k_2 &= f\left(x_j + \frac{h}{2}, y_j + \frac{h}{2} k_1\right),
\end{aligned}
$$
(7.18)

then the method is determined by the recursion

$$
y_{j+1} = y_j + hk_2.
$$
(7.19)

This is a Runge–Kutta method with $t = 2$ and

$$
\alpha_1 = 0, \quad \alpha_2 = 1, \quad \mu_2 = \tfrac{1}{2}, \quad \lambda_{21} = \tfrac{1}{2}.
$$

If we apply the corrected Euler's formula with $h = 0.1$ to our test equation

$$
y' = x^2 + y^2, \qquad y(0) = 0,
$$

for $j = 0$, our calculation proceeds as follows:

$$
k_1 = x_0^2 + y_0^2 = 0^2 + 0^2 = 0
$$
$$
k_2 = \left(0 + \frac{0.1}{2}\right)^2 + \left(0 + \frac{0.1}{2} \cdot 0\right)^2 = 0.00250.
$$

Consequently,

$$
y_1 = y_0 + hk_2 = 0 + 0.1 \cdot 0.00250 = 0.000250.
$$

Updating x_j and y_j and continuing this recursion, we obtain the results given in Table 7.5.

TABLE 7.5 Output for the Corrected Euler's Computation

x_j	y_j	Exact Value $y(x_j)$
0.0	0.000000	0.000000
0.1	0.000250	0 000333
0.2	0.002500	0.002667
0.3	0.008752	0.009003
0.4	0.021020	0.021359
0.5	0.041354	0.041791
0.6	0.071895	0.072448
0.7	0.114958	0.115660
0.8	0.173171	0.174080
0.9	0.249692	0.250907
1.0	0.348545	0.350232

━━━━━ **EXAMPLE 7.6** ━━━

The *Heun method* is a popular two-stage ($t = 2$) rule. It is determined by the parameters

$$\alpha_1 = \tfrac{1}{4}, \qquad \alpha_2 = \tfrac{3}{4}$$

$$\mu_2 = \tfrac{2}{3}, \qquad \lambda_{21} = \tfrac{2}{3}.$$

Thus in the Heun method,

$$
\begin{array}{l}
k_1 = f(x_j, y_j) \\[4pt]
k_2 = f(x_j + \tfrac{2}{3}h, \; y_j + \tfrac{2}{3}hk_1) \\[4pt]
\text{and} \\[4pt]
y_{j+1} = y_j + h(\tfrac{1}{4}k_1 + \tfrac{3}{4}k_2).
\end{array}
\tag{7.20}
$$

Consider again the differential equation

$$y' = x^2 + y^2, \qquad y(0) = 0,$$

and let $h = 0.1$. Then for $j = 0$,

$$k_1 = 0^2 + 0^2 = 0,$$

$$k_2 = (0 + \tfrac{2}{3} \cdot 0.1)^2 + 0^2 = 0.004444$$

$$y_1 = 0 + 0.1(\tfrac{1}{4} \cdot 0 + \tfrac{3}{4} \cdot 0.004444) = 0.000333.$$

Table 7.6 gives a listing of further solution values obtained by the Heun method.

TABLE 7.6 Computational Example for Heun's Method

x_j	y_j	Exact Value, $y(x_j)$
0.0	0.000000	0.000000
0.1	0.000333	0 000333
0.2	0.002667	0.002667
0.3	0.009002	0.009003
0.4	0.021355	0.021359
0.5	0.041776	0.041791
0.6	0.072411	0.072448
0.7	0.115577	0.115660
0.8	0.173913	0.174080
0.9	0.250586	0.250907
1.0	0.349640	0.350232

■

━━━━ **EXAMPLE 7.7** ━━━━

The most popular four-stage ($t = 4$) Runge–Kutta method is the *classical* Runge–Kutta formula, which is determined by the relations

$$
\begin{aligned}
k_1 &= f(x_j, y_j) \\
k_2 &= f(x_j + \tfrac{1}{2}h, \; y_j + \tfrac{1}{2}hk_1) \\
k_3 &= f(x_j + \tfrac{1}{2}h, \; y_j + \tfrac{1}{2}hk_2) \\
k_4 &= f(x_j + h, \; y_j + hk_3) \\
y_{j+1} &= y_j + \frac{h}{6}(k_1 + 2k_2 + 2k_3 + k_4).
\end{aligned}
$$

(7.21)

TABLE 7.7   Subroutine for the Classical Runge–Kutta Method

```
      SUBROUTINE RUKU(X,Y,N,B)
C
C     ****************************************************************
C     *  FUNCTION: THIS SUBROUTINE COMPUTES THE SOLUTION OF A       *
C     *            DIFFERENTIAL EQUATION Y'=F(X,Y) OVER             *
C     *            THE INTERVAL [X(1),B] USING THE CLASSICAL        *
C     *            4-TH ORDER RUNGE-KUTTA METHOD                    *
C     *  USAGE:                                                     *
C     *       CALL SEQUENCE: CALL RUKU(X,Y,N,B)                     *
C     *       EXTERNAL FUNCTIONS/SUBROUTINES: FUNCTION F(U,V)       *
C     *  PARAMETERS:                                                *
C     *     INPUT:                                                  *
C     *       X(1)=INDEPENDENT VARIABLE INITIAL VALUE               *
C     *       Y(1)=DEPENDENT VARIABLE INITIAL VALUE                 *
C     *       N=NUMBER OF STEPS                                     *
C     *       B=SOLUTION INTERVAL ENDPOINT (LAST X VALUE)           *
C     *     OUTPUT:                                                 *
C     *       X=N+1 BY 1 ARRAY OF INDEPENDENT VARIABLE              *
C     *         VALUES                                              *
C     *       Y=N+1 BY 1 ARRAY OF DEPENDENT VARIABLE                *
C     *         SOLUTION VALUES                                     *
C     ****************************************************************
C
      DIMENSION X(N+1),Y(N+1)
      H=(B-X(1))/N
C     *** TJ IS AS K(J) IN THE TEXT ***
      DO 1 I=2,N+1
         X(I)=X(I-1)+H
         U=X(I-1)
         V=Y(I-1)
         T1=F(U,V)
         U=U+0.5*H
         V=V+0.5*H*T1
         T2=F(U,V)
         V=Y(I-1)+0.5*H*T2
         T3=F(U,V)
         U=U+0.5*H
         V=Y(I-1)+H*T3
         T4=F(U,V)
         Y(I)=Y(I-1)+H*(T1+2.0*T2+2.0*T3+T4)/6.0
    1 CONTINUE
      RETURN
      END
```

TABLE 7.8 Program Illustrating the Classical Runge–Kutta Method

```
C       PROGRAM RKMETH
C
C       **********************************************************************
C       THIS PROGRAM WILL SET UP AND SOLVE THE DIFFERENTIAL EQUATION
C       Y'=X**2+Y**2 WITH THE INITIAL CONDITION Y(0.)=0.
C       THE SOLUTION IS OBTAINED AT 11 GRID POINTS IN THE INTERVAL
C       [0,1] USING THE CLASSICAL RUNGE-KUTTA METHOD
C       CALLS:    RUKU
C       OUTPUT:
C                 X(I)=VALUE OF X FOR I=1,11
C                 Y(I)=APPROXIMATED VALUE OF Y AT X(I)
C       **********************************************************************
C
        DIMENSION X(11),Y(11)
C
C       *** FIRST, THE INITIAL CONDITION AND ENDPOINT ARE ESTABLISHED***
C
        X(1)=0.
        Y(1)=0.
        B=1.
        N=10
        N1=N+1
C
C       *** SUBROUTINE RUKU WILL APPROXIMATE THE SOLUTION          ***
C       *** RETURNING N1 VALUES IN ARRAYS X AND Y                 ***
C
        CALL RUKU(X,Y,N,B)
        WRITE(10,1)(X(I),Y(I),I=1,N1)
      1 FORMAT(5X,F4.1,5X,F10.6)
        STOP
        END
C
C       *** FUNCTION F SPECIFIES THE DIFFERENTIAL EQUATION. IT IS  ***
C       *** CALLED BY SUBROUTINE RUKU                             ***
C
        FUNCTION F(X,Y)
        F=X**2+Y**2
        RETURN
        END
```

TABLE 7.9 Output for Example 7.7

x_j	Classical Runge–Kutta, y_j	Exact Value, $y(x_j)$
0.0	0.000000	0.000000
0.1	0.000333	0 000333
0.2	0.002667	0.002667
0.3	0.009003	0.009003
0.4	0.021359	0.021359
0.5	0.041791	0.041791
0.6	0.072448	0.072448
0.7	0.115660	0.115660
0.8	0.174081	0.174080
0.9	0.250908	0.250907
1.0	0.350234	0.350232

In applying this method for our test equation

$$y' = x^2 + y^2, \qquad y(0) = 0$$

with $h = 0.1$ and $j = 0$, we get the values

$$k_1 = 0^2 + 0^2 = 0$$

$$k_2 = \left(0 + \frac{0.1}{2}\right)^2 + \left(0 + \frac{0.1}{2} \cdot 0\right)^2 = 0.0025$$

$$k_3 = \left(0 + \frac{0.1}{2}\right)^2 + \left(0 + \frac{0.1}{2} \cdot 0.0025\right)^2 \approx 0.002500$$

$$k_4 = (0 + 0.1)^2 + (0 + 0.1 \cdot 0.002500)^2 \approx 0.010000.$$

Consequently,

$$y_1 = 0 + \frac{0.1}{6}(0 + 0.005000 + 0.005000 + 0.010000)$$

$$\approx 0.000333.$$

The subroutine RUKU for this method is listed in Table 7.7. This subroutine is called by the program listed in Table 7.8, which applies the classical Runge–Kutta method to our test problem. The numerical results are presented in Table 7.9. ∎

─────── **ENGINEERING EXAMPLE** ───────

Assume that $N(t)$ denotes population size of wild rabbits, say, in an isolated area at time t. By "isolated" we mean that migration does not occur. Further, we presume that the population is relatively large so that whereas, strictly speaking, $N(t)$ is integer, nevertheless, to nearsighted eyes, a plot would look continuous. So with this argument, $N(t)$ will be viewed as a continuous function. For an animal such as a rabbit, it is reasonable to assume that if food is plentiful, the population increase depends only on the reproductive capacity of the population. On the other hand, if the population becomes too large, lack of food becomes a factor, and the rate of increase declines or even becomes negative because malnutrition enhances susceptibility to diseases and reduces reproductive capacity.

An elementary model quantifying these effects (e.g., Haberman, 1977, Sec. 34) is based on a differential equation of the form

$$N' = R(N)N.$$

The function $R(N)$ is called the population *growth rate*. The most elementary model assumes that $R(N) = R$, a constant, and the solution is the exponential growth function,

$$N(t) = N(t_0) \exp [R(t - t_0)].$$

Here $N(t_0)$ is the population at initial time t_0.

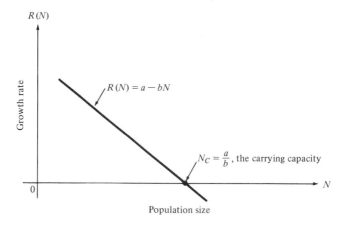

FIGURE 7–2   Linear Growth Rate Function

This model implies that the population increases ever more rapidly and without bound. This seems unreasonable, and the next level of complexity is the postulation that $R(N)$ is a line with negative slope. The root of this line (sketched in Figure 7.2) would be a point at which the derivative is zero, that is, a point at which the population remains constant. This point $N_C = a/b$ is sometimes spoken of appropriately as the "carrying capacity" of the population. The implication, then, is that the population grows according to the differential equation

$$N' = (a - bN)N.$$

This model is known as the *logistic* equation. For $N \ll N_C$, the population does grow nearly exponentially, since $R(N) \approx a$ is nearly constant. As N increases the derivative converges to 0, and the population tends to the constant, $N_C = a/b$. If somehow N were larger than the carrying capacity, then the growth rate would be negative and the population would decline toward the carrying capacity.

Suppose that at time 0, there are $N(0) = 500$ rabbits, and $a = 1.0$ and $b = 10^{-3}$. We wish to know how many rabbits there will be at time $t = 1$. Our plan will be to apply the classical Runge–Kutta rule [equation (7.21)] by means of subroutine RUKU (Table 7.7). Now the key question, of course, is: What size should step size h be? An intuitive answer is, small enough that if we halve the step size from h to $h/2$, the solution does not change much. (In Section 7.7, more sophisticated notions for step-size choice are offered.) Our computation plan began by starting with $h = 1$ and successively solving the population equation with $h = 1, \frac{1}{2}, \frac{1}{4}, \ldots$ until to five places, the computed approximation of $N(1)$ remained unchanged from the last run. Our findings are summarized as follows:

Step Size, h	Population at $t = 1$, $N(1)$
1	730.9103
1/2	731.0474
1/4	731.0579
1/8	731.0585
1/16	731.0586

Intuitively, from the regularity of these numbers, we feel confident that to six or seven significant decimals, the exact solution is 731.0586. It turns out (Haberman, 1977) that the logistics equation has an analytic solution which agrees with our computed solution to the accuracy shown.　■

*7.3.2 Parameter Selection for Runge–Kutta Formulas

Later in this section we give some rationale for selection of the number t of stages. For now, assume that t has already been selected. Attention focuses on choice of the parameters α_i, μ_i, and λ_{im} in (7.15) and (7.17) for our fixed number t.

The central principle of the Runge–Kutta approach is that the parameters should be chosen so that a power series expansion in h of the right side of (7.15) agrees with the Taylor's series expansion of $y(x_j + h)$ for as high a power of h as possible. Let $y(x)$ denote the solution of the differential equation $y' = f(x, y)$, which passes through the point (x_j, y_j). Let us denote the heady expression on the right side of (7.15) by $g(x_j, y_j, h)$, where we have presumably accounted for the dependence of the k_i's on h through (7.17). Then

$$y_{j+1} = y_j + hg(x_j, y_j, h).$$

As we have mentioned above, the idea behind the Runge–Kutta approach is to choose the parameters so that for as large an integer p as possible, the pth-degree Taylor's polynomial in h of the solution $y(x_{j+1}) = y(x_j + h)$ coincides with the Taylor's polynomial of $y_j + hg(x_j, y_j, h)$. That is, we choose parameters so that for maximal p,

$$\sum_{q=0}^{p} \frac{y^{(q)}(x_j)}{q!} h^q = y_j + h \sum_{q=0}^{p-1} \frac{\dfrac{d^q}{dh^q} g(x_j, y_j, h)}{q!} h^q. \tag{7.22}$$

Equating like coefficients of the qth power of h in (7.22) and replacing derivatives of $g(x_j, y_j, h)$ with those of the defining expression (7.15), we obtain the criterion for determining the Runge–Kutta parameters; namely, for $q = 0$ we get the relation $y(x_j) = y_j$, and for $q = 1, 2, \ldots, p$,

$$\boxed{\; y^{(q)}(x_j) = q \frac{d^{q-1}}{dh^{q-1}}\left[\sum_{i=1}^{t} \alpha_i f\left(x_j + h\mu_i, y_j + h \sum_{m=1}^{i-1} \lambda_{im} k_m\right)\right]\Bigg|_{h=0}. \;} \tag{7.23}$$

The parameter design criterion has us choose parameters α_i, μ_i, and λ_{im} so that (7.23) holds for as large an integer p as possible over all sufficiently many times differentiable functions $y(x)$ and $f(x, y)$, with $y'(x)$ and $f(x, y)$ related by $y' = f(x, y)$. Observe that (7.16) implies that in (7.23), $\mu_1 = 0$. As we will see in the example to follow, the procedure involves expressing $y^{(q)}(x_j)$ explicitly in terms of $f(x_j, y_j)$ and its derivatives, through (7.4), and then equating like coefficients of these derivatives.

The criterion (7.23) appears formidable, and indeed for moderate to large values of t, its detailed implementation is messy. But the next pair of examples will help to illustrate that the idea behind the use of (7.23) for constructing Runge–Kutta parameters is straightforward, if tedious.

The criterion (7.23) is appealing in the following sense: For any method of type (7.15) not satisfying (7.23), we may anticipate that there is some Runge–Kutta formula with the same number t of stages which is ultimately more accurate as h gets smaller.

────── **EXAMPLE 7.8** ──────────────────────────────────────

Let $t = 1$. Then in view of (7.15), the only parameter to be selected is α_1. From (7.23), with $q = 1$, we have

$$y'(x_j) = \alpha_1 f(x_j, y_j).$$

But necessarily, $y'(x_j) = f(x_j, y_j)$, so to satisfy (7.23), with $q = 1$, we must set $\alpha_1 = 1$. Thus the one-stage Runge–Kutta rule coincides with the Euler method. ■

────── **EXAMPLE 7.9** ──────────────────────────────────────

Let $t = 2$. Then

$$y_{j+1} = y_j + h[\alpha_1 f(x_j, y_j) + \alpha_2 f(x_j + h\mu_2, y_j + h\lambda_{21}f(x_j, y_j))].$$

From (7.23), with $q = 1$, we have

$$y'(x_j) = f(x_j, y_j) = (\alpha_1 + \alpha_2)f(x_j, y_j). \tag{7.24}$$

For $q = 2$, and by use of the chain rule (Appendix B, item 36) as in (7.4),

$$
\begin{aligned}
y''(x_j) &= f_x(x_j, y_j) + f_y(x_j, y_j)f(x_j, y_j) \\
&= 2\alpha_2[\mu_2 f_x(x_j, y_j) + \lambda_{21}f_y(x_j, y_j)f(x_j, y_j)].
\end{aligned}
\tag{7.25}
$$

In the above, subscripts of f denote partial derivatives with respect to the indicated variable. Equations (7.24) and (7.25) must hold for every possible $f(x_j, y_j)$, $f_x(x_j, y_j)$, and $f_y(x_j, y_j)$. So, for example, from (7.24) we must have $\alpha_1 + \alpha_2 = 1$. By equating like coefficients of $f_x(x_j, y_j)$ in (7.25), we have that $2\alpha_2\mu_2 = 1$, and from the coefficients of $f_y(x_j, y_j)f(x_j, y_j)$, we see that $2\alpha_2\lambda_{21} = 1$. Collecting these relations, we may assert that every two-stage Runge–Kutta formula satisfies

$$
\boxed{
\begin{aligned}
\alpha_1 + \alpha_2 &= 1 \\
2\alpha_2\mu_2 &= 1 \\
2\alpha_2\lambda_{21} &= 1.
\end{aligned}
}
\tag{7.26}
$$

The two-stage Heun method (see Example 7.6) with $\alpha_1 = \frac{1}{4}$, $\alpha_2 = \frac{3}{4}$, $\mu_2 = \frac{2}{3}$, and $\lambda_{21} = \frac{2}{3}$ satisfies these conditions, as does the corrected Euler method, with parameters $\alpha_1 = 0$, $\alpha_2 = 1$, $\mu_2 = \frac{1}{2}$, and $\lambda_{21} = \frac{1}{2}$. Since there are more unknowns than equations, system (7.26) has infinitely many solutions. Thus there are infinitely many two-stage rules satisfying (7.23) for $p = 2$. One may confirm that for $p = 3$ in (7.23) there are more conditions than unknowns. Consequently, no parameter set for a two-stage rule satisfies (7.23) for $p = 3$. ∎

In the introductory section of Chapter 3 we said that a function $g(h)$ was $O(h^q)$ if for some positive number C,

$$|g(h)| \leq C|h|^q$$

whenever $|h|$ is small enough. We say that a Runge–Kutta method is of order p if after one step of size h, and with $y_0 = y(x_0)$,

$$|y_1 - y(x_1)| = |y_1 - y(x_0 + h)| = O(h^{p+1}).$$

If two power series in h agree through the pth power, their difference is $O(h^{p+1})$, as the reader may check. Therefore, if a Runge–Kutta rule satisfies (7.23) for a certain number p, it is pth order. By construction, the single-stage rule (Euler's method) is first order, and the two-stage rules given in Examples 7.5 and 7.6 are second order. The classical four-stage Runge–Kutta rule is known to be fourth order. Above four stages, the order is less than the number of stages. Up to 11 stage rules, the relation between the order and number of stages (see Butcher, 1964) is as summarized in Table 7.10.

To illustrate forcefully that the order of a Runge–Kutta method has practical merit, we mention a result (derived in Henrici, 1962, for example) that for a differential equation having a solution on a finite interval $[x_0, b]$, and for the sequence $y_0, y_1, y_2, \ldots$ determined by a pth-order Runge–Kutta method, and $y(x)$ the solution of (7.3),

$$\max_{0 \leq k \leq N} |y(x_0 + kh) - y_k| = O(h^p), \qquad (7.27)$$

where $x_0 + Nh = b$. This result can be interpreted as follows. In one step the error of the approximating solution is $O(h^{p+1})$, but these errors are accumulating during the entire calculation. Inequality (7.27) guarantees that the accumulated error is $O(h^p)$. We may summarize by saying that the order of the accumulated error equals the order of the method.

TABLE 7.10   Attainable Runge–Kutta Method Order p as Function of Number of Stages t

Number of Stages, t	1	2	3	4	5	6	7	8	9	10	11
Attainable Order, p	1	2	3	4	4	5	6	6	7	8	9

EXAMPLE 7.10

We used subroutine RUKU (Table 7.7) with various step sizes h on our test equation

$$y' = x^2 + y^2, \qquad y(0) = 0,$$

to indicate a consequence of (7.27) that $|y(1) - y_{N(h)}| = O(h^4)$, where $N(h) = 1/h$, 4 being the order of the classical Runge–Kutta method, as mentioned. By undertaking our calculations in double precision, Table 7.11 was compiled. In the rightmost column we see that the error, when divided by h^4, tends to a constant as h decreases.

TABLE 7.11   Illustration of Order of Error for the Classical Runge–Kutta Method

h	Error	Error/h^4
0.500000	-0.5550E-03	-0.008881
0.250000	-0.5745E-04	-0.014707
0.125000	-0.4449E-05	-0.018225
0.062500	-0.3070E-06	-0.020120
0.031250	-0.2005E-07	-0.021025
0.015625	-0.1214E-08	-0.020367

■

In theory, for any differential equation with a "smooth" solution, as the step size h becomes smaller, the higher the value of p in (7.23) is, the more accurate the Runge–Kutta solution approximation will be. Experience will confirm that for a fixed step size, higher-order Runge–Kutta rules do enhance the accuracy of the computation. For instance, in comparing Examples 7.3, 7.5, and 7.7, we see that by successively increasing t from 1 to 2, and then to 4, the estimates of $y(1)$ have increased in accuracy from one significant place to five, without much change in programming or computational effort. Computation of the solution of the differential equation in our examples to five-place accuracy by Euler's method would require on the order of 1000-fold increase in computational effort over that required by Table 7.9.

In practice, specialists prefer Runge–Kutta methods with stage numbers between 4 and 6. These families of Runge–Kutta methods provide sufficient accuracy for well-behaved problems. In order for a greater number of stages (say 10 or higher) to be advantageous, we would have to use large step sizes h; but for large step sizes, the Taylor's polynomial approximations motivating the Runge–Kutta method become unreliable.

7.4
ADAMS AND LINEAR MULTISTEP METHODS

Runge–Kutta methods are referred to as *single-step methods* because, as is readily seen by inspection of (7.15), y_{j+1} depends on y_j, but not on "steps" y_{j-1}, y_{j-2}, . . ., further back in the past. The methods to be considered next are *multistep methods*: y_{j+1} will depend on at least one of the past values y_{j-1}, y_{j-2}, ...

Assume again that the solution of the initial-value problem (7.3) is to be calculated at node points in an interval $[x_0, b]$. Define step size h by $h = (b - x_0)/N$, where $N > 1$ is a given integer. As in the case of the single-step methods, the solution will be approximated at the set of node points $x_j = x_0 + jh$ ($j = 0, 1, 2, \ldots, N$). Assume that the approximations $y_0, y_1, \ldots, y_j$ of $y(x_0)$, $y(x_1), \ldots, y(x_j)$ are already available. Then the subsequent value y_{j+1} can be calculated recursively according to the *linear multistep method*, which has the general form

$$y_{j+1} = -\sum_{m=1}^{i} \alpha_m y_{j+1-m} + h \sum_{m=0}^{i} \beta_m f_{j+1-m}, \tag{7.28}$$

where

$$f_{j+1-m} = f(x_{j+1-m}, y_{j+1-m}).$$

The integer i and the coefficients α_m and β_m are the parameters of the method. The rule (7.28) is explicit only if $\beta_0 = 0$. Otherwise, the function value $f_{j+1} = f(x_{j+1}, y_{j+1})$ appears on the right-hand side. This depends on the unknown value y_{j+1}. In such a case (the "implicit" case), a nonlinear equation must be solved for y_{j+1}.

The literature of multistep methods concentrates on a certain subfamily, the *Adams methods*. These methods are characterized by the constraints that

$$\alpha_1 = -1 \text{ and } \alpha_m = 0, \qquad \text{for } m > 1. \tag{7.29}$$

The Adams method coefficients β_m are chosen so that the truncation error $y(x_j + h) - y_{j+1}$ is of as high an order as possible in step size h. We defer details to (Szidarovszky and Yakowitz, 1978, Sec. 8.1.4). But a procedure is outlined in Problem 10.

One reason for specialization in Adams methods is that outside this class, the linear multistep algorithms are often found to be unstable, in the sense that the error increases exponentially as it propagates with increasing j in (7.28). But even among stable linear multistep rules, the Adams methods have certain attractive theoretical features with respect to error attenuation. Another reason for their popularity is that the implementation of the Adams methods is particularly easy.

In view of the constraint (7.29), Adams methods are characterized by the formula

$$y_{j+1} = y_j + h \sum_{m=0}^{i} \beta_m f_{j+1-m}. \tag{7.30}$$

In the case that β_0 in (7.30) is constrained to be zero, we have what are known as the *Adams–Bashforth* methods. These methods have the feature that they are explicit in the sense that the right side of (7.30) will not depend on the unknown y_{j+1}. The coefficients of the first several Adams–Bashforth formulas are given in Table 7.12 and subroutine ADAMS for implementing the three-step Adams–Bashforth formula is provided by Table 7.13.

TABLE 7.12   Coefficients for Adams–Bashforth Formulas

Number of Steps i	β_1	β_2	β_3	β_4	β_5
2	$\dfrac{3}{2}$	$-\dfrac{1}{2}$			
3	$\dfrac{23}{12}$	$-\dfrac{16}{12}$	$\dfrac{5}{12}$		
4	$\dfrac{55}{24}$	$-\dfrac{59}{24}$	$\dfrac{37}{24}$	$-\dfrac{9}{24}$	
5	$\dfrac{1901}{720}$	$-\dfrac{2774}{720}$	$\dfrac{2616}{720}$	$-\dfrac{1276}{720}$	$\dfrac{251}{720}$

TABLE 7.13   Subroutine for the Third-Order Adams–Bashforth Method

```
      SUBROUTINE ADAMS(X,Y,N,B)
C
C     ****************************************************************
C     *  FUNCTION: THIS SUBROUTINE COMPUTES THE SOLUTION OF A       *
C     *            DIFFERENTIAL EQUATION OVER THE INTERVAL          *
C     *            [X(1),B] USING THE DEFINING FUNCTION F(U,V)      *
C     *            AND THE EXPLICIT 3-STEP ADAMS-BASHFORTH          *
C     *            METHOD                                           *
C     *  USAGE:                                                     *
C     *      CALL SEQUENCE: CALL ADAMS(X,Y,N,B)                     *
C     *      EXTERNAL FUNCTIONS/SUBROUTINES: FUNCTION F(U,V)        *
C     *  PARAMETERS:                                                *
C     *      INPUT:                                                 *
C     *          N=NUMBER OF ITERATIONS (<1000)                    *
C     *      X(1),X(2),X(3)                                         *
C     *          =INDEPENDENT VARIABLE INITIAL VALUES               *
C     *          B=SOLUTION INTERVAL ENDPOINT (LAST X VALUE)       *
C     *      Y(1),Y(2),Y(3)                                         *
C     *          =DEPENDENT VARIABLE INITIAL SOLUTION VALUES        *
C     *      OUTPUT:                                                *
C     *          X=N+1 BY 1 ARRAY OF INDEPENDENT VARIABLE           *
C     *              VALUES                                         *
C     *          Y=N+1 BY 1 ARRAY OF DEPENDENT VARIABLE            *
C     *              SOLUTION VALUES                                *
C     ****************************************************************
C
      DIMENSION X(N+1),Y(N+1),T(1001)
C     *** INITIALIZATION ***
      H=(B-X(1))/N
      DO 1 I=1,3
         U=X(I)
         V=Y(I)
         T(I)=F(U,V)
    1 CONTINUE
C     *** COMPUTE ITERATIVE SOLUTION ***
      DO 2 I=4,N+1
         X(I)=X(I-1)+H
         Y(I)=Y(I-1)+H*(23.0*T(I-1)-16.0*T(I-2)+5.0*T(I-3))/12.0
         U=X(I)
         V=Y(I)
         T(I)=F(U,V)
    2 CONTINUE
      RETURN
      END
```

─────── **EXAMPLE 7.11** ───────────────────────────────────

By the use of the three-step Adams–Bashforth formula, we solve our standard test equation

$$y' = x^2 + y^2, \qquad y(0) = 0.$$

From Table 7.12 we see that the formula to be used has the form

$$y_{j+1} = y_j + h\left(\frac{23}{12} f_j - \frac{16}{12} f_{j-1} + \frac{5}{12} f_{j-2}\right).$$

We first illustrate the use of this formula for $h = 0.1$ and $j = 2$ (since this is the smallest value of j for which the method can be applied). The values y_1 and y_2

TABLE 7.14   Program to Exemplify the Adams Method

```
C       PROGRAM ABMETH
C
C       ********************************************************************
C       THIS PROGRAM WILL SET UP AND SOLVE NUMERICALLY THE DIFFERENTIAL
C       EQUATION Y'=X**2+Y**2 WITH THE INITIAL CONDITION Y(0.)=0.
C       THE SOLUTION IS OBTAINED AT 11 POINTS OVER THE INTERVAL [0,1]
C       USING THE THREE STEP ADAMS-BASHFORTH METHOD
C       CALLS:    ADAMS
C       OUTPUT:
C               X(I)=VALUE OF X FOR I=1,11
C               Y(I)=APPROXIMATED VALUE OF Y AT X(I)
C       ********************************************************************
C
        DIMENSION X(11),Y(11)
C
C       *** FIRST, THE INITIAL VALUES AND ENDPOINT ARE DEFINED      ***
C
        H=.1
        X(1)=0.
        X(2)=H
        X(3)=2*H
        Y(1)=0.
        Y(2)=.000333
        Y(3)=.002667
        B=1.
        N=10
C
C       *** SUBROUTINE ADAMS WILL APPROXIMATE THE SOLUTION         ***
C       *** RETURNING 11 VALUES IN ARRAYS X AND Y                  ***
C
        CALL ADAMS(X,Y,N,B)
        WRITE(10,1)(X(I),Y(I),I=1,11)
      1 FORMAT(5X,F4.1,5X,F10.6)
        STOP
        END
C
C       *** FUNCTION F SPECIFIES THE DIFFERENTIAL EQUATION. IT IS  ***
C       *** CALLED BY SUBROUTINE ADAMS                            ***
C
        FUNCTION F(X,Y)
        F=X**2+Y**2
        RETURN
        END
```

**TABLE 7.15   Output for
Adams–Bashforth Example**

x_j	y_j	Exact Value, $y(x_j)$
0.0	0.000000	0.000000
0.1	0.000333	0 000333
0.2	0.002667	0.002667
0.3	0.009002	0.009003
0.4	0.021350	0.021359
0.5	0.041760	0.041791
0.6	0.072370	0.072448
0.7	0.115494	0.115660
0.8	0.173758	0.174080
0.9	0.250318	0.250907
1.0	0.349191	0.350232

are taken as the values obtained in the classical Runge–Kutta computation (Example 7.7). Thus to start the computation, we have the values

$$y_0 = 0, \qquad y_1 = 0.000333, \qquad y_2 = 0.002667.$$

Then

$$f_0 = x_0^2 + y_0^2 = 0^2 + 0^2 = 0$$

$$f_1 = x_1^2 + y_1^2 = (0.1)^2 + (0.000333)^2 \approx 0.010000$$

$$f_2 = x_2^2 + y_2^2 = (0.2)^2 + (0.002667)^2 \approx 0.040007,$$

and consequently

$$y_3 = 0.002667 + 0.1\left(\frac{23}{12} \cdot 0.040007 - \frac{16}{12} \cdot 0.010000 + \frac{5}{12} \cdot 0\right)$$

$$\approx 0.009002.$$

We have performed the complete calculations by means of the calling program listed in Table 7.14, for obtaining the values $y_3, y_4, \ldots, y_{10}$ of the solution. The results are reported in Table 7.15.   ∎

When in (7.30), $\beta_0 \neq 0$, the right-hand side depends on the unknown value of y_{j+1} [since $f_{j+1} = f(x_{j+1}, y_{j+1})$], and we have an *implicit* Adams formula. This class of Adams formulas is known as the *Adams–Moulton method*. Of course, such implicit equations entail solving a nonlinear equation problem for y_{j+1}. Nevertheless, for certain types of differential equations (notably what are known as "stiff" differential equations), the Adams–Moulton class has certain theoretical and practical appeal in comparison to Adams–Bashforth formulas. For that reason, they are useful despite the technical complications arising from their being im-

TABLE 7.16   Coefficients for Adams–Moulton Formulas

Number of Steps i	β_0	β_1	β_2	β_3	β_4
1	$\dfrac{1}{2}$	$\dfrac{1}{2}$	0	0	0
2	$\dfrac{5}{12}$	$\dfrac{8}{12}$	$-\dfrac{1}{12}$	0	0
3	$\dfrac{9}{24}$	$\dfrac{19}{24}$	$-\dfrac{5}{24}$	$\dfrac{1}{24}$	0
4	$\dfrac{251}{720}$	$\dfrac{646}{720}$	$-\dfrac{264}{720}$	$\dfrac{106}{720}$	$-\dfrac{19}{720}$

plicit. The coefficients of the first few Adams–Moulton formulas are summarized in Table 7.16.

A common procedure for implementing an Adams–Moulton method is to use it conjunctively with an Adams–Bashforth formula of like number of steps, as follows. Assume that values $y_j, \ldots, y_{j+1-i}$ have already been obtained. Let y^I_{j+1} denote a "predictor" of y_{j+1} obtained from the (explicit) Adams–Bashforth formula. Then define f^I_{j+1} to be $f(x_{j+1}, y^I_{j+1})$ and construct the "corrected" y^{II}_{j+1} according to the Adams–Moulton formula, but with f^I_{j+1} replacing f_{j+1} in (7.30). Some authors recommend continuing the "correction" process by computing f^{II}_{j+1} from y^{II}_{j+1}, and substituting this new corrector f^{II}_{j+1} into the Adams–Moulton formula and continuing such iterations until magnitudes of successive corrections are less than some chosen threshold. Other authors argue that refinement of corrections is not computationally tenable. Regardless of whether subsequent correction iterations are used, conjunctive application of Adams–Bashforth and Adams–Moulton methods are known as *predictor–corrector methods*.

7.5
SIMULTANEOUS AND HIGHER-ORDER DIFFERENTIAL EQUATIONS

In many applications we are required to solve a system of several first-order differential equations with several unknown functions. The initial value problems of such equations may be expressed as

$$
\begin{aligned}
y_1'(x) &= f_1(x, y_1(x), \ldots, y_n(x)), \ y_1(x_0) = y_{10} \\
y_2'(x) &= f_2(x, y_1(x), \ldots, y_n(x)), \ y_2(x_0) = y_{20} \\
&\ \vdots \qquad\qquad\qquad \vdots \qquad\qquad\qquad \vdots \\
y_n'(x) &= f_n(x, y_1(x), \ldots, y_n(x)), \ y_n(x_0) = y_{n0}.
\end{aligned}
\tag{7.31}
$$

It is convenient to write this system in a vector form by introducing the notation

$$\mathbf{y}(x) = (y_1(x), \ldots, y_n(x))^T, \quad \mathbf{f}(x, \mathbf{y}(x)) = (f_1(x, \mathbf{y}(x)), \quad \ldots, \quad f_n(x, \mathbf{y}(x))^T$$

and

$$\mathbf{y}_0 = (y_{10}, \ldots, y_{n0})^T.$$

In this notation (7.31) becomes

$$\boxed{\mathbf{y}'(x) = \mathbf{f}(x, \mathbf{y}(x)), \qquad \mathbf{y}(x_0) = \mathbf{y}_0.} \qquad (7.32)$$

The point of this section is that our earlier Runge–Kutta and linear multistep methods can readily be applied to solve simultaneous first-order differential equations.

Before exploring details of the solution of (7.32), we first present a standard device for converting higher-order differential equations into first-order simultaneous equations. This extends the domain of application of the techniques to be described. The initial value problem of a kth-order differential equation has the form

$$\boxed{\begin{aligned} y^{(k)}(x) &= f(x, y(x), y'(x), \ldots, y^{(k-1)}(x)), \\ y^{(j)}(x_0) &= y_0^{(j)}, \qquad (0 \le j \le k - 1) \end{aligned}} \qquad (7.33)$$

relating the kth order derivative of $y(x)$ to derivatives of lower order. Here y and f are real-valued functions of the indicated variables, and $y_0^{(j)}$ $(0 \le j \le k - 1)$ are given numbers: Equation (7.33) can be reduced to a simultaneous system of first-order equations by introducing the new variables

$$y_1(x) = y(x), \quad y_2(x) = y'(x), \quad \ldots, \quad y_k(x) = y^{(k-1)}(x).$$

For then, (7.33) can be written as

$$\boxed{\begin{aligned} y_1'(x) &= y_2(x), & y_1(x_0) &= y_0^{(0)} \\ y_2'(x) &= y_3(x), & y_2(x_0) &= y_0^{(1)} \\ &\cdots & &\cdots \\ y_{k-1}'(x) &= y_k(x), & y_{k-1}(x_0) &= y_0^{(k-2)} \\ y_k'(x) &= f(x, y_1(x), \ldots, y_k(x)), & y_k(x_0) &= y_0^{(k-1)}. \end{aligned}} \qquad (7.34)$$

That is, in this case, (7.34) has the structure (7.32), where

$$\mathbf{y}(x) = \begin{bmatrix} y_1(x) \\ \vdots \\ y_{k-1}(x) \\ y_k(x) \end{bmatrix}, \qquad \mathbf{y}_0 = \begin{bmatrix} y_0^{(0)} \\ \vdots \\ y_0^{(k-2)} \\ y_0^{(k-1)} \end{bmatrix}$$

and

$$\mathbf{f}(x, \mathbf{y}(x)) = \begin{bmatrix} y_2(x) \\ \vdots \\ y_k(x) \\ f(x, \mathbf{y}(x)) \end{bmatrix}.$$

───── **EXAMPLE 7.12** ─────

Consider the second-order differential equation

$$y''(x) = y(x)y'(x) + x^2 + 1$$

with initial conditions

$$y(0) = y'(0) = 0.$$

Then by introducing the new variables

$$y_1(x) = y(x), \qquad y_2(x) = y'(x)$$

we get

$$\begin{aligned} y_1'(x) &= y_2(x) & y_1(0) &= 0 \\ y_2'(x) &= y_1(x)y_2(x) + x^2 + 1 & y_2(0) &= 0, \end{aligned} \tag{7.35}$$

which is an initial-value problem for a system of two first-order differential equations with two unknown functions, $y_1(x)$ and $y_2(x)$. ∎

Although there are numerical methods specifically devoted to the simultaneous differential equation problem, it turns out that the obvious vector generalizations of the univariate techniques described earlier in this chapter are usually adequate and effective. If the reader will inspect the Runge–Kutta- and Adams-type formulas, he or she will see that they continue to have meaning if the variables y and the derivative functions $f(x, y)$ are presumed to be vector valued. Thus with this understanding, our methodology for single differential equations is directly applicable.

The subroutine MRUKU listed in Table 7.17 is an obvious multivariable generalization of the classical Runge–Kutta rule RUKU (Table 7.7). The number of simultaneous equations is entered as calling parameter M, and the derivative func-

TABLE 7.17   Subroutine MRUKU for Multivariate Runge–Kutta Method

```
      SUBROUTINE MRUKU(X,Y,N,B,M)
C
C     ****************************************************************
C     *  FUNCTION: THIS SUBROUTINE COMPUTES THE SOLUTION TO A SET*
C     *            OF SIMULTANEOUS DIFFERENTIAL EQUATIONS OVER    *
C     *            THE INTERVAL [X(1),B] GIVEN THE J-TH COORDIN-  *
C     *            ATE OF THE DEFINING FUNCTION F(U,V,J) AND      *
C     *            USING THE 4-TH ORDER RUNGE-KUTTA METHOD        *
C     *  USAGE:                                                   *
C     *       CALL SEQUENCE: CALL MRUKU(X,Y,N,B,M)               *
C     *       EXTERNAL FUNCTIONS/SUBROUTINES: FUNCTION F(U,V,J)   *
C     *  PARAMETERS:                                              *
C     *       INPUT:                                              *
C     *           M=NUMBER OF DIFFERENTIAL EQUATIONS              *
C     *             (MAXIMUM OF 100)                              *
C     *           X(1)=INDEPENDENT VARIABLE INITIAL VALUE         *
C     *           Y(1,1),Y(2,1),...,Y(M,1)                        *
C     *             =DEPENDENT VARIABLE INITIAL VALUES            *
C     *           B=SOLUTION INTERVAL ENDPOINT (LAST X VALUE)     *
C     *       OUTPUT:                                             *
C     *           X=N+1 BY 1 ARRAY OF INDEPENDENT VARIABLE        *
C     *             VALUES                                        *
C     *           Y=M BY N+1 ARRAY OF DEPENDENT VARIABLE SOLU-    *
C     *             TION VALUES (EACH ARRAY ROW IS THE SOLUTION   *
C     *             FOR ONE OF THE M DIFFERENTIAL EQUATIONS)      *
C     ****************************************************************
C
      DIMENSION X(N+1),Y(M,N+1),V(100),T(100,4)
      H=(B-X(1))/N
      DO 1 I=2,N+1
        X(I)=X(I-1)+H
        U=X(I-1)
        DO 2 J=1,M
          V(J)=Y(J,I-1)
    2   CONTINUE
        DO 3 J=1,M
          T(J,1)=F(U,V,J)
    3   CONTINUE
        U=U+0.5*H
        DO 4 J=1,M
          V(J)=V(J)+0.5*H*T(J,1)
    4   CONTINUE
        DO 5 J=1,M
          T(J,2)=F(U,V,J)
    5   CONTINUE
        DO 6 J=1,M
          V(J)=Y(J,I-1)+0.5*H*T(J,2)
    6   CONTINUE
        DO 7 J=1,M
          T(J,3)=F(U,V,J)
    7   CONTINUE
        U=U+0.5*H
        DO 8 J=1,M
          V(J)=Y(J,I-1)+H*T(J,3)
    8   CONTINUE
        DO 9 J=1,M
          T(J,4)=F(U,V,J)
    9   CONTINUE
        DO 10 J=1,M
          Y(J,I)=Y(J,I-1)+H*(T(J,1)+2.0*T(J,2)+2.0*T(J,3)+T(J,4))/6.0
   10   CONTINUE
    1 CONTINUE
      RETURN
      END
```

tion $\mathbf{f}(x, \mathbf{y})$ in (7.32) is presumed supplied by an external function program F(X,Y,J), where Y is an array of dimension M, and J denotes the index of the coordinate of $\mathbf{f}(x, \mathbf{y})$ to be obtained at that evaluation.

EXAMPLE 7.13

The two dimensional "predator–prey" model

$$y_1' = \gamma_1 y_1 - \gamma_2 y_1 y_2 \tag{7.36}$$

$$y_2' = -\gamma_3 y_2 + \gamma_4 y_1 y_2 \tag{7.37}$$

has appeared prominently in recent texts on mathematical modeling and numerical methods for differential equations. Here $y_1(t)$ represents the prey, and the first term on the right in (7.36) tells us that the population growth per unit time is proportional to the prey population, and the second, that the decrease rate (per unit time) is proportional to the product of the population of prey and predator. The right side of (7.37) implies that the predator growth is negatively proportional to the predator population (presumably reflecting competition) and is positively proportional to the product of the two populations. The prey population influences the growth of the number of predators by way of nutrient.

From the computational standpoint, an interesting aspect of the predator–prey equation is that its solution is known (e.g., Haberman, 1977, Sec. 48) to be periodic. That is, for some positive number T, and all t,

$$y_j(t + T) = y_j(t), \qquad j = 1, 2.$$

If one plots the points $(y_1(t), y_2(t))$ in the y_1–y_2-plane, the solution curve must therefore be closed. An inadequate numerical solution will manifest itself by producing "phase-plane" y_1–y_2 curves that do not close on themselves.

By means of the calling program listed in Table 7.18, we have utilized MRUKU to produce phase-plane plots. Following Ortega and Poole (1981), we have chosen as parameters and initial condition, the numbers

$$\gamma_1 = 0.25, \qquad \gamma_2 = 0.01, \qquad \gamma_3 = 1.0$$

$$\gamma_4 = 0.01, \qquad y_1(0) = 80, \qquad y_2(0) = 30.$$

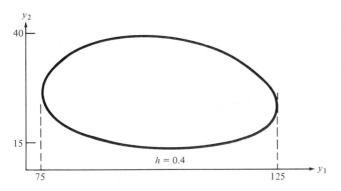

FIGURE 7–3 Computed Two-Dimensional Predator–Prey Trajectory with $h = 0.4$

TABLE 7.18   Calling Program for Example 7.13

```
C       PROGRAM MRKMETH
C
C       *********************************************************************
C       THIS PROGRAM WILL SET UP AND SOLVE NUMERICALLY THE PAIR OF
C       DIFFERENTIAL EQUATIONS Y1'=.25*Y1-.01*Y1*Y2 AND Y2'=-Y2+.01*Y1*Y2
C       WITH THE INITIAL CONDITION Y1(0.)=80., Y2(0.)=30.
C       THE SOLUTION IS OBTAINED AT 25 POINTS OVER THE INTERVAL [0,80]
C       USING MULTIVARIATE 4-TH ORDER RUNGE-KUTTA METHOD
C       CALLS:    MRUKU
C       OUTPUT:
C               X(I)=VALUE OF X FOR I=1,25
C               Y(J,I)=APPROXIMATED VALUE OF Y1 AND Y2 AT X(I)
C       *********************************************************************
C
        DIMENSION X(100),Y(2,101)
C
C       *** FIRST, THE INITIAL CONDITION AND ENDPOINT ARE ESTABLISHED ***
C
        Y(1,1)=80.
        Y(2,1)=30.
        X(1)=0.
        B=80.
        N=100
C
C       *** SUBROUTINE MRUKU WILL APPROXIMATE THE SOLUTION         ***
C       *** RETURNING N+1 VALUES IN ARRAYS X AND Y                 ***
C
        CALL MRUKU(X,Y,N,B,2)
        WRITE(10,1)(X(I),Y(1,I),Y(2,I),I=1,N+1)
      1 FORMAT(5X,F4.1,5X,F10.6,5X,F10.6)
        STOP
        END
C
C       *** FUNCTION F SPECIFIES THE DIFFERENTIAL EQUATION. IT IS  ***
C       *** CALLED BY SUBROUTINE MRUKU                             ***
C
        FUNCTION F(X,Y,M)
        DIMENSION Y(2)
        IF(M.EQ.1) THEN
            F=.25*Y(1)-.01*Y(1)*Y(2)
        ELSE
            F=-Y(2)+.01*Y(1)*Y(2)
        ENDIF
        RETURN
        END
```

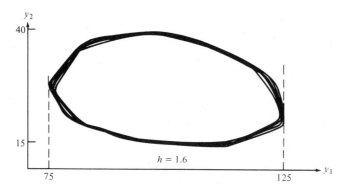

FIGURE 7–4   Computed Two-Dimensional Predator–Prey Trajectory with $h = 1.6$

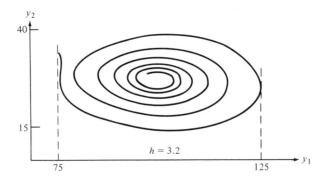

FIGURE 7–5 Computed Two-Dimensional Predator–Prey Trajectory with $h = 3.2$

In Figures 7.3 to 7.5, the (connected) plots of the solution points are presented for various step sizes h. The time interval was taken as $[0, 80]$ (i.e., $b = 80$). Through these plots we see the deterioration of performance as the step size increases. To the accuracy of the plotter resolution, Figure 7.3 is correct: Six successive cycles are essentially coincident in the phase plane. ∎

*7.6
TWO-POINT BOUNDARY-VALUE PROBLEMS

Let a and b ($a < b$) be given numbers and $g(u, v)$ a given real- or vector-valued function of two variables. Two-point boundary-value problems of first-order differential equations require the solution of the differential equation

$$y' = f(x, y) \tag{7.38}$$

subject to a two-point boundary condition, usually expressible as

$$g(y(a), y(b)) = 0, \tag{7.39}$$

the numbers a and b being the "two points." Here f, g, and y can be either real or vector valued.

Many two-point boundary-value problem studies concentrate on real second-order differential equations, which are equations of the form

$$y'' = f(x, y, y'). \tag{7.40}$$

Such equations can, of course, be converted into a system of the form (7.38) by the procedure at the beginning of Section 7.5. On the other hand, the exposition of techniques is sometimes a bit clearer in the real variable case. We will follow the latter path in the discussion of the linear equation case (Section 7.6.1), but in Section 7.6.2, reduction to first-order equations is undertaken.

Typically, (7.39), as well as (7.38), is a system of equations. A common instance of (7.39) requires that $y(a) = A$ and $y(b) = B$, A and B being given constants. This can be phrased as

$$g(y(a), y(b)) = \begin{bmatrix} y(a) - A \\ y(b) - B \end{bmatrix} = \begin{bmatrix} 0 \\ 0 \end{bmatrix}. \qquad (7.41)$$

If the boundary conditions are specified as $f(a) = A$, $f'(b) = B$, this can be rewritten as

$$g(y(a), y'(b)) = \begin{bmatrix} y(a) - A \\ y'(b) - B \end{bmatrix} = \begin{bmatrix} 0 \\ 0 \end{bmatrix}. \qquad (7.42)$$

───── **EXAMPLE 7.14** ─────

Take the points $a = 0$ and $b = 1$. Our test differential equation

$$y' = x^2 + y^2$$

leads to a two-point boundary-value problem if the conditions

$$g(y(0), y(1)) = y(0) + y(1) - 1 = 0$$

are imposed. Solution of this problem will exemplify the "shooting method" discussed in Section 7.6.2.

■

*7.6.1   Linear Two-Point Boundary-Value Problems

If the differential equation and the boundary conditions are linear in y, the problem may be converted to an initial-value problem and solved relatively conveniently by the technique introduced in Sections 7.3 and 7.5. We illustrate this conversion in the second-order case (7.40) with condition (7.41). In the linear case, the problem takes the form

$$y'' + p(x)y' + q(x)y = r(x), \qquad y(a) = A, \quad y(b) = B, \qquad (7.43)$$

where the functions $p(x)$, $q(x)$, and $r(x)$ are assumed to be continuous on the interval $[a, b]$. The reduction to initial-value problems proceeds in two stages.

First, solve the homogeneous initial-value problem

$$u'' + p(x)u' + q(x)u = 0, \qquad u(a) = 0, \quad u'(a) = 1, \qquad (7.44)$$

and then solve the inhomogeneous problem

$$z'' + p(x)z' + q(x)z = r(x), \qquad z(a) = A, \quad z'(a) = 0. \qquad (7.45)$$

Let $u(x)$ and $z(x)$ denote the solutions of the initial-value problems (7.44) and (7.45), respectively. As we shall verify, a suitable linear combination of functions $u(x)$ and $z(x)$ gives the solution of the original boundary-value problem.

Consider the function

$$y(x) = \alpha u(x) + z(x), \tag{7.46}$$

where the number α is to be determined. Regardless of α, the function $y(x)$ satisfies the differential equation (7.43), since by substitution we have

$$
\begin{aligned}
y'' + p(x)y' + q(x)y &= (\alpha u'' + z'') + p(x)(\alpha u' + z') + q(x)(\alpha u + z) \\
&= \alpha[u'' + p(x)u' + q(x)u] + [z'' + p(x)z' + q(x)z] \\
&= \alpha \cdot 0 + r(x) = r(x).
\end{aligned}
$$

The boundary condition $y(a) = A$ is also satisfied since

$$y(a) = \alpha u(a) + z(a) = \alpha \cdot 0 + A = A.$$

Coefficient α will now be determined so that $y(x)$ satisfies the remaining boundary condition $y(b) = B$. By substitution we obtain

$$y(b) = \alpha u(b) + z(b),$$

which equals B if and only if

$$\alpha = \frac{B - z(b)}{u(b)}.$$

Thus the function

$$\boxed{y(x) = z(x) + \frac{B - z(b)}{u(b)} u(x)} \tag{7.47}$$

is the solution of the original boundary-value problem.

───── **EXAMPLE 7.15** ─────────────────────────────────────

Consider the boundary-value problem

$$y'' - 3y' + 2y = 2, \qquad y(0) = 1, \quad y(1) = 4.$$

In this case the initial-value problem (7.44) has the form

$$u'' - 3u' + 2u = 0, \qquad u(0) = 0, \quad u'(0) = 1,$$

which has a unique solution

$$u(x) = e^{2x} - e^x.$$

The initial-value problem (7.45) can be written as

$$z'' - 3z' + 2z = 2, \qquad z(0) = 1, \quad z'(0) = 0,$$

and

$$z(x) = 1$$

is the only solution. Since

$$u(b) = u(1) = e^2 - e \text{ and } z(b) = 1,$$

(7.47) implies that the function

$$y(x) = 1 + \frac{4 - 1}{(e^2 - e)} \cdot (e^{2x} - e^x) = 1 + \frac{3(e^{2x} - e^x)}{e^2 - e}$$

gives the solution of the original boundary-value problem. ∎

*7.6.2   The Shooting Method for Nonlinear Two-Point Boundary-Value Problems

If either the differential equation or the boundary condition is nonlinear in y, we have a nonlinear two-point boundary-value problem. Consider first the first-order representation (7.38). If the value of $y(a)$ were known, it would give a complete initial condition through the differential equation (7.38), and under the usual regularity conditions, it uniquely determines $y(b)$. In the shooting method, we regard $y(a)$ as a variable and seek to determine the value of $y(a)$ that satisfies the two-points boundary condition

$$F(y(a)) = g(y(a), y(b)) = 0. \tag{7.48}$$

The rationale here is that $y(b)$ is a uniquely determined function (through the differential equation) of $y(a)$. We remark that in the second-order case (7.40), the initial condition is the vector $\mathbf{y}(a) = (y(a), y'(a))^T$, and in this case, in the shooting method, we seek the root of $F(y(a), y'(a))$. The shooting method is not really a "method" in the sense that it leaves unspecified how we are supposed to solve the resulting nonlinear equation.

The nomenclature "shooting" can be motivated by the ballistics problem illustrated in Figure 7.6. Here we have a cannon located at position a on the x-axis, and a target at $x = b$. If an artillery shell with known initial velocity V is subject only to Newton's law of motion and viscous damping, then its trajectory $y(x)$ satisfies a second-order differential equation of the form (7.40). The natural boundary condition is that the height $y(x)$ of the shell should be 0 at a and b, or $y(a) = y(b) = 0$. Under the shooting method, one adjusts $y'(a) = \tan(\theta)$ until $y(b) = 0$.

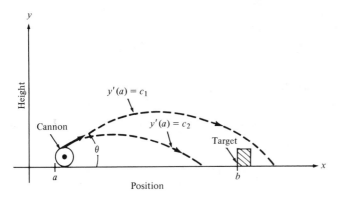

FIGURE 7–6 Shooting Method Application

Toward solving (7.48), one may employ techniques of Chapter 5. If, as in the ballistics problem, the initial condition allows only one free variable z, then (7.48) has the form

$$F(z) = 0.$$

The secant method leads, in this case, to the iteration scheme

$$z_{k+1} = z_k - F(z_k) \frac{z_k - z_{k-1}}{F(z_k) - F(z_{k-1})}, \qquad (7.49)$$

where z_0 and z_1 are two initial numbers to be specified or guessed. Then in this case, the shooting method can be implemented by conjunctive use of the subroutine SECA (Table 5.5) for secant iterations on successive initial conditions, and the classical Runge–Kutta routine RUKU (Table 7.7) to calculate the requisite terminal condition $y(b)$ needed to evaluate $F(z)$.

───── **EXAMPLE 7.16** ─────

We illustrate the shooting method implementation using secant iterations to update the initial condition and Runge–Kutta computation of the terminal state. Consider Example 7.14, which we restate:

$$y' = x^2 + y^2, \qquad y(0) + y(1) = 1. \qquad (7.50)$$

By means of the calling program listed in Table 7.19, we have obtained initial state iterations (Table 7.20) which do indeed converge to a solution of (7.50). The starting parameters z_0 and z_1 for the routine SECA were selected as 0.0 and 0.1, respectively. We call the reader's attention to our redefining the external function for SECA as FUNCTION $Y1(X)$, instead of FUNCTION $F(X)$, as described in the SECA comment lines. The reason for this modification, which must be accounted for in the two instructions at which SECA calls for $F(X)$, is that RUKU also requires an external function designated by F, and this refers to the derivative function $f(x, y)$ of the differential equation, not the boundary condition.

TABLE 7.19   Program Implementing the Shooting Method

```
C       PROGRAM SMETH
C
C       ****************************************************************
C       THIS PROGRAM WILL SET UP AND SOLVE NUMERICALLY THE DIFFERENTIAL
C       EQUATION Y'=X**2+Y**2 WITH THE BOUNDARY CONDITION Y(0.)+Y(1.)=1.
C       THE SOLUTION IS OBTAINED USING THE SHOOTING METHOD.
C       SUBROUTINE SECA IS USED TO UPDATE THE INITIAL CONDITIONS
C       AND RUKU WILL CALCULATE THE TERMINAL STATE AT EACH ITERATION
C       CALLS:    SECA (MODIFIED AS DESCRIBED IN THE TEXT),RUKU
C       OUTPUT:
C                 X(I)=VALUE OF X FOR I=1,21
C                 Y(I)=APPROXIMATED VALUE OF Y AT X(I)
C       ****************************************************************
C
C       *** FIRST, ASSIGN STARTING PARAMETERS FOR SUBROUTINE SECA   ***
C
        Z0=0.
        Z1=.1
        EPS=1.E-5
C
C       *** SUBROUTINE SECA WILL UPDATE INITIAL CONDITIONS          ***
C       *** UNTIL THE BOUNDARY CONDITIONS ARE SATISFIED            ***
C
        CALL SECA(Z0,Z1,XX,EPS)
        STOP
        END
C
C       *** FUNCTION Y1 WILL SOLVE THE INITIAL VALUE PROBLEM USING  ***
C       *** SUBROUTINE RUKU AND THE INITIAL CONDITION PASSED BY SECA ***
C
        FUNCTION Y1(C)
        DIMENSION X(21),Y(21)
C
C       *** SET UP INITIAL CONDITIONS                              ***
C
        X(1)=0.
        Y(1)=C
        CALL RUKU(X,Y,20,1.,21)
        Y1=1-(C+Y(21))
        WRITE(10,1) C,Y(21),Y1
      1 FORMAT(5X,3(F10.6,5X))
        RETURN
        END
C
C       *** FUNCTION F SPECIFIES THE DIFFERENTIAL EQUATION. IT IS   ***
C       *** CALLED BY SUBROUTINE RUKU                              ***
C
        FUNCTION F(X,Y)
        F=X**2+Y**2
        RETURN
        END
```

**TABLE 7.20   Results of Shooting
Method Iteration**

j	$y(0) = z_j$	$y(1)$	$F(z_j)$
0	0.100000	0.482560	0.417440
1	0.000000	0.350232	0.649768
2	0.279677	0.817244	-0.096922
3	0.245821	0.741406	0.012774
4	0.249763	0.749868	0.000369
5	0.249880	0.750121	-0.000002
6	0.249880	0.750120	0.000000

*7.7
ADAPTIVE STEP-SIZE SELECTION AND ERROR CONTROL

Up to this point we have not discussed how the step size h of the preceding methods is to be chosen. Obviously, there is a trade-off to be made: If the step size is too small, then computer time is needlessly wasted and accumulation of arithmetic roundoff errors can become a hazard. A large step size invites large truncation error associated with higher-order terms neglected in the construction of the methods. For simplicity, our developments will be concerned only with Runge–Kutta rules.

Techniques for automatic step-size selection are based on estimating the local error at each step and then choosing the step size to keep this estimated error within some tolerance bound. Thus step-size selection hinges on estimation of the *local error,* which at the jth step is defined to be

$$\hat{y}(x_{j+1}) - y_{j+1}.$$

Here y_{j+1} is, of course, the computed approximation of $y(x_{j+1})$, and $\hat{y}(x_{j+1})$ we define to be the exact value at x_{j+1} of the differential equation solution that passes through the point (x_j, y_j). That is, $\hat{y}(x)$ solves the initial-value problem

$$\hat{y}' = f(x, \hat{y}), \qquad \hat{y}(x_j) = y_j.$$

In contrast to local errors, the *global error* at x_{j+1} is defined to be

$$y(x_{j+1}) - y_{j+1},$$

where $y(x)$ is the exact solution of the original initial-value problem (7.3). Figure 7.7 illustrates the relationships between $y(x)$, $\hat{y}(x)$, and local and global errors. Intuitively, the local error is the additional truncation error arising from inexact solution at a given step. The global error gives the accumulated total error propagating from the entire sequence of steps.

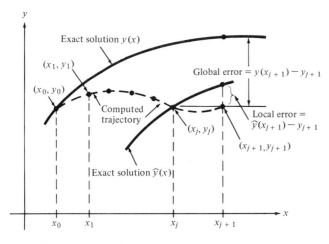

FIGURE 7–7 Relationship Between $y(x)$, $\hat{y}(x)$, Local and Global Errors

Assume that some Runge–Kutta procedure has been selected. We let y_0, y_1, y_2, . . . denote the computed solution values at the arguments x_0, x_1, x_2, ... The local error estimation techniques at each stage apply a higher-order technique to compute an additional approximation, say z_{j+1}, of $y(x_{j+1})$. Since a higher-order technique is used, if the solution is "well behaved" and the step size h is small enough that neglected terms really are negligible, then one may anticipate that the local error of the higher-order method is much less than that of the selected Runge–Kutta procedure. That is,

$$|\hat{y}(x_{j+1}) - z_{j+1}| << |\hat{y}(x_{j+1}) - y_{j+1}|. \tag{7.51}$$

If the approximation above indeed holds, then

$$z_{j+1} - y_{j+1} \simeq \hat{y}(x_{j+1}) - y_{j+1}. \tag{7.52}$$

and we take $z_{j+1} - y_{j+1}$ as the estimate of local error.

Of course, computation of z_{j+1} is typically more expensive than that of y_{j+1} itself, since z_{j+1} must be more accurate. Here, as in other walks of life, information must be paid for. A popular idea toward making this expense as small as possible has been offered by Fehlberg (1964). For a given order, say $p + 1$, the corresponding member of the Fehlberg family computes z_{j+1} with a minimum number of function calls, according to the limitations in Table 7.10, and then provides the pth-order estimate y_{j+1} without any additional function calls. A particularly popular Fehlberg rule is given in Table 7.21, which gives a fifth-order estimate z_{j+1} for a fourth-order rule y_{j+1}.

Subroutine RKF (Table 7.22) implements a single step of this Runge–Kutta–Fehlberg formula, outputting y_{j+1} and z_{j+1} as the parameters YOUT and ZOUT. In view of (7.52), the difference of these values provides a local error estimate. Subroutine ARUKU (Table 7.23) utilizes RKF to update the step size as the computation progresses. If the absolute value of ZOUT-YOUT is less than

TABLE 7.21 Runge–Kutta–Fehlberg Formula

$$k_1 = f(x_j, y_j)$$

$$k_2 = f\left(x_j + \frac{1}{4}h, y_j + \frac{1}{4}hk_1\right)$$

$$k_3 = f\left(x_j + \frac{3}{8}h, y_j + h\left(\frac{3}{32}k_1 + \frac{9}{32}k_2\right)\right)$$

$$k_4 = f\left(x_j + \frac{12}{13}h, y_j + h\left(\frac{1932}{2197}k_1 - \frac{7200}{2197}k_2 + \frac{7296}{2197}k_3\right)\right)$$

$$k_5 = f\left(x_j + h, y_j + h\left(\frac{439}{216}k_1 - 8k_2 + \frac{3680}{513}k_3 - \frac{845}{4104}k_4\right)\right)$$

$$k_6 = f\left(x_j + \frac{1}{2}h, y_j + h\left(-\frac{8}{27}k_1 + 2k_2 - \frac{3544}{2565}k_3 + \frac{1859}{4104}k_4 - \frac{11}{40}k_5\right)\right)$$

$$y_{j+1} = y_j + h\left(\frac{25}{216}k_1 + \frac{1408}{2565}k_3 + \frac{2197}{4104}k_4 - \frac{1}{5}k_5\right)$$

$$z_{j+1} = y_j + h\left(\frac{16}{135}k_1 + \frac{6656}{12825}k_3 + \frac{28561}{56430}k_4 - \frac{9}{50}k_5 + \frac{2}{55}k_6\right)$$

TABLE 7.22 Subroutine RKF for the Runge–Kutta–Fehlberg Formula

```
      SUBROUTINE RKF(XI,YI,H,YOUT,ZOUT)
C
C     *************************************************************
C     *  FUNCTION: A CALL TO THIS SUBROUTINE COMPUTES ONE STEP OF*
C     *            THE SOLUTION AND A GUESS OF THE ERROR FOR A   *
C     *            DIFFERENTIAL EQUATION Y'=F(X,Y) WITH INITIAL  *
C     *            VALUES XI, YI. THIS SOLUTION IS OBTAINED      *
C     *            USING A 4-TH ORDER RUNGE-KUTTA FEHLBERG STEP  *
C     *            METHOD IMBEDDED IN A 5-TH ORDER STEP SOLUTION *
C     *  USAGE:                                                  *
C     *        CALL SEQUENCE: CALL RKF(XI,YI,H,YOUT,ZOUT)        *
C     *        EXTERNAL FUNCTIONS/SUBROUTINES: FUNCTION F(U,V)   *
C     *  PARAMETERS:                                             *
C     *        INPUT:                                            *
C     *            XI=INDEPENDENT VARIABLE INITIAL VALUE         *
C     *            YI=DEPENDENT VARIABLE INITIAL VALUE           *
C     *            H=INTERVAL STEP SIZE                          *
C     *        OUTPUT:                                           *
C     *          YOUT=4-TH ORDER SOLUTION ESTIMATE              *
C     *          ZOUT=5-TH ORDER SOLUTION ESTIMATE              *
C     *             (ZOUT-YOUT=LOCAL ERROR ESTIMATE)            *
C     *************************************************************
C
      REAL K1,K2,K3,K4,K5,K6
      K1=F(XI,YI)
      U=XI+0.25*H
      V=YI+0.25*H*K1
      K2=F(U,V)
      U=XI+(3./8.)*H
      V=YI+H*((3./32.)*K1+(9./32.)*K2)
      K3=F(U,V)
      U=XI+H*(12./13.)
      V=YI+(H/2197.)*(1932.*K1-7200.*K2+7296.*K3)
      K4=F(U,V)
      U=XI+H
      V=YI+H*((439./216.)*K1-8.*K2+(3680./513.)*K3-
     1(845./4104.)*K4)
      K5=F(U,V)
      U=XI+0.5*H
      V=-(8./27.)*K1+2.*K2-(3544./2565.)*K3+
     1(1859./4104.)*K4-(11./40.)*K5
      V=YI+H*V
      K6=F(U,V)
      YOUT=(25./216.)*K1+(1408./2565.)*K3+
     1(2197./4104.)*K4-K5/5.
      YOUT=YI+H*YOUT
      ZOUT=(16./135.)*K1+(6656./12825.)*K3+
     1(28561./56430.)*K4-(9./50.)*K5
      ZOUT=ZOUT+(2./55.)*K6
      ZOUT=YI+H*ZOUT
      RETURN
      END
```

the user-specified value TOL (for tolerance), the value ZOUT is accepted for y_{j+1}, and a larger step size (by a factor of 3) is chosen for the next step. Otherwise, h is reduced by a factor of 10, and the computation is repeated from the same condition x_j and y_j. Strictly speaking, YOUT, rather than ZOUT, should be chosen for y_{j+1}, but since in principle the higher-order estimate ZOUT should be more accurate, and since it is available, we adopt the pragmatic viewpoint that it should be used. The reader will note that AKUKU is an obvious modification of subroutine ASIMP for adaptive quadrature (Section 3.8.2).

**TABLE 7.23   Subroutine AKUKU for the Adaptive
Runge–Kutta Method**

```
        SUBROUTINE ARUKU(X,Y,B,M,TOL)
C
C  ****************************************************************
C  *  FUNCTION: THIS SUBROUTINE COMPUTES THE SOLUTION OF A        *
C  *            DIFFERENTIAL EQUATION BY ADAPTIVELY CHOOSING      *
C  *            THE STEP SIZE TO LIMIT THE LOCAL ERROR EST-       *
C  *            IMATE WITHIN A GIVEN TOLERANCE. A 4-TH ORDER      *
C  *            RUNGE-KUTTA-FEHLBERG METHOD IS USED               *
C  *  USAGE:                                                      *
C  *      CALL SEQUENCE: CALL ARUKU(X,Y,B,M,TOL)                  *
C  *      EXTERNAL FUNCTIONS/SUBROUTINES:                         *
C  *                SUBROUTINE RKF(XI,YI,H,YOUT,ZOUT)             *
C  *  PARAMETERS:                                                 *
C  *      INPUT:                                                  *
C  *        X(1)=INDEPENDENT VARIABLE INITIAL VALUE               *
C  *        Y(1)=DEPENDENT VARIABLE INITIAL VALUE                 *
C  *          B=SOLUTION INTERVAL ENDPOINT (LAST X VALUE)         *
C  *          M=MAXIMUM NUMBER OF ITERATIONS                      *
C  *      OUTPUT:                                                 *
C  *          X=M BY 1 ARRAY OF INDEPENDENT VARIABLE VALUES       *
C  *          Y=M BY 1 ARRAY OF DEPENDENT VARIABLE SOLUTION       *
C  *            VALUES                                            *
C  ****************************************************************
C
        DIMENSION X(M),Y(M)
C  *** INITIALIZATION ***
        H=.10E-02
        I=1
        N=0
C  *** COMPUTE SOLUTION ITERATIVELY ***
        DO WHILE(X(I).LE.B)
          N=N+1
          CALL RKF(X(I),Y(I),H,YOUT,ZOUT)
          ERR=ZOUT-YOUT
C  *** TEST IF THE NUMBER OF ITERATIONS EXCEEDED ***
          IF(N.GT.M) THEN
            WRITE(6,1)
   1        FORMAT(1X,'PROGRAM STOPPED TOO MANY ITERATIONS')
            STOP
          END IF
C  *** TEST STEP SIZE ***
          IF(ABS(ERR).LT.TOL) THEN
            I=I+1
            X(I)=X(I-1)+H
            H=3.0*H
            Y(I)=ZOUT
          ELSE
            H=H/10.0
          END IF
        END DO
        M=I
        H=B-X(I-1)
        X(M)=X(I-1)+H
        CALL RKF(X(I-1),Y(I-1),H,Y(M),ZOUT)
        RETURN
        END
```

We serve notice that the code ARUKU is intended only to illustrate the principles of automatic error control. It is inefficient and does not have the safeguards
of a professional differential equation program package. More will be said about
this matter after the following computational example.

━━━━━ **EXAMPLE 7.17** ━━━

By means of the calling program given in Table 7.24, the automatic step-size routine ARUKU is called on to solve the differential equation

$$y' = y, \qquad y(0) = 1 \tag{7.53}$$

over the interval [0, 1]. We chose this over our "usual" differential equation because in the present case it is easy to compute the exact local error $\hat{y}(x_{j+1}) - y_{j+1}$ and thereby see how well the RKF error estimator is doing. Specifically, the solution of (7.53) that passes through points (x_j, y_j) is

$$\hat{y}(x) = y_j \exp (x - x_j),$$

and if h is the current step size, then the exact local error is given by

$$y_j \exp (h) - \text{YOUT}.$$

TABLE 7.24 Calling Program for the Subroutine ARUKU

```
C       PROGRAM RKFMETH
C
C       ****************************************************************
C       THIS PROGRAM WILL SET UP AND SOLVE NUMERICALLY THE DIFFERENTIAL
C       EQUATION Y'=Y WITH THE INITIAL CONDITION Y(0.)=1.
C       THE SOLUTION IS OBTAINED USING THE AUTOMATIC STEPSIZE ROUTINE
C       USING 4-TH ORDER RUNGE-KUTTA FEHLBERG METHOD
C       CALLS:    ARUKU,RKF (BOTH MODIFIED FOR DOUBLE PRECISION)
C       OUTPUT:
C                 X(I)=VALUE OF X FOR I=1,...  (MAX=50)
C                 Y(I)=APPROXIMATED VALUE OF Y AT X(I)
C       ****************************************************************
C
        IMPLICIT DOUBLE PRECISION(A-H,O-Z)
        DIMENSION X(50),Y(50)
C
C       *** FIRST, THE INITIAL CONDITION AND ENDPOINT ARE ESTABLISHED***
C       *** THE MAXIMUM NUMBER OF ITERATIONS IS SET TO 50          ***
C
        X(1)=0.D0
        Y(1)=1.D0
        B=1.D0
        MAX=50
        TOL=1.D-4
C
C       *** SUBROUTINE ARUKU WILL APPROXIMATE THE SOLUTION         ***
C       *** RETURNING AT MOST 50 VALUES IN ARRAYS X AND Y          ***
C
        CALL ARUKU(X,Y,B,MAX,TOL)
        WRITE(10,*)(X(I),Y(I),I=1,50)
        STOP
        END
C
C       *** FUNCTION F SPECIFIES THE DIFFERENTIAL EQUATION. IT IS  ***
C       *** CALLED BY SUBROUTINE RKF                               ***
C
        FUNCTION F(X,Y)
        IMPLICIT DOUBLE PRECISION(A-H,O-Z)
        F=Y
        RETURN
        END
```

TABLE 7.25   Computation Using ARUKU

x_j	y_j	h	Computed Local Error	Actual Local Error
0.0000000	1.0000000	0.001	-0.278D-16	-0.278D-16
0.0010000	1.0010005	0.003	-0.305D-15	-0.305D-15
0.0040000	1.0040080	0.009	-0.757D-13	-0.753D-13
0.0130000	1.0130849	0.027	-0.184D-10	-0.181D-10
0.0400000	1.0408108	0.081	-0.451D-08	-0.424D-08
0.1210000	1.1286249	0.243	-0.111D-05	-0.892D-06
0.3640000	1.4390740	0.729	-0.276D-03	-0.455D-04
0.3640000	1.4390740	0.073	-0.369D-08	-0.350D-08
0.4369000	1.5479010	0.219	-0.911D-06	-0.750D-06
0.6556000	1.9262975	0.656	-0.226D-03	-0.651D-04
0.6556000	1.9262975	0.066	-0.293D-08	-0.279D-08
0.7212100	2.0569201	0.197	-0.722D-06	-0.608D-06
0.9180400	2.5043763	0.590	-0.179D-03	-0.696D-04
0.9180400	2.5043763	0.059	-0.225D-08	-0.216D-08
0.9770890	2.6567105	0.177	-0.555D-06	-0.477D-06
1.0000000	2.7182811	0.023	-0.555D-06	-0.477D-06

The bound TOL was set to 10^{-4}. In addition to converting the subroutines and program functions to double precision, a print statement was added to ARUKU to obtain output (Table 7.25) giving the estimated and actual local error and other information. The first line of Table 7.25 corresponds to the starting step-size value of $h = 0.001$. Since the computed local error ($-0.278E-16$) is less than TOL $= 10^{-4}$, in the next step h is multiplied by 3. Since the local error remains again under 10^{-4}, h is multiplied again by 3, and so on. The consecutive increased values of h, from 0.001 to 0.729, can be found in the first seven lines of the table. In the seventh line, the computed local error becomes greater than 10^{-4} (actually it equals $-0.276E-3$). Consequently, we divide h by 10 and repeat the step. In this case the new value of h becomes $0.729E-1$, and the new value of x becomes $0.3640000 + 0.072900 = 0.436900$. Along this path, we obtain the further lines of Table 7.25. The computed estimate of $y(1)$ is 2.7182811. Note that to the accuracy shown, the exact value $y(1)$ is 2.7182818. We see that in this case, the Runge–Kutta–Fehlberg technique is successful in giving "ballpark" error estimates. As is to be expected, the smaller the step size h, the better the estimate. The computation required 16 calls to RKF, of which 13 were accepted for new y_j values.

We emphasize that routine ARUKU is intended for illustrative use only. It does not have sophistication and safeguards one would expect of a professional program package. A popular package, the code RKF45 by Shampine and Watts, is presented in Forsythe et al. (1977, Chap. 6). This algorithm, like ours, is based on the formula given in Table 7.21, but it has much more stringent tests for step-size adjustment and initialization. It also includes provision for approximating the "optimal" step size—the largest step that does not lead to violation of the local error bound. The price paid for the greater sophistication is that the code occupies 14 pages of fairly small print.

7.8
SUPPLEMENTARY NOTES AND DISCUSSIONS

Numerical solution of differential equations can, in some respects, be regarded as a generalization of numerical integration, or quadrature, discussed in Chapter 3. For example, if $f(x, y)$ in (7.3) really depends only on x, that is, if $f(x, y) = f(x)$, then the initial-value problem has the form $y'(x) = f(x)$, $y(x_0) = y_0$ and the solution is

$$y(x) = y(x_0) + \int_{x_0}^{x} f(x)\, dx.$$

For such functions, some of the differential equation methods coincide with standard quadrature formulas. For example, the corrected Euler's method presented in Example 7.5 specializes to the midpoint rule.

Another similarity between the major integration and differential equation solution methods are that they tend to provide evidence of the value of mathematical analysis. Sophisticated but unintuitive methods are far more accurate than commonsense approaches. For example, in integration, common sense would seem to suggest that the best one could do with n points is to uniformly sample and average the functional values over the domain of integration. This should result in something like the midpoint or trapezoidal rule. But, for well-behaved functions, Gaussian quadrature, the points and weights of which it is fair to say are unlikely to be guessed by anyone, is typically more accurate by many orders of magnitude. Similarly, in this chapter we saw that Runge–Kutta rules, about which we have no heuristic explanation to offer except in terms of Taylor's expansions and numerical differentiation, are much more successful, for a given computational effort, than the intuitively obvious Euler's method.

It is a mathematical fact that the values of x_j and y_j uniquely determine the solution values at x greater than x_j. With this fact in mind, without allusion to interpolation polynomials, it is hard to explain why Adams methods should successfully use earlier information (x_{j-1}, y_{j-1}), (x_{j-2}, y_{j-2}), . . . from the more distant past. It is in this sense that we assert that the demonstrated success of Gauss quadrature, Runge–Kutta, and linear multistep methods are a tribute to the value of series expansions in particular and mathematical analysis in general.

In the examples of this chapter we were able to obtain fairly accurate answers to our test differential equation problem, without great programming effort or expenditure of running time. In fact, hand-held programmable calculators often suffice for textbook sized problems. As further evidence of the manageability of initial-value problems for ordinary differential equations, we mention that there are a number of good computer packages available. For example, the IMSL library, resident at many computer centers, has three packages, one using a Runge–Kutta method, another based on linear-multistep techniques, and a third on an approach not studied here. These and other sophisticated codes provide inexpensive, reliable answers to well-behaved initial-value differential equation problems. Such packages have error control features that incorporate adaptive step-size selection in order to maintain an error tolerance specified by the user.

Two classes of numerical differential equation problems require considerable sophistication. The first of these is the nonlinear two-point boundary problem. By "nonlinear" we mean that $f(x, y)$ in (7.3) cannot be expressed in the special form

$$f(x, y(x)) = g_1(x)y(x) + g_2(x). \tag{7.54}$$

By "two-point" we mean constraints such as those discussed in Section 7.6. For such problems readers are advised to seek professional help or continue their studies in numerical methods. The other important class of truculent problems is initial-value differential equations that are "stiff." The notion of "stiff" is somewhat imprecise, but it embodies the idea that the solution has various components, some of which change much more rapidly than others. Quite often, it is hard to tell in advance whether a given differential equation is stiff. Among the methods discussed in this chapter, implicit Adams formulas have the most attractive theoretical features for stiff systems. But implicit Runge–Kutta formulas (which we have not discussed) are also of value in this context. The IMSL library program DGEAR has a stiff differential equation option based on the Adams multistep approach. Recent editions of IMSL also contain two-point boundary-value problem codes, one based on the shooting method; the other, for linear problems, uses difference equation techniques. The basic idea of difference equation methods can be found in Szidarovszky and Yakowitz (1978, Sec. 8.2.3).

PROBLEMS

Section 7.2

1. Approximate the solutions of the following initial-value problems by the Taylor's series technique (7.5), with $m = 3$. Show your derivations.
   (a) $y' = x^2 y$, $y(0) = 1$.
   (b) $y' = x^2 + y$, $y(0) = 1$.
   (c) $y' = \cos(x)y$, $y(0) = 1$.
   (d) $y' = x^2 y$, $y(1) = 1$.

2. Apply the recursive Taylor's series technique (7.6) to the differential equations in Problem 1. Use a computer and steps of size $h = 0.1$. Evaluate for Taylor's series orders $m = 1, 2$, and 3. Compute the solution. Below we list exact solutions to the different parts of Problem 1. Compare your computed solution with these.
   (a) $y(x) = \exp(x^3/3)$.
   (b) $y(x) = e^x + 2e^x[1 - e^{-x}(1 + x + x^2/2)]$.
   (c) $y(x) = \exp[\sin(x)]$.
   (d) $y(x) = \exp\left(\dfrac{x^3}{3} - \dfrac{1}{3}\right)$.

*3. Extend the theory behind the Taylor's series technique (7.5) to approximate a second-order differential equation

   $$y'' = f(x, y, y'), \qquad y(x_0) = y_0, \quad y'(x_0) = y_0'.$$

   Apply your method to the problem

   $$y'' = y' + y^2 + x, \qquad y(0) = 1, \quad y'(0) = 1.$$

   Find the fifth-degree Taylor's polynomial for $y(x)$. [**HINT:** $y(x_0)$ and $y'(x_0)$ are given by the initial condition, $y''(x_0)$ can be obtained from the differential equation, and the derivatives $y^{(3)}(x_0)$, $y^{(4)}(x_0)$, . . . can be determined by repeated use of the chain rule.]

4. Show that the Taylor's series approach is not effective for the initial-value problem

$$y' = y\sqrt{x}, \qquad y(0) = 1.$$

[**HINT:** Show that $y''(0)$ does not exist.]

5. Solve the initial-value problems in Problem 1 over the interval $[0, 1]$ by Euler's method. Repeat using step sizes $h = (0.1)^k$, $k = 1, 2, 3$. Compare with exact solutions, given in Problem 2.

Section 7.3
*6. Solve the differential equations of Problem 1 over the interval $[0, 1]$ by the corrected Euler's formula (7.18)–(7.19). Take step size $h = (0.1)^k$, $k = 1, 2, 3$. Compare with the results of Problem 5, if you did it, and with exact solutions (Problem 2).

7. Redo Problem 5 with the classical Runge–Kutta formula. (**HINT:** Subroutine RUKU may be used.)

*8. Find a two-stage Runge–Kutta rule in which $\alpha_1 = \frac{1}{3}$. Is it unique?

Section 7.4
9. Solve Problem 1 by the Adams–Bashforth formulas, with number of steps 2 and 4 (see Table 7.12). Obtain the initial values by the classical Runge–Kutta method. Compare with exact solutions (Problem 2).

*10. Whereas we noted that the coefficients β_m in the Adams methods (7.30) are defined so as to give a maximal-order truncation error, it turns out that this can be achieved by integrating the interpolation polynomial for the data $\{(x_{j+1-m}, f_{j+1-m})\}$. Specifically, after integrating the differential equation

$$y'(x) = f(x, y(x))$$

we conclude that

$$y(x_{j+1}) = y(x_j) + \int_{x_j}^{x_{j+1}} f(x, y(x)) \, dx.$$

Let $p(x) = \sum_{m=1}^{i} f_{j+1-m} l_m(x)$ be the interpolation polynomial for $f(x, y(x))$, based on the values at (x_{j+1-m}, f_{j+1-m}), $1 \le m \le i$. Thus

$$l_m(x) = \prod_{\substack{v=1 \\ v \ne m}}^{i} \frac{x - x_v}{x_m - x_v}.$$

Then the interpolatory quadrature formula for $\int_{x_j}^{x_{j+1}} f(x, y(x)) \, dx$ is, in view of Section 3.3, given by $\sum_{m=1}^{i} h\beta_m f_{j+1-m}$, where

$$h\beta_m = \int_{x_j}^{x_{j+1}} l_m(x) \, dx.$$

Use this insight to verify the Adams–Bashforth coefficients for $i = 2$ in Table 7.12.

*11. Solve Problem 1(a) by the predictor–corrector technique. Take $h = 0.1$, use $i = 2$ steps for both the predictor and corrector, and use only one corrector iteration at each step.

Section 7.5

12. Modify subroutine EULER (Table 7.2) to solve simultaneous initial-value problems of any input dimension M.

13. Solve the following initial-value problems by use of the vectorial version of the Euler's method developed in Problem 12. Take $h = 0.1$ and perform 10 steps.

(a) $y_1' = x^2 + y_1^2 + y_2$ $\qquad$ $y_1(0) = 0$

$\qquad$ $y_2' = x - y_1 + y_2^2$ $\qquad$ $y_2(0) = 1.$

(b) $y_1' = xy_1y_2$ $\qquad\qquad$ $y_1(1) = 1$

$\qquad$ $y_2' = x + y_1 + y_2$ $\qquad$ $y_2(1) = 2.$

(c) $y_1' = xy_1 + y_2^2 + y_3$ $\qquad$ $y_1(0) = 0$

$\qquad$ $y_2' = y_1 - y_2 + y_3$ $\qquad$ $y_2(0) = 1$

$\qquad$ $y_3' = y_1y_2y_3$ $\qquad\qquad$ $y_3(0) = 2.$

14. Solve the simultaneous ordinary differential equations in Problem 13 using the subroutine MRUKU (Table 7.17).

15. Reduce the following higher-order equations to systems of first-order differential equations.

(a) $y^{(3)} = x + y + 2y' + y''^2$ $\qquad$ $y(0) = y'(0) = y''(0) = 0.$

(b) $y'' = (x + y + y'^2)^3$ $\qquad\qquad$ $y(1) = 1, \quad y'(1) = -2.$

(c) $y_1'' = xy_1y_2'$ $\qquad\qquad\qquad$ $y_1(0) = y_1'(0) = 0$

$\qquad$ $y_2'' = x - y_1' + y_2'^2$ $\qquad$ $y_2(0) = y_2'(0) = 1.$

(**Hint:** For part (c) introduce the new variables $z_1 = y_1, z_2 = y_1', z_3 = y_2, z_4 = y_2'.$)

Section 7.6

16. Reduce the following boundary-value problems to initial-value problems.

(a) $y'' + xy' + e^x y = 1$ $\qquad\qquad\qquad\qquad\qquad$ $y(0) = 0, \quad y(1) = 1.$

(b) $y'' + y = e^x$ $\qquad\qquad\qquad\qquad\qquad\qquad\quad$ $y(1) = 2, \quad y(2) = 4.$

(c) $y'' + \sin(x) \cdot y' + \cos(x) \cdot y = \tan(x) + 1$ $\qquad$ $y(0) = 0, \quad y(1) = 1.$

17. Solve the boundary-value problems given in Problem 16 using the shooting method. Compare the results with those obtained by the reduction to an initial-value problem.

*18. The boundary-value problem

$$y'' = 2y' - y, \qquad y(0) = -1, \quad y(1) = 0, \quad y(2) = e^2$$

is called overdetermined, since the number of boundary conditions is greater than the order of the differential equation.

(a) Prove that the problem has a unique solution.

(b) If the value of $y(2)$ is specified to be 0 instead of e^2, prove that the problem has no solution. [**HINT:** Find first the unique solution of the problem when the third condition $y(2) = e^2$ is dropped.]

Section 7.7

*19. Repeat the calculations of Example 7.17 with TOL $= 10^{-4}$, $b = 1$, in the case of the following initial-value problems.

(a) $y' = x^2 y$ $\qquad\qquad$ $y(0) = 1$.
(b) $y' = x^2 + y$ $\qquad\quad$ $y(0) = 1$.
(c) $y' = \cos(x)y$ $\qquad\;$ $y(0) = 1$.
(d) $y' = x^2 + y^2$ $\qquad$ $y(0) = 0$ (our test equation).

Compare with exact solutions (given in Problem 2). (**HINT:** Modify the program given in Table 7.24.)

References ▬▬▬▬▬▬▬▬

Abramowitz, M., and I. A. Stegun, eds. (1965), *Handbook of Mathematical Functions with Formulas, Graphs and Mathematical Tables,* Dover Publications, Inc., New York.

Apostol, T. M. (1957), *Mathematical Analysis: A Modern Approach to Advanced Calculus,* 3rd printing, Addison-Wesley Publishing Co., Inc., Reading, Mass.

Bellman, R. (1970), *Introduction to Matrix Analysis,* 2nd ed., McGraw-Hill Book Company, New York.

Bickel, P. J., and K. Doksum (1977), *Mathematical Statistics,* Holden-Day, Inc., San Francisco.

Bloomfield, P. (1976), *Fourier Analysis of Time Series: An Introduction,* John Wiley & Sons, Inc., New York.

Butcher, J. C. (1964), On Runge–Kutta Processes of Higher Order, *J. Austral. Math. Soc., 4,* 179–194.

Dahlquist, G., and A. Bjorck (1974), *Numerical Methods,* Prentice-Hall, Inc., Englewood Cliffs, N.J.

Davis, P., and P. Rabinowitz (1975), *Methods of Numerical Integration,* Academic Press, Inc., New York.

de Boor, C. (1978), *A Practical Guide to Splines,* Springer-Verlag New York, Inc., New York.

Dennis, J. E., and R. B. Schnabel (1983), *Numerical Methods for Unconstrained Optimization,* Prentice-Hall, Inc., Englewood Cliffs, N.J.

Fehlberg, E. (1964), New High-Order Runge–Kutta Formulas with Stepsize Controls for Systems of First and Second Order Differential Equations, *Z. Angew. Math. Mech., 44,* 83–88.

Forsythe, G., M. Malcolm, and C. Moler (1977), *Computer Methods for Mathematical Computations,* Prentice-Hall, Inc., Englewood Cliffs, N.J.

Golub, G., and C. Van Loan (1983), *Matrix Computations,* The Johns Hopkins University Press, Baltimore, Md.

Haberman, R. (1977), *Mathematical Models,* Prentice-Hall, Inc., Englewood Cliffs, N.J.

Hart, J. F., et al. (1968), *Computer Approximations,* John Wiley & Sons, Inc., New York.

Henrici, P. (1962), *Discrete Variable Methods in Ordinary Differential Equations,* John Wiley & Sons, Inc., New York.

Henrici, P. (1982), *Essentials of Numerical Analysis,* John Wiley & Sons, Inc., New York.

Herstein, I. (1964), *Topics in Algebra,* Blaisdell Publishing Company, New York.

Huff, D. (1954), *How to Lie with Statistics,* W. W. Norton & Company, Inc., New York.

Isaacson, E., and H. Keller (1966), *Analysis of Numerical Methods,* John Wiley & Sons, Inc., New York.

Johnson, L. W., and R. Riess (1977), *Numerical Analysis,* Addison-Wesley Publishing Co., Inc., Reading, Mass.

Kaplan, W. (1952), *Advanced Calculus*, Addison-Wesley Publishing Co., Inc., Reading, Mass.

Knuth, D. (1969), *Seminumerical Algorithms*, Addison-Wesley Publishing Co. Inc., Reading, Mass.

Lancaster, P. (1969), *Theory of Matrices*, Academic Press, Inc., New York.

Lawson, C. L., and H. J. Hanson (1974), *Solving Least Squares Problems*, Prentice-Hall, Inc., Englewood Cliffs, N.J.

Luenberger, D. (1973), *Introduction to Linear and Nonlinear Programming*, Addison-Wesley Publishing Co., Inc., Reading, Mass.

Metropolis, N., J. Howlett, and G. C. Rota, eds. (1980), *A History of Computing in the Twentieth Century*, Academic Press, Inc., New York.

Nering, E. D. (1963), *Linear Algebra and Matrix Theory*, John Wiley & Sons, Inc., New York.

Ortega, J., and W. Poole (1981), *Numerical Methods of Differential Equations*, Pitman Publishing, Inc., Marshfield, Mass.

Ortega, J., and W. Rheinboldt (1970), *Iterative Solution of Nonlinear Equations in Several Variables*, Academic Press, Inc., New York.

Ralston, A., and P. Rabinowitz (1978), *A First Course in Numerical Analysis*, 2nd ed., McGraw-Hill Book Company, New York.

Rice, J. (1983), *Numerical Methods, Software, and Analysis*, IMSL Reference Edition, McGraw-Hill Book Company, New York.

Savage, L. J. (1954), *The Foundations of Statistics*, John Wiley & Sons, Inc., New York.

Stegun, I., and M. Abramowitz (1956), Pitfalls in Computation, *J. SIAM, 4*, 207–219.

Stewart, G. (1973), *Introduction to Matrix Computations*, Academic Press, Inc., New York.

Stoer, J., and R. Bulirsch (1980), *Introduction to Numerical Analysis*, Springer-Verlag New York, Inc., New York.

Szidarovszky F, and S. Yakowitz (1978), *Principles and Procedures of Numerical Analysis*, Plenum Press, New York.

U.S. Bureau of the Census (1984), *Statistical Abstract of the United States 1984*, U.S. Government Printing Office, Washington, D.C.

Wilf, H. (1960), The Numerical Solution of Polynomial Equations, in *Mathematical Methods for Digital Computers*, H. Ralston and H. Wilf, eds., John Wiley & Sons, Inc., New York.

Yakowitz, S. (1977), *Computational Probability and Simulation*, Addison-Wesley Publishing Co., Inc., Reading, Mass.

Fundamentals of Matrix Theory

Consider the following simultaneous linear equations

$$2x_1 + x_2 - x_3 = 5$$
$$x_1 + x_2 + x_3 = 5$$
$$-x_1 \qquad + x_3 = -1,$$

where x_1, x_2, and x_3 are the unknowns. The coefficients of the unknowns are conveniently written as a rectangular or, in this special case, a square array,

$$\mathbf{A} = \begin{bmatrix} 2 & 1 & -1 \\ 1 & 1 & 1 \\ -1 & 0 & 1 \end{bmatrix}, \tag{A.1}$$

which is called a matrix of coefficients. For purposes of this book, a *matrix* is any rectangular array of numbers.

$$\mathbf{A} = \begin{bmatrix} a_{11} & a_{12} & \cdots & a_{1n} \\ a_{21} & a_{22} & \cdots & a_{2n} \\ \vdots & \vdots & & \vdots \\ a_{m1} & a_{m2} & \cdots & a_{mn} \end{bmatrix} \tag{A.2}$$

A matrix with m rows and n columns is said to be of *order $m \times n$* (pronounced "m by n"). One often designates the coefficient in the ith row and jth column by a subscripted lowercase letter such as a_{ij}. Thus the coefficient a_{32} of the matrix in (A.1) is 0. When we write "the (i, j) coefficient (or coordinate or element) of $\mathbf{A}$," we refer to a_{ij}.

Matrices are denoted in this book by boldface capital letters, such as $\mathbf{A}$, $\mathbf{B}$, $\mathbf{C}$. The matrix having coefficients a_{ij} is sometimes denoted by $\mathbf{A} = (a_{ij})$ or by

$$\mathbf{A} = [a_{ij}]_{i,j=1}^{m,n}$$

An $m \times 1$ matrix is called a *column vector*, and a $1 \times n$ matrix is called a *row vector*. Thus vectors are matrices of a special form. In matrix algebra, single numbers are called *scalars*. Vectors are commonly designated by boldface lowercase letters such as **a**, **b**, **x**, and scalars by regular letters such as *a, b,* and *x*.

A.1
SPECIAL FORMS AND ALGEBRA
OF MATRICES AND VECTORS

In this section some definitions and properties of vectors and matrices are discussed.

1. For an arbitrary matrix of order $m \times n$, m, of course, need not be equal to n. In the important case that $m = n$, the matrix is called a *square matrix*. In this case, m is the *order* of the matrix. The coefficients $a_{11}, a_{22}, a_{33}, \ldots$ of a square matrix are called *diagonal elements,* and they form the main diagonal or simply the *diagonal* of the matrix.

2. A square matrix with all off-diagonal coefficients equal to zero is a *diagonal matrix*.

Figure A.1 and Example A.1 will help to illustrate the notion of square and diagonal matrices, as well as other special forms given below.

3. A square matrix in which all coefficients below the diagonal are equal to zero is called *upper triangular;* and similarly a square matrix with all zero coefficients above the diagonal is called *lower triangular.*

EXAMPLE A.1

Consider the following matrices:

$$
\mathbf{A} = \begin{bmatrix} 1 & 2 \\ 2 & 3 \\ 4 & 5 \end{bmatrix}, \qquad \mathbf{b} = [1 \quad 1 \quad 0 \quad 1], \qquad \mathbf{c} = \begin{bmatrix} 1 \\ 2 \\ 3 \end{bmatrix},
$$

$$
\mathbf{D} = \begin{bmatrix} 1 & 2 & 3 \\ 1 & 1 & 0 \\ 0 & 1 & -1 \end{bmatrix}, \qquad \mathbf{E} = \begin{bmatrix} 1 & 0 & 0 \\ 0 & 2 & 0 \\ 0 & 0 & -1 \end{bmatrix}, \qquad \mathbf{F} = \begin{bmatrix} 1 & 2 & 1 \\ 0 & 2 & 4 \\ 0 & 0 & -1 \end{bmatrix},
$$

$$
\mathbf{G} = \begin{bmatrix} 1 & 0 & 0 \\ 1 & 0 & 0 \\ -1 & -5 & 4 \end{bmatrix}.
$$

The matrix **A** is a 3×2 matrix with no special properties, **b** is a row vector, **c** is a column vector, **D** is a square matrix, **E** is diagonal, **F** is upper triangular, and **G** is lower triangular.

Observe that any matrix that is both lower triangular and upper triangular is necessarily diagonal.

4. A matrix composed entirely of zeros is called a *zero* (or *null*) *matrix,* and a vector of zeros is called a *zero* (or *null*) *vector*. We denote them by **O** and **0**, respectively.

$$\begin{bmatrix} \times & \cdots & \times \\ \cdot & & \cdot \\ \cdot & & \cdot \\ \cdot & & \cdot \\ \times & \cdots & \times \end{bmatrix} \quad \begin{array}{l} n \\ \text{rows} \end{array} \qquad \begin{bmatrix} \times & 0 & \cdots & 0 \\ 0 & \times & \cdots & 0 \\ \cdot & \cdot & \cdot & \cdot \\ \cdot & \cdot & \cdot & \cdot \\ 0 & 0 & \cdots & \times \end{bmatrix}$$

n columns

(a) Square matrix   (b) Diagonal matrix
                                    ($a_{ij} = 0$ if $i \neq j$)

$$\begin{bmatrix} \times & \times & \cdots & \times \\ 0 & \times & \cdots & \times \\ \cdot & \cdot & \cdot & \cdot \\ \cdot & \cdot & \cdot & \cdot \\ 0 & 0 & \cdots & \times \end{bmatrix} \qquad \begin{bmatrix} \times & 0 & \cdots & 0 \\ \times & \times & \cdots & 0 \\ \cdot & \cdot & & \cdot \\ \cdot & \cdot & & \cdot \\ \times & \times & \cdots & \times \end{bmatrix}$$

(c) Upper triangular matrix   (d) Lower triangular matrix
      ($a_{ij} = 0$ if $i > j$)   ($a_{ij} = 0$ if $i < j$)

FIGURE A–1   Some Special Matrix Forms

5. A diagonal matrix with all diagonal coefficients equal to 1 is called the *identity* (or *unit*) *matrix* and is denoted by **I**.

EXAMPLE A.2

Consider the matrices

$$\mathbf{A} = \begin{bmatrix} 0 & 0 & 0 \\ 0 & 0 & 0 \end{bmatrix}, \qquad \mathbf{B} = \begin{bmatrix} 1 & 0 & 0 \\ 0 & 1 & 0 \\ 0 & 0 & 1 \end{bmatrix}, \qquad \mathbf{c} = \begin{bmatrix} 0 \\ 0 \\ 0 \end{bmatrix}, \qquad \mathbf{d} = [0 \quad 0 \quad 0].$$

Here **A** is a zero matrix, **c** and **d** are zero vectors, and **B** is the identity matrix.

6. Matrices **A** and **B** are *equal* if they are of the same order $m \times n$ and the corresponding coefficients in the two matrices are equal. That is, $a_{ij} = b_{ij}$ for all i and j. If **A** and **B** are equal, we write **A** = **B**.

7. The *transpose* of an $m \times n$ matrix **A** is an $n \times m$ matrix, denoted by $\mathbf{A}^T$, and defined so that the (i, j) coefficient of $\mathbf{A}^T$ equals a_{ji}, the (j, i) coefficient of **A**.

8. If a square matrix **A** satisfies the relation $\mathbf{A}^T = \mathbf{A}$, then **A** is called *symmetric*. A symmetric matrix **A** is characterized by the property that $a_{ij} = a_{ji}$ for all i and j.

EXAMPLE A.3

An arbitrary diagonal matrix is symmetric. The matrices

$$\mathbf{A} = \begin{bmatrix} 1 & 3 & -4 \\ 3 & 0 & -1 \\ -4 & -1 & 1 \end{bmatrix} \qquad \text{and} \qquad \mathbf{B} = \begin{bmatrix} 1 & -1 \\ -1 & 1 \end{bmatrix}$$

are also symmetric. Also note that

$$
\begin{bmatrix} 1 \\ 1 \\ 1 \end{bmatrix}^T = [1 \quad 1 \quad 1], \qquad
\begin{bmatrix} 0 & 0 & 1 \\ 1 & 1 & 4 \end{bmatrix}^T = \begin{bmatrix} 0 & 1 \\ 0 & 1 \\ 1 & 4 \end{bmatrix}.
$$

9. The sum of matrices **A** and **B** is defined whenever **A** and **B** have the same number of rows and columns. Each coefficient of the sum **A** + **B** equals the sum of the two corresponding coefficients of **A** and **B**. In other words, **A** + **B** is the matrix the (i, j) coefficient of which is $a_{ij} + b_{ij}$, a_{ij} and b_{ij} being the (i, j) coefficients of **A** and **B**, respectively. The difference matrix **A** − **B** is analogously defined to be the matrix $(a_{ij} - b_{ij})$.

10. A product of a scalar α and a matrix **A** is the matrix denoted by α**A** and defined so that each coefficient is obtained by multiplying the corresponding coefficient of **A** by the constant α. Thus α**A** = $(\alpha \cdot a_{ij})$.

EXAMPLE A.4

$$
\begin{bmatrix} 1 & 1 & 0 \\ 2 & 3 & 4 \end{bmatrix} + \begin{bmatrix} 2 & -1 & 0 \\ 1 & 4 & 5 \end{bmatrix} = \begin{bmatrix} 3 & 0 & 0 \\ 3 & 7 & 9 \end{bmatrix}
$$

$$
\begin{bmatrix} 1 & 1 & 1 \\ 2 & 3 & 4 \end{bmatrix} - \begin{bmatrix} 2 & -1 & 0 \\ 1 & 4 & 5 \end{bmatrix} = \begin{bmatrix} -1 & 2 & 1 \\ 1 & -1 & -1 \end{bmatrix}
$$

$$
2 \begin{bmatrix} 1 & 0 & 0 \\ 2 & 3 & -1 \end{bmatrix} = \begin{bmatrix} 2 & 0 & 0 \\ 4 & 6 & -2 \end{bmatrix}.
$$

11. The matrix operations above satisfy the following relations:

(a) **A** + **B** = **B** + **A**   (commutative)
(b) (**A** + **B**) + **C** = **A** + (**B** + **C**)   (associative)
(c) **A** + **O** = **A**
(d) $(\mathbf{A}^T)^T = \mathbf{A}$
(e) $(\mathbf{A} + \mathbf{B})^T = \mathbf{A}^T + \mathbf{B}^T$
(f) $\alpha(\mathbf{A} + \mathbf{B}) = \alpha\mathbf{A} + \alpha\mathbf{B}$ (distributive)
(g) $(\alpha + \beta)\mathbf{A} = \alpha\mathbf{A} + \beta\mathbf{A}$ (distributive)
(h) For positive integers n,

$$
\mathbf{A} + \mathbf{A} + \cdots + \mathbf{A} = n\mathbf{A},
$$

where on the left-hand side we presume **A** to be added n times. From (b), the order in which addition is done is immaterial. The symbols α and β denote scalars.

12. The product of two matrices **A** and **B** is defined only when the number of columns of **A** equals the number of rows of **B**. If **A** and **B** are $m \times n$ and $n \times p$, respectively, the product **C** = **A** · **B** is $m \times p$, and coefficient c_{ij} of

C is defined by

$$c_{ij} = \sum_{k=1}^{n} a_{ik}b_{kj}. \qquad (A.3)$$

In other words, the coefficient c_{ij} of **C** is obtained by multiplying each coefficient in the ith row of **A** by the corresponding element in the jth column of **B**, and adding these products.

───── **EXAMPLE A.5** ─────────────────────────────────────

$$[1 \quad 1 \quad 3] \begin{bmatrix} 4 \\ 4 \\ 3 \end{bmatrix} = 1 \cdot 4 + 1 \cdot 4 + 3 \cdot 3 = 17$$

$$\begin{bmatrix} 1 \\ 1 \\ 3 \end{bmatrix} [4 \quad 4 \quad 3] = \begin{bmatrix} 4 & 4 & 3 \\ 4 & 4 & 3 \\ 12 & 12 & 9 \end{bmatrix}$$

$$\begin{bmatrix} 1 & 0 \\ 2 & 2 \end{bmatrix} \begin{bmatrix} 1 & 0 & 3 \\ 1 & 1 & 2 \end{bmatrix} = \begin{bmatrix} 1 & 0 & 3 \\ 4 & 2 & 10 \end{bmatrix}.$$

■

13. The multiplication operation satisfies the following properties:

(a) $(\mathbf{AB})\mathbf{C} = \mathbf{A}(\mathbf{BC})$ (associative)
(b) $(\mathbf{A} + \mathbf{B})\mathbf{C} = \mathbf{AC} + \mathbf{BC}$ (distributive)
(c) $\mathbf{A}(\mathbf{B} + \mathbf{C}) = \mathbf{AB} + \mathbf{AC}$ (distributive)
(d) $\mathbf{A} \cdot \mathbf{I} = \mathbf{A}, \quad \mathbf{I} \cdot \mathbf{A} = \mathbf{A}$
(e) $(\mathbf{AB})^T = \mathbf{B}^T\mathbf{A}^T$
(f) $\mathbf{AO} = \mathbf{O}$

14. Note that even if **AB** is defined, the product **BA** may not be defined. Moreover, if both **AB** and **BA** exist, *they are not necessarily the same*. These properties are illustrated next.

───── **EXAMPLE A.6** ─────────────────────────────────────

Let

$$\mathbf{A} = \begin{bmatrix} 1 & 1 & 0 \\ 1 & 0 & 0 \\ 0 & 0 & 0 \end{bmatrix}, \qquad \mathbf{B} = \begin{bmatrix} 1 & 0 \\ 0 & 1 \\ 0 & 0 \end{bmatrix};$$

then

$$\mathbf{AB} = \begin{bmatrix} 1 & 1 & 0 \\ 1 & 0 & 0 \\ 0 & 0 & 0 \end{bmatrix} \begin{bmatrix} 1 & 0 \\ 0 & 1 \\ 0 & 0 \end{bmatrix} = \begin{bmatrix} 1 & 1 \\ 1 & 0 \\ 0 & 0 \end{bmatrix},$$

but the product

$$\mathbf{BA} = \begin{bmatrix} 1 & 0 \\ 0 & 1 \\ 0 & 0 \end{bmatrix} \begin{bmatrix} 1 & 1 & 0 \\ 1 & 0 & 0 \\ 0 & 0 & 0 \end{bmatrix}$$

does not exist. Let

$$\mathbf{C} = \begin{bmatrix} 1 & 1 \\ 1 & 1 \end{bmatrix}, \quad \mathbf{D} = \begin{bmatrix} 1 & -1 \\ 1 & -1 \end{bmatrix};$$

then

$$\mathbf{CD} = \begin{bmatrix} 1 & 1 \\ 1 & 1 \end{bmatrix} \begin{bmatrix} 1 & -1 \\ 1 & -1 \end{bmatrix} = \begin{bmatrix} 2 & -2 \\ 2 & -2 \end{bmatrix},$$

but

$$\mathbf{DC} = \begin{bmatrix} 1 & -1 \\ 1 & -1 \end{bmatrix} \begin{bmatrix} 1 & 1 \\ 1 & 1 \end{bmatrix} = \begin{bmatrix} 0 & 0 \\ 0 & 0 \end{bmatrix}.$$

The last result shows that the product of two matrices may be the zero matrix, even if neither of the matrices is zero. ∎

A.2
LINEAR EQUATIONS AND MATRIX INVERSES

15. Linear equations may be rewritten in a compact form by using matrix and vector notation. Let us consider the linear equations

$$\begin{aligned} a_{11}x_1 + a_{12}x_2 + \cdots + a_{1n} x_n &= b_1 \\ a_{21}x_2 + a_{22}x_2 + \cdots + a_{2n} x_n &= b_2 \\ \vdots \qquad \vdots \qquad\qquad \vdots \quad\ \ \vdots \\ a_{m1}x_1 + a_{m2}x_2 + \cdots + a_{mn}x_n &= b_m. \end{aligned} \qquad (A.4)$$

Introduce the following notation:

$$\mathbf{A} = (a_{ij})_{i,j=1}^{m,n} = \begin{bmatrix} a_{11} & a_{12} & \cdots & a_{1n} \\ a_{21} & a_{22} & \cdots & a_{2n} \\ \vdots & \vdots & & \vdots \\ a_{m1} & a_{m2} & \cdots & a_{mn} \end{bmatrix},$$

$$\mathbf{x} = (x_j)_{j=1}^{n} = \begin{bmatrix} x_1 \\ x_2 \\ \vdots \\ x_n \end{bmatrix}, \quad \mathbf{b} = (b_i)_{i=1}^{m} = \begin{bmatrix} b_1 \\ b_2 \\ \vdots \\ b_m \end{bmatrix}.$$

Then equation (A.4) can be written in the compact form

$$\mathbf{Ax} = \mathbf{b}. \qquad (A.5)$$

EXAMPLE A.7 ──────────────────────────────────────

In the case of the equations given at the beginning of this appendix, we have $m = n = 3$, and

$$\mathbf{A} = \begin{bmatrix} 2 & 1 & -1 \\ 1 & 1 & 1 \\ -1 & 0 & 1 \end{bmatrix}, \qquad \mathbf{x} = \begin{bmatrix} x_1 \\ x_2 \\ x_3 \end{bmatrix}, \qquad \mathbf{b} = \begin{bmatrix} 5 \\ 5 \\ -1 \end{bmatrix}.$$

■

---

16. The *powers* or *exponents* of square matrices are defined by the relations $\mathbf{A}^1 = \mathbf{A}$ and $\mathbf{A}^n = \mathbf{A} \times \mathbf{A}^{n-1}(n \geq 2)$. Thus $\mathbf{A}^n = \mathbf{A} \times \mathbf{A} \times \cdots \times \mathbf{A}$, where on the right-hand side the number of factors equals n. Observe that if $\mathbf{A}$ is not a square matrix, then $\mathbf{A} \times \mathbf{A}$ is not defined. Therefore, exponents of matrices are meaningful only for square matrices.

17. The *inverse* of a square matrix $\mathbf{A}$ is that matrix, denoted by $\mathbf{A}^{-1}$, which satisfies the relation

$$\mathbf{A} \cdot \mathbf{A}^{-1} = \mathbf{A}^{-1} \cdot \mathbf{A} = \mathbf{I}.$$

EXAMPLE A.8 ──────────────────────────────────────

Not all squares matrices have inverses. For example, the zero matrix has no inverse, since for any arbitrary matrix $\mathbf{X}$ of the same order, $\mathbf{O} \cdot \mathbf{X} = \mathbf{X} \cdot \mathbf{O} = \mathbf{O} \neq \mathbf{I}$. Consider next the matrix

$$\mathbf{A} = \begin{bmatrix} 2 & 1 & -1 \\ 1 & 1 & 1 \\ -1 & 0 & 1 \end{bmatrix};$$

then one can verify that

$$\mathbf{A}^{-1} = \begin{bmatrix} -1 & 1 & -2 \\ 2 & -1 & 3 \\ -1 & 1 & -1 \end{bmatrix},$$

since

$$\mathbf{A}^{-1}\mathbf{A} = \begin{bmatrix} -1 & 1 & -2 \\ 2 & -1 & 3 \\ -1 & 1 & -1 \end{bmatrix} \begin{bmatrix} 2 & 1 & -1 \\ 1 & 1 & 1 \\ -1 & 0 & 1 \end{bmatrix} = \begin{bmatrix} 1 & 0 & 0 \\ 0 & 1 & 0 \\ 0 & 0 & 1 \end{bmatrix}$$

and

$$\mathbf{AA}^{-1} = \begin{bmatrix} 2 & 1 & -1 \\ 1 & 1 & 1 \\ -1 & 0 & 1 \end{bmatrix} \begin{bmatrix} -1 & 1 & -2 \\ 2 & -1 & 3 \\ -1 & 1 & -1 \end{bmatrix} = \begin{bmatrix} 1 & 0 & 0 \\ 0 & 1 & 0 \\ 0 & 0 & 1 \end{bmatrix}.$$

■

18. If **A** is a square matrix having an inverse, then (A.5) has a unique solution, which can be written as $\mathbf{x} = \mathbf{A}^{-1}\mathbf{b}$.

This relation can be verified by premultiplying both sides of equation $\mathbf{Ax} = \mathbf{b}$ by the inverse matrix $\mathbf{A}^{-1}$ to get

$$\mathbf{A}^{-1}\mathbf{Ax} = \mathbf{A}^{-1}\mathbf{b},$$

where the left-hand side equals $\mathbf{Ix} = \mathbf{x}$.

EXAMPLE A.9

The inverse of the matrix of coefficients of the linear equations discussed in Example A.7 has been determined in Example A.8. In view of item 18, this implies that

$$\mathbf{x} = \mathbf{A}^{-1}\mathbf{b} = \begin{bmatrix} -1 & 1 & -2 \\ 2 & -1 & 3 \\ -1 & 1 & -1 \end{bmatrix} \begin{bmatrix} 5 \\ 5 \\ -1 \end{bmatrix} = \begin{bmatrix} 2 \\ 2 \\ 1 \end{bmatrix}.$$

That is, the solutions of the linear equations at the beginning of the appendix are $x_1 = x_2 = 2$ and $x_3 = 1$.

■

19. Let **A** be a square matrix. If $\mathbf{A}^{-1}$ exists, **A** is called *nonsingular*. Otherwise, **A** is said to be *singular*.

EXAMPLE A.10

Consider the matrix

$$\mathbf{A} = \begin{bmatrix} 0 & 1 \\ 0 & 2 \end{bmatrix},$$

and let

$$\mathbf{X} = \begin{bmatrix} x_{11} & x_{12} \\ x_{21} & x_{22} \end{bmatrix}$$

be an arbitrary matrix. Then

$$\mathbf{XA} = \begin{bmatrix} x_{11} & x_{12} \\ x_{21} & x_{22} \end{bmatrix} \begin{bmatrix} 0 & 1 \\ 0 & 2 \end{bmatrix} = \begin{bmatrix} 0 & x_{11} + 2x_{12} \\ 0 & x_{21} + 2x_{22} \end{bmatrix} \neq \mathbf{I}$$

and consequently, no matrix **X** satisfies the relation

$$\mathbf{XA} = \mathbf{I}$$

since the first column of **I** is not zero. Thus $\mathbf{A}^{-1}$ does not exist. That is, **A** is singular.

■

20. Let **A** and **B** be nonsingular $n \times n$ matrices. Then **AB** is nonsingular and

$$(\mathbf{AB})^{-1} = \mathbf{B}^{-1}\mathbf{A}^{-1}. \tag{A.6}$$

For notice that in view of the associative law (item 13),

$$(\mathbf{AB})(\mathbf{B}^{-1}\mathbf{A}^{-1}) = \mathbf{A}(\mathbf{BB}^{-1})\mathbf{A}^{-1} = \mathbf{AIA}^{-1} = (\mathbf{AI})\mathbf{A}^{-1} = \mathbf{AA}^{-1} = \mathbf{I}.$$

21. Consider the matrix

$$\mathbf{B} = (\mathbf{b}^{(1)}, \mathbf{b}^{(2)}, \ldots, \mathbf{b}^{(n)}), \tag{A.7}$$

where $\mathbf{b}^{(1)}, \mathbf{b}^{(2)}, \ldots, \mathbf{b}^{(n)}$ are the columns of **B**, and let **A** be a matrix such that **AB** is defined. Then

$$\mathbf{AB} = (\mathbf{Ab}^{(1)}, \mathbf{Ab}^{(2)}, \ldots, \mathbf{Ab}^{(n)}). \tag{A.8}$$

That is, the columns of **AB** can be obtained by multiplying the matrix **A** by each of the columns of **B**, and then assembling a new matrix column by column from these products. This property is illustrated in the next example.

EXAMPLE A.11

Define the matrices

$$\mathbf{A} = \begin{bmatrix} 1 & 2 \\ -1 & 1 \end{bmatrix}, \quad \mathbf{B} = \begin{bmatrix} 2 & 0 \\ 1 & 3 \end{bmatrix}.$$

Then

$$\mathbf{AB} = \begin{bmatrix} 1 & 2 \\ -1 & 1 \end{bmatrix} \begin{bmatrix} 2 & 0 \\ 1 & 3 \end{bmatrix} = \begin{bmatrix} 4 & 6 \\ -1 & 3 \end{bmatrix}.$$

But

$$\mathbf{Ab}^{(1)} = \begin{bmatrix} 1 & 2 \\ -1 & 1 \end{bmatrix} \begin{bmatrix} 2 \\ 1 \end{bmatrix} = \begin{bmatrix} 4 \\ -1 \end{bmatrix}$$

$$\mathbf{Ab}^{(2)} = \begin{bmatrix} 1 & 2 \\ -1 & 1 \end{bmatrix} \begin{bmatrix} 0 \\ 3 \end{bmatrix} = \begin{bmatrix} 6 \\ 3 \end{bmatrix}.$$

From inspection we see that $\mathbf{Ab}^{(1)}$ and $\mathbf{Ab}^{(2)}$ are indeed the columns of **AB**. ∎

22. For any matrices $\mathbf{V}$ and $\mathbf{A}$ such that the product $\mathbf{VA}$ is defined, if $\mathbf{x}$ is a solution of the simultaneous linear equation

$$\mathbf{Ax} = \mathbf{b}, \tag{A.9}$$

then by multiplying both sides on the left by matrix $\mathbf{V}$, we see that it also satisfies

$$\mathbf{VAx} = \mathbf{Vb}. \tag{A.10}$$

If $\mathbf{A}$ and $\mathbf{V}$ are nonsingular, then we may conclude that $\mathbf{x}$ satisfies (A.9) if and only if it satisfies (A.10).

23. The notion of row operation plays a central role in numerical computation procedures for solving simultaneous linear equations. We now consider the operation of adding a multiple α of the jth row of matrix $\mathbf{A}$ to the ith row. If $\mathbf{A} = (a_{ij})$, the resulting matrix will be

$$\tilde{\mathbf{A}} = \begin{bmatrix} a_{11} & a_{12} & \cdots & a_{in} \\ \vdots & \vdots & & \vdots \\ a_{i-1,1} & a_{i-1,2} & \cdots & a_{i-1,n} \\ a_{i1} + \alpha a_{j1} & a_{i2} + \alpha a_{j2} & \cdots & a_{in} + \alpha a_{jn} \\ a_{i+1,1} & a_{i+1,2} & \cdots & a_{i+1,n} \\ \vdots & \vdots & & \vdots \\ a_{n1} & a_{n2} & \cdots & a_{nn} \end{bmatrix}$$

This operation is achieved by $\tilde{\mathbf{A}} = \mathbf{VA}$, where

$$\mathbf{V} = \begin{bmatrix} 1 & & & & & & & & & & & \\ & \ddots & & & & & & & & & & \\ & & \ddots & & & & & & & & & \\ & & & 1 & & & & & & & & \\ \cdots & \cdots & \cdots & \cdot & 1 & \cdot & \cdots & \cdots & \cdot\alpha\cdot & \cdots & \cdots & i \\ & & & & 1 & & & & & & & \\ & & & & & \ddots & & & & & & \\ & & & & & & \ddots & & & & & \\ & & & & & & & 1 & & & & \\ \cdots & \cdots & \cdots & \cdots & \cdots & \cdots & \cdot & 1 & \cdots & \cdots & \cdot & j \\ & & & & & & & \cdot 1 & & & & \\ & & & & & & & & \ddots & & & \\ & & & & & & & & & \ddots & & \\ & & & & & & & & & & 1 \end{bmatrix}$$

That is, $\mathbf{V}$ is the same as the identity matrix $\mathbf{I}$, except that it has an additional element α in the (i, j) position. It is easy to verify that $\mathbf{V}^{-1}$ exists, and, in fact,

V^{-1} equals V, except that α is replaced by $-\alpha$. The reader may check this assertion by evaluating VV^{-1} and comparing it to I.

24. Another important row operation, the interchange of the ith and jth rows, is obtained from A by the product WA, where

$$
W = \begin{bmatrix}
1 & & & & & & & & & & & & & \\
& 1 & & & & & & & & & & & & \\
& & \ddots & & & & & & & & & & & \\
& & & \ddots & & & & & & & & & & \\
& & & & 1 & & & & & & & & & \\
\cdot & \cdot & \cdot & \cdot & \cdot 0 \cdot & \cdot & \cdot & \cdot & \cdot 1 \cdot & \cdot & \cdot & \cdot & \cdot & i \\
& & & & 1 & & & & & & & & & \\
& & & & & \ddots & & & & & & & & \\
& & & & & & \ddots & & & & & & & \\
& & & & & & & \ddots & & & & & & \\
& & & & & & & & 1 & & & & & \\
\cdot & \cdot & \cdot & \cdot & \cdot 1 \cdot & \cdot & \cdot & \cdot & \cdot 0 \cdot & \cdot & \cdot & \cdot & \cdot & j \\
& & & & & & & & \cdot 1 & & & & & \\
& & & & & & & & & \ddots & & & & \\
& & & & & & & & & & \ddots & & & \\
& & & & & & & & & & & \ddots & & \\
& & & & & & & & & & & & 1 &
\end{bmatrix}
$$

The coefficients not shown are presumed to be 0. W is diagonal, except for 1's in the (i, j) and (j, i) locations. The diagonal coefficients are all 1's, except for 0's at (i, i) and (j, j). The matrix W is its own inverse. An important conclusion we need from this discussion is that if matrix $\tilde{A}$ differs from a nonsingular matrix A only in that a multiple of one row of A is added to another, or that rows of A are interchanged, then x is a solution to

$$Ax = b$$

if and only if it is a solution to

$$\tilde{A}x = \tilde{b},$$

where $\tilde{b}$ is determined from b by the same row operation by which $\tilde{A}$ was constructed.

EXAMPLE A.12

Let

$$
A = \begin{bmatrix}
1 & 2 & 3 \\
4 & 5 & 6 \\
7 & 8 & 9
\end{bmatrix}.
$$

The matrix $\tilde{\mathbf{A}}$ obtained by adding $2 \times$ (row 2) to row 1 is

$$\tilde{\mathbf{A}} = \begin{bmatrix} 9 & 12 & 15 \\ 4 & 5 & 6 \\ 7 & 8 & 9 \end{bmatrix}.$$

$\tilde{\mathbf{A}}$ can be obtained by $\tilde{\mathbf{A}} = \mathbf{VA}$, where

$$\mathbf{V} = \begin{bmatrix} 1 & 2 & 0 \\ 0 & 1 & 0 \\ 0 & 0 & 1 \end{bmatrix}.$$

∎

Calculus Background

This reference guide summarizes the major definitions, concepts, and results of calculus that are used in this book.

B.1
CONVERGENCE AND CONTINUITY

1. *Convergence* of a sequence is defined as follows: The sequence $\{x_i\}$ of real numbers converges to the limit x^* if, for arbitrary $\varepsilon > 0$, there exists an index $N = N(\varepsilon)$ such that for all $i \geq N$ we have $|x_i - x^*| < \varepsilon$. In other words, for any $i \geq N$ the term x_i approximates the limit x^* with error less than ε. This relation between $\{x_i\}$ and x^* is denoted as

$$\lim_{i \to \infty} x_i = x^*.$$

Convergent sequences have the following properties:

2. $\lim_{i \to \infty} (x_i + y_i) = \lim_{i \to \infty} x_i + \lim_{i \to \infty} y_i.$

3. $\lim_{i \to \infty} (x_i - y_i) = \lim_{i \to \infty} x_i - \lim_{i \to \infty} y_i.$

4. $\lim_{i \to \infty} (x_i y_i) = \lim_{i \to \infty} x_i \cdot \lim_{i \to \infty} y_i.$ $\hspace{2em}$ (B.1)

5. $\lim_{i \to \infty} (x_i/y_i) = \lim_{i \to \infty} x_i / \lim_{i \to \infty} y_i.$

where in (B.1) we have assumed that sequences $\{x_i\}$ and $\{y_i\}$ are convergent, and in property 5, all y_i's as well as $\lim_{i \to \infty} y_i$ differ from zero.

6. *Convergence of a function* at a point is defined in the following way: The limit of function $f(x)$ at the point x_0 equals A if for arbitrary $\varepsilon > 0$ there exists a $\delta = \delta(\varepsilon) > 0$ such that $|x - x_0| < \delta$ implies that $|f(x) - A| < \varepsilon$. This relation is denoted as

$$\lim_{x \to x_0} f(x) = A.$$

A function $f(x)$ is *continuous* at x_0 if and only if

$$\lim_{x \to x_0} f(x) = f(x_0). \tag{B.2}$$

Continuous real-valued functions have the following properties:

7. Let $f(x)$ be continuous on $[a, b]$ (i.e., $f(x)$ is continuous at each point $x_0 \in [a, b]$). Then there are points x_1, $x_2 \in [a, b]$ such that for all $x \in [a, b]$,

$$f(x_1) \leq f(x) \leq f(x_2). \tag{B.3}$$

8. Let $f(x)$ be continuous on $[a, b]$, and let c be any value between $f(a)$ and $f(b)$. Then there exists a point $\xi \in [a, b]$ such that $f(\xi) = c$.

The following consequence is central to the bisection method of Chapter 5: If $f(x)$ is continuous on the interval $[a, b]$, and $f(a)$ and $f(b)$ have different signs, then $f(x)$ has at least one zero in $[a, b]$.

B.2
DERIVATIVES

9. The *derivative* of $f(x)$ at point x_0 is defined by

$$\frac{d}{dx} f(x) \bigg|_{x=x_0} = f'(x_0) = \lim_{h \to 0} \frac{f(x_0 + h) - f(x_0)}{h}, \tag{B.4}$$

and it gives the slope of the tangent line of function $f(x)$ at x_0 (see Figure B.1). One-sided derivatives, denoted by $f'(x_0 + 0)$ or $f'(x_0 - 0)$ are defined as in (B.4) but with h restricted to positive or negative values, respectively.

The derivatives of functions satisfy the following properties:

10. If $f(x) \equiv c$, c being a constant, then $f'(x) = 0$.

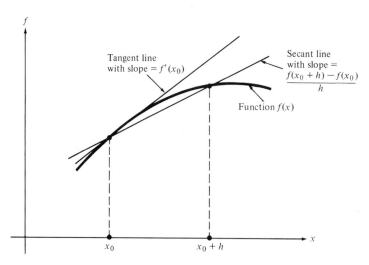

FIGURE B–1 The Derivative and Its Approximation

11. $[cf(x)]' = cf'(x)$.

12. $[f(x) + g(x)]' = f'(x) + g'(x)$.

13. $[f(x) - g(x)]' = f'(x) - g'(x)$.   (B.5)

14. $[f(x)g(x)]' = f'(x)g(x) + f(x)g'(x)$.

15. $\left[\dfrac{f(x)}{g(x)}\right]' = \dfrac{f'(x)g(x) - f(x)g'(x)}{(g(x))^2}$.

16. $[f(g(x))]' = f'(g(x))g'(x)$.

Property 16 is called the *chain rule* of calculus. The relations listed above are illustrated in the following examples.

───── **EXAMPLE B.1** ─────────────────────────────

By using the definition of derivatives, we shall verify that $(x^3)' = 3x^2$. Note that

$$(x^3)' = \lim_{h \to 0} \frac{(x + h)^3 - x^3}{h} = \lim_{h \to 0} \frac{x^3 + 3x^2h + 3xh^2 + h^3 - x^3}{h}$$

$$= \lim_{h \to 0} (3x^2 + 3xh + h^2) = \lim_{h \to 0} 3x^2 + \lim_{h \to 0} 3xh + \lim_{h \to 0} h^2.$$

In the limit we have

$$(x^3)' = 3x^2 + 0 + 0 = 3x^2.$$

───── **EXAMPLE B.2** ─────────────────────────────

By using relations 12 and 16, we get

$$[\exp (x^2 + x + 1)]' = \exp (x^2 + x + 1)[x^2 + x + 1]'$$
$$= \exp (x^2 + x + 1)(2x + 1)$$
$$[\sin (x^2)]' = \cos (x^2)[x^2]' = \cos (x^2)2x$$
$$[(x + 1)^3]' = 3(x + 1)^2[x + 1]' = 3(x + 1)^2;$$

and by using relations 14 and 15, we have

$$[\sin (x) \cos (x)]' = \cos (x) \cos (x) + \sin (x)[-\sin (x)] = \cos^2(x) - \sin^2(x)$$

$$[\tan (x)]' = \left[\frac{\sin (x)}{\cos (x)}\right]' = \frac{\cos (x) \cos (x) - \sin (x) [-\sin (x)]}{\cos^2(x)} = \frac{1}{\cos^2(x)}.$$

17. We say that function $f(x)$ is *differentiable* in the interval $[a, b]$ if for each $x \in (a, b)$ the derivative $f'(x)$ exists, and furthermore, the one-sided derivatives $f'(a + 0)$ and $f'(b - 0)$ exist.

Differentiable functions have the following properties:

18. (Rolle's Theorem)   Let $f(x)$ be differentiable on $[a, b]$ and let $f(a) =$

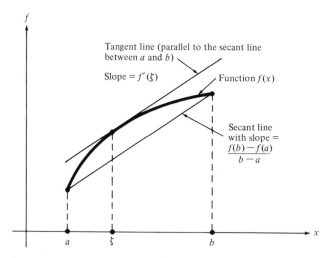

FIGURE B–2   Illustration of Fact 19

$f(b)$. Then there exists a point ζ between a and b such that

$$f'(\zeta) = 0. \tag{B.6}$$

19. (Mean-Value Theorem)   Let $f(x)$ be differentiable on $[a, b]$. Then there exists a point ζ between a and b such that

$$f'(\zeta) = \frac{f(b) - f(a)}{b - a}. \tag{B.7}$$

This result is illustrated in Figure B.2.

20. (The Cauchy Mean-Value Theorem)   Let $f(x)$ and $g(x)$ be differentiable on $[a, b]$. Then there exists a point ζ between a and b such that

$$\frac{f'(\zeta)}{g'(\zeta)} = \frac{f(b) - f(a)}{g(b) - g(a)}. \tag{B.8}$$

21. The *higher-order derivatives* are defined recursively by

$$f^{(n)}(x) = [f^{(n-1)}(x)]', \tag{B.9}$$

where $f^{(n-1)}(x)$ and $f^{(n)}(x)$ denote the $(n - 1)$st-order and nth-order derivatives of function $f(x)$, respectively.

This recursive relation is illustrated in the following example.

EXAMPLE B.3

$$[\sin (x)]' = \cos (x)$$
$$[\sin (x)]'' = [\sin (x)']' = [\cos (x)]' = -\sin (x) \tag{B.10}$$
$$[\sin (x)]^3 = [\sin (x)'']' = [-\sin (x)]' = -\cos (x),$$

and so on.

There are important relations between the signs of the first two derivatives and the fundamental properties of the graphs of functions, which are discussed next.

22. A function $f(x)$ is *locally increasing* (or *decreasing*) at point x_0 if there exists a positive ε such that for all $x_0 - \varepsilon < x < x_0$, $f(x) < f(x_0)$ [or $f(x) > f(x_0)$] and for all $x_0 < x < x_0 + \varepsilon$, $f(x_0) < f(x)$ [or $f(x_0) > f(x)$].

23. The point x_0 is a local *minimum* (or *maximum*) point of function $f(x)$ if there exists a positive ε such that for all $x_0 - \varepsilon < x < x_0 + \varepsilon$, $f(x_0) \leq f(x)$ [or $f(x_0) \geq f(x)$].

The relations between the derivative and the foregoing properties are as follows:

24. If $f(x)$ is differentiable in the neighborhood of x_0, [i.e., for $|x - x_0|$ sufficiently small, $f'(x)$ exists], and $f'(x_0) > 0$ [or $f'(x_0) < 0$], then $f(x)$ is locally increasing (or decreasing at x_0).

The converse of this statement is not true in general, as the function $f(x) = x^3$ shows at $x_0 = 0$, but the following statement holds:

25. If $f(x)$ is differentiable in the neighborhood of x_0 and $f(x)$ is locally increasing (or decreasing) at x_0, then $f'(x_0) \geq 0$ [or $f'(x_0) \leq 0$].

26. If $f(x)$ is differentiable in the neighborhood of x_0 and x_0 is a local minimum or maximum point of $f(x)$, then $f'(x_0) = 0$.

27. The converse of this statement is not true in general, as the case of the function $f(x) = x^3$ shows at $x_0 = 0$, but the following holds:

28. Assume that $f(x)$ is twice differentiable in the neighborhood of x_0. If $f'(x_0) = 0$ and $f''(x_0) > 0$ [or $f''(x_0) < 0$], then x_0 is a local minimum (or maximum) point of function $f(x)$.

———— **EXAMPLE B.4** ————

At $x_0 = 0$;

(a) Function $f_1(x) = 2x$ is locally increasing and function $f_2(x) = -2x + x^2$ is locally decreasing, since $f_1'(0) = 2$ and $f_2'(0) = -2$.

(b) Function $f_3(x) = x^2$ has a local minimum and $f_4(x) = -x^2$ has a local maximum, since $f_3''(0) = 2 > 0$ and $f_4''(0) = -2 < 0$, and $f_3'(0) = f_4'(0) = 0$. ■

B.3
INFINITE SERIES

29. An infinite series

$$s = a_1 + a_2 + \cdots + a_i + \cdots \qquad (B.11)$$

is said to be *convergent* if the sequence $\{s_i\}$ of the partial sums defined by

$$s_i = a_1 + a_2 + \cdots + a_i$$

is convergent, and the sum of the infinite series is defined as the limit of the sequence $\{s_i\}$.

Some important properties of infinite series are summarized below.

30. $\sum_{i=1}^{\infty} (1/i^{\alpha})$ is convergent for $\alpha > 1$ and divergent for $\alpha \leq 1$.

31. Assume that there exists a convergent infinite series

$$\sum_{i=1}^{\infty} b_i \text{ such that } b_i \geq 0 \ (i \geq 1), \text{ and for all } i, \ |a_i| \leq b_i.$$

Then the series (B.11) is convergent.

32. Assume that in the infinite series

$$s = a_1 - a_2 + a_3 - a_4 + \cdots + (-1)^{i+1} a_i + \cdots$$

all terms a_i have the same sign. Furthermore, suppose that

$$\lim_{i \to \infty} a_i = 0$$

$$|a_{i+1}| \leq |a_i| \qquad (i \geq 1).$$

Then the infinite series is convergent. Also,

$$|s - s_i| \leq |a_{i+1}|, \tag{B.12}$$

where

$$s_i = a_1 - a_2 + a_3 - a_4 + \cdots + (-1)^{i+1} a_i$$

denotes the ith partial sum and s is the limit.

EXAMPLE B.5

(a) $\sum_{i=1}^{\infty} (1/i^2)$ is convergent and $\sum_{i=1}^{\infty} (1/i)$ is divergent, since in these cases $\alpha = 2$ and $\alpha = 1$, respectively.

(b) The alternating series

$$1 - \tfrac{1}{2} + \tfrac{1}{3} - \tfrac{1}{4} + \tfrac{1}{5} - \tfrac{1}{6} + \cdots$$

is convergent, since it satisfies the conditions of property 32.

Furthermore, in this case

$$|s - s_i| \leq \frac{1}{i+1},$$

since

$$(-1)^{i+2} \frac{1}{i+1}$$

is the first neglected term in constructing s_i.

33. An infinite series of functions

$$s(x) = f_1(x) + f_2(x) + \cdots + f_i(x) + \cdots$$

is said to be converged for a given value $x = x_0$ if the sequence $\{s_i(x_0)\}$ is convergent, where for $i = 1, 2, \ldots$, and all x,

$$s_i(x) = f_1(x) + f_2(x) + \cdots + f_i(x).$$

The sum of the series at x_0 is defined as the limit of the sequence $\{s_i(x_0)\}$.

Taylor's polynomials and Taylor's series play important roles in calculus. Their definitions and properties are omitted from this appendix because they are discussed in Section 2.3.

B.4
PARTIAL DERIVATIVES

34. The *partial derivative* of a multivariable function $f(x_1, \ldots, x_n)$, with respect to x_k, is defined as

$$\frac{\partial f}{\partial x_k}(x_1, \ldots, x_n) = f_{x_k}(x_1, \ldots, x_n) \tag{B.13}$$

$$= \lim_{h \to 0} \frac{f(x_1, \ldots, x_{k-1}, x_k + h, x_{k+1}, \ldots, x_n) - f(x_1, \ldots, x_{k-1}, x_k, x_{k+1}, \ldots, x_n)}{h}.$$

35. *Higher-order partial derivatives* are defined by the recurrent relation

$$\frac{\partial^k f}{\partial x_{i_1} \partial x_{i_2} \cdots \partial x_{i_k}} = \frac{\partial}{\partial x_{i_k}} \frac{\partial^{k-1} f}{\partial x_{i_1} \partial x_{i_2} \cdots \partial x_{i_{k-1}}}. \tag{B.14}$$

It is known that if higher-ordered partial derivatives are continuous, their values are independent of the order in which differentiation is done.

────── **EXAMPLE B.6** ──────

Consider the bivariable function

$$f(x_1, x_2) = 2x_1^2 - 2x_1 x_2 + x_2^2 - 2x_1 + 1.$$

Then

$$\frac{\partial f}{\partial x_1}(x_1, x_2) = 4x_1 - 2x_2 - 2,$$

$$\frac{\partial f}{\partial x_2}(x_1, x_2) = -2x_1 + 2x_2.$$

36. The *chain rule* for multivariable functions can be stated in the following way: Let $f(x_1, \ldots, x_n)$ be an n-variable differentiable function, and let $g_1(x), \ldots, g_n(x)$ be differentiable real functions. Consider now the real function

$$h(x) = f(g_1(x), \ldots, g_n(x)).$$

Then function $h(x)$ is differentiable; furthermore,

$$h'(x) = \frac{\partial f}{\partial g_1}(g_1(x), \ldots, g_n(x))g_1'(x) + \cdots + \frac{\partial f}{\partial g_n}(g_1(x), \ldots, g_n(x))g_n'(x).$$

$$(B.15)$$

───── **EXAMPLE B.7** ─────

Consider the bivariate function

$$f(x_1, x_2) = x_1^2 + x_2^2,$$

and

$$g_1(x) = \sin(x), \qquad g_2(x) = e^x.$$

Then

$$h(x) = f(g_1(x), g_2(x)) = \sin^2(x) + (e^x)^2.$$

Simple differentiation shows that

$$h'(x) = 2\sin(x)\cos(x) + 2e^{2x}.$$

On the other hand, by using the chain rule, the same result can be obtained:

$$h'(x) = 2x_1\big|_{x_1 = g_1(x)}\, g_1'(x) + 2x_2\big|_{x_2 = g_2(x)}\, g_2'(x)$$

$$= 2g_1(x)g_1'(x) + 2g_2(x)g_2'(x) = 2\sin(x)\cos(x) + 2e^x e^x$$

$$= 2\sin(x)\cos(x) + 2e^{2x}.$$ ∎

B.5
INTEGRATION

37. The *primitive function* (or *indefinite integral*) $F(x)$ of a real function $f(x)$ is defined by the relation

$$f(x) = F'(x) \qquad (B.16)$$

and is denoted as

$$F(x) = \int^X f(z)\, dz.$$

Some principal properties of primitive functions are summarized as follows:
38. Let $F_1(x)$ and $F_2(x)$ be two primitive functions of $f(x)$. Then $F_1(x) - F_2(x)$ is a constant.
39. If $F(x)$ is a primitive function of $f(x)$, then for arbitrary constant c, $F(x) + c$ is also a primitive function of $f(x)$.
The following relations hold:
40. $\int (f(x) + g(x))\, dx = \int f(x)\, dx + \int g(x)\, dx.$
41. $\int (f(x) - g(x))\, dx = \int f(x)\, dx - \int g(x)\, dx.$
42. $\int cf(x)\, dx = c \int f(x)\, dx.$
43. $\int f'(x)g(x)\, dx = f(x)g(x) - \int f(x)g'(x)\, dx$ (integration by parts).

44. $$\int f'(x)f^n(x)\, dx = \frac{f^{n+1}(x)}{n+1} + c. \tag{B.17}$$

45. $$\int \frac{f'(x)}{f(x)}\, dx = \log |f(x)| + c.$$

46. If $F(x)$ is a primitive function of $f(x)$, then

$$\int f(g(x))g'(x)\, dx = F(g(x)) + c,$$

and as a special case,

$$\int f(ax + b)\, dx = \frac{F(ax + b)}{a} + c.$$

EXAMPLE B.8

By using the properties above, we get the following relations:

(a) $$\int \frac{\log (x)}{x}\, dx = \int \frac{1}{x} [\log (x)]\, dx = \frac{[\log (x)]^2}{2} + c,$$

since the derivative of $\log (x)$ equals $1/x$.

(b) $$\int \log (x)\, dx = \int 1 \log (x)\, dx = x \log (x) - \int x \frac{1}{x}\, dx = x \log (x)$$
$- x + c$ since we can choose $f(x) = x$ and $g(x) = \log (x)$ in property 43.

(c) $$\int \frac{x}{1 + x^2}\, dx = \frac{1}{2} \int \frac{2x}{1 + x^2}\, dx = \frac{1}{2} \log (1 + x^2) + c,$$

since we may define $f(x) = 1 + x^2$ and use property 45. ∎

---

47. Let $f(x)$ be a bounded function on an interval $[a, b]$, let $a = x_0 < x_1 < \cdots < x_n = b$ be given domain points, and let $\zeta_k \in [x_{k-1}, x_k]$ be arbitrary points for $k = 1, \ldots, n$. Then

$$\sum_{k=1}^{n} f(\zeta_k)(x_k - x_{k-1})$$

is called a *Riemann sum* of function $f(x)$. If the Riemann sums are convergent for any selection of points x_j such that max $\{x_{j+1} - x_j; 0 \le j < n\} \to 0$ as $n \to \infty$, and the limit is always the same, then we say that function $f(x)$ is *integrable* on the interval $[a, b]$, and the *definite integral* of function $f(x)$ on $[a, b]$ is defined by the common limit of the Riemann sums and it is denoted by

$$\int_a^b f(x) \, dx.$$

The relation between primitive functions and definite integrals is given by the following fact:

48. Let $F(x)$ be a primitive function of $f(x)$. Then

$$\int_a^b f(x) \, dx = F(b) - F(a) = F(x)|_a^b. \tag{B.18}$$

49. In applied mathematics there are many applications of the *mean-value theorem for integrals*, which is: Assume that $f(x)$ and $g(x)$ are integrable on $[a, b]$, and in this interval

$$f(x) \ge 0.$$

Then there exists a point $\zeta \in (a, b)$ such that

$$\int_a^b f(x)g(x) \, dx = g(\zeta) \int_a^b f(x) \, dx. \tag{B.19}$$

50. If either function $f(x)$ is not bounded on $[a, b]$, and/or the interval $[a, b]$ is not finite, then the above definition of definite integrals is inapplicable. These integrals are called *improper*. Some improper integrals can be defined in the following way:

(a) $\int_a^\infty f(x) \, dx = \lim\limits_{b \to \infty} \int_a^b f(x) \, dx.$

(b) $\int_{-\infty}^\infty f(x) \, dx = \lim\limits_{a \to -\infty} \int_a^c f(x) \, dx + \lim\limits_{b \to \infty} \int_c^b f(x) \, dx$, where c is an arbitrary number.

───── **EXAMPLE B.9** ─────────────────────────────

(a) $\int_0^\pi \sin(x) \, dx = [-\cos(x)]_0^\pi = [-\cos(\pi)] - [-\cos(0)] = 1 + 1 = 2.$

(b) $\int_0^1 x^2 \, dx = \left[\dfrac{x^3}{3}\right]_0^1 = \dfrac{1}{3} - \dfrac{0}{3} = \dfrac{1}{3}.$

(c) $\int_1^\infty \dfrac{1}{x^2} \, dx = \lim\limits_{b \to \infty} \int_1^b \dfrac{1}{x^2} \, dx = \lim\limits_{b \to \infty} \left[-\dfrac{1}{x}\right]_1^b = \lim\limits_{b \to \infty} \left(-\dfrac{1}{b} + 1\right) = 0 + 1$
$= 1.$ ■

Simplex Subroutine
for Linear Programming

In Chapter 6 our construction of the best approximating polynomial required solution of a linear programming problem. Specifically, the subroutine BAP, which computes the best approximating polynomial (see Table 6.7), calls subroutine LINPRO, which is presented in Table C.1. Actually, LINPRO finds the solution of a general form of the linear programming problem.

Because LINPRO is a general-purpose simplex algorithm for linear programming, the reader may find this subroutine useful in other contexts. In this appendix we discuss this algorithm and provide an illustrative example of its use. The principles of LINPRO stem directly from the plan of the simplex method given in Dahlquist and Bjorck (1974, Chap. 10). A more general discourse on linear programming and the simplex method is to be found in Luenberger (1973, Part 1).

TABLE C.1   Simplex Subroutine LINPRO

```
      SUBROUTINE LINPRO(N,M1,M2,M3,EPS,X,KOD)
C
C ****************************************************************
C *    FUNCTION: THIS SUBROUTINE COMPUTES THE SOLUTION OF A      *
C *              LINEAR PROGRAMMING PROBLEM BY USING THE         *
C *              SIMPLEX METHOD. THE OBJECTIVE FUNCTION IS       *
C *              MAXIMIZED. PARAMETERS A,B, AND C ARE PRESUMED   *
C *              TRANSMITTED THROUGH A COMMON DECLARATION IN     *
C *              THE CALLING PROGRAM                             *
C *    USAGE:                                                    *
C *        CALL SEQUENCE: CALL LINPRO(N,M1,M2,M3,EPS,X,KOD)      *
C *    PARAMETERS:                                               *
C *        INPUT:                                                *
C *            N  = NUMBER OF VARIABLES                          *
C *            M1 = NUMBER OF CONSTRAINTS OF THE LE TYPE         *
C *            M2 = NUMBER OF CONSTRAINTS OF THE EQ TYPE         *
C *            M3 = NUMBER OF CONSTRAINTS OF THE GE TYPE         *
C *           EPS = TOLERANCE, COEFFICIENT WITH ABSOLUTE         *
C *                 VALUE LESS THEN EPS IS CONSIDERED TO BE      *
C *                 ZERO                                         *
C *             A = MATRIX OF COEFFICIENTS (IN COMMON BLOCK)     *
C *             B = RIGHT HAND SIDE VECTOR (IN COMMON BLOCK)     *
C *             C = COEFFICIENTS OF THE OBJECTIVE FUNCTION       *
C *                 (IN COMMON BLOCK)                            *
C *        OUTPUT:                                               *
C *          X(I) = I-TH COMPONENT OF THE OPTIMAL SOLUTION       *
C *                 FOR I = 1, ... ,N                            *
C *        X(N+1) = OPTIMAL VALUE OF THE OBJECTIVE FUNCTION      *
C *           KOD = KEY SHOWING THE CATEGORY OF THE LP PROBLEM   *
C *                 IF KOD = 1 THEN AN OPTIMAL SOLUTION EXISTS   *
C *                 IF KOD = 2 THEN NO FEASIBLE SOLUTION EXISTS  *
C *                 IF KOD = 3 THEN THE OBJECTIVE FUNCTION IS    *
C *                 NOT BOUNDED                                  *
C *    REMARKS:   IN CONSTRUCTING THE SIMPLEX TABLE, FIRST       *
C *               CONSTRAINTS OF THE LE TYPE ARE PLACED,         *
C *               NEXT CONSTRAINTS OF THE EQ TYPE, AND FINALLY   *
C *               CONSTRAINTS OF THE GE TYPE                     *
C *               THERE IS NO NEED FOR INTRODUCING SLACK AND     *
C *               ARTIFICIAL VARIABLES                           *
C *    CONDITIONS:                                               *
C *               M1+M2+M3 MUST BE LESS THEN OR EQUAL TO 50      *
C *               N MUST NOT BE LARGER THEN 50 UNLESS THE        *
C *               SUBROUTINE IS REDIMENSIONED                    *
C ****************************************************************
C
      COMMON A(52,151),B(50),C(50)
      DIMENSION X(51),KK(150)
C     *** LOAD UP THE INITIAL TABLEAU WITH SLACK AND ***
C     *** ARTIFICIAL VARIABLES, ORIGIONAL AND        ***
C     *** SECONDARY OBJECTIVE FUNCTIONS              ***
      L1=N+1
      L2=M1+M2+M3+1
      N1=N+1
C     *** FIX THE SIZE OF THE APPENDED MATRIX ***
      MM=M1+M2+M3+2
      NN=N+M1+M2+(2*M3)+1
C     *** INITIALIZE BOTH OBJECTIVES TO ZERO ***
      DO 10 I=L2,MM
        DO 10 J=1,NN
          A(I,J)=0.0
   10 CONTINUE
      L3=M1+M2+M3
```

TABLE C.1 (Continued)

```
C      *** INITIALIZE ALL SLACK AND ARTIFICIAL COEFFICIENTS ***
C      *** TO ZERO                                          ***
       DO 20 I=1,L3
          DO 20 J=L1,NN
             A(I,J) = 0.0
   20 CONTINUE
       L=N
       IF(M1.GT.0) THEN
C      *** FIX SLACK COEFFICIENTS FOR LE CONSTRAINTS ***
          DO 40 I=1,M1
             A(I,L+I)=1.0
   40     CONTINUE
          L=L+M1
       END IF
       IF(M3.GT.0) THEN
C      *** FIX SLACK COEFFICIENTS FOR GE CONSTRAINTS ***
          DO 70 I=1,M3
             I1=M1+M2+I
             A(I1,L+I)=-1.0
   70     CONTINUE
          L=L+M3
       END IF
       M4=M2+M3
       IF(M4.GT.0) THEN
C      *** FIX ARTIFICIAL COEFFICIENTS FOR EQ AND GE ***
C      *** CONSTRAINTS                                ***
          DO 100 I=1,M4
             A(M1+I,L+I)=1.0
  100     CONTINUE
       END IF
       M5=MM-1
C      *** PLACE THE OBJECTIVE FUNCTION ***
       DO 120 J=1,N
          A(M5,J)=C(J)
  120 CONTINUE
       I2=M1+M2+M3
       IF(M4.GT.0) THEN
          I1=M1+1
C      *** CONSTRUCT THE SECONDARY OBJECTIVE FUNCTION ***
          DO 140 J=1,N
             DO 140 I=I1,I2
                A(MM,J)=A(MM,J)+A(I,J)
  140     CONTINUE
          J1=N+M1+1
          J2=N+M1+M3
          IF(M3.GT.0) THEN
             DO 160 J=J1,J2
                A(MM,J) = -1.0
  160        CONTINUE
          END IF
          DO 180 I=I1,I2
             A(MM,NN)=A(MM,NN)+B(I)
  180     CONTINUE
       END IF
       DO 210 I = 1, I2
          A(I,NN)=B(I)
  210 CONTINUE
       A(M5,NN)=0.0
```

TABLE C.1   (Continued)

```
C       *** LOAD UP THE INITIAL BASIC SOLUTION ***
        IF(M1.GT.0) THEN
            DO 220 I=1,M1
                KK(I)=I+N
 220        CONTINUE
        END IF
        IF(M4.GT.0) THEN
            I1=M1+1
            DO 240 I=I1,I2
                KK(I)=N+M1+M3+I-I1+1
 240        CONTINUE
        ELSE
C       *** IF NO CONSTRAINT OF EQ AND GT TYPE, THEN   ***
C       *** NO SECONDARY OBJECTIVE FUNCTION IS NEEDED ***
            MM=MM-1
        END IF
        IFIRST=0
        U=0.0
        DO 500 WHILE(IFIRST.EQ.0.OR.U.GT.EPS)
            IFIRST=1
            N5=NN-1
C       *** CHECK FOR THE PIVOT ELEMENT                     ***
C       *** FIND THE LARGEST COEFFICIENT OF THE OBJECTIVE ***
            U=A(MM,1)
            J0=1
            DO 280 J=1,N5
                IF(A(MM,J).GT.U) THEN
                    U=A(MM,J)
                    J0=J
                ENDIF
 280        CONTINUE
            IF(U.GT.EPS) THEN
                KEY=0
                I0=0
                DO 300 WHILE(KEY.EQ.0.AND.I0.LT.I2)
                    I0=I0+1
                    IF(A(I0,J0).GE.EPS) KEY=1
 300            CONTINUE
C       *** TEST IF THERE IS A POSITIVE A(I,J) ***
C       *** IN THIS COLUMN, IF NOT THEN THE    ***
C       *** OBJECTIVE FUNCTION IS NOT BOUNDED  ***
                IF (KEY.EQ.0) THEN
                    KOD=3
                    RETURN
                ENDIF
C       *** FIND THE PIVOT ELEMENT FOR ELIMINATION ***
                IF(I0.LT.I2) THEN
                    I1=I0+1
                    U=A(I0,NN)/A(I0,J0)
                    DO 350 I=I1,I2
                        IF (A(I,J0).GE.EPS) THEN
                            IF((A(I,NN)/A(I,J0)).LT.U) THEN
                                I0=I
                                U=A(I,NN)/A(I,J0)
                            END IF
                        END IF
 350                CONTINUE
                END IF
```

TABLE C.1   (Continued)

```
C                    *** PERFORM THE ELIMINATION STEP ***
               U=A(I0,J0)
               DO 370 J=1,NN
                  A(I0,J)=A(I0,J)/U
      370      CONTINUE
               DO 410 I=1,MM
                  IF(I.NE.I0) THEN
                     DO 400 J=1,NN
                        IF(J.NE.J0) THEN
                           A(I,J)=A(I,J)-A(I0,J)*A(I,J0)
                        END IF
      400            CONTINUE
                  END IF
      410      CONTINUE
               DO 430 I=1,MM
                  IF(I.NE.I0) THEN
                     A(I,J0)=0.0
                  END IF
      430      CONTINUE
C                    *** REGISTER THE NEWEST BASIS VECTOR AND GO       ***
C                    *** BACK TO PERFORM THE NEXT STEP OF ELIMINATION ***
               KK(I0)=J0
            ELSE
               IF(M4.GT.0) THEN
                  L=N+M1+M3
C                    *** CHECK FOR FEASIBLE SOLUTION FROM THE   ***
C                    *** SECONDARY OBJECTIVE                    ***
                  IF(A(MM,NN).GT.0.0) THEN
                     DO 460 I=1,I2
                        IF(KK(I).GT.L) THEN
                           KOD = 2
                           RETURN
                        END IF
      460            CONTINUE
                  END IF
                  MM=MM-1
C                    *** REMOVE THE ARTIFICIAL VARIABLES AND THE ***
C                    ***         SECONDARY OBJECTIVE FUNCTION    ***
                  M4=0
                  NM=NN-M2-M3
                  DO 490 I=1,MM
                     A(I,NM)=A(I,NN)
      490            CONTINUE
                  NN=NM
                  IFIRST=0
               END IF
            END IF
C                 *** FROM HERE WE GO BACK TO CONTINUE ELIMINATION ***
      500 CONTINUE
            KOD = 1
C          *** SET UP THE OPTIMAL SOLUTION ***
         DO 510 I=1,N
            X(I)=0.0
      510 CONTINUE
         DO 520 I=1,I2
            J = KK(I)
            X(J)=A(I,NN)
      520 CONTINUE
         X(N+1)=-A(MM,NN)
         RETURN
         END
```

C.1
CALLING PARAMETERS

It is assumed that the linear programming problem is formulated as follows:

Maximize $c_1 x_1 + c_2 x_2 + \cdots + c_n x_n$

subject to

$$a_{11} x_1 \quad + a_{12} x_2 \quad + \cdots + a_{1n} x_n \quad \leq b_1$$

$$\vdots \qquad\qquad \vdots \qquad\qquad\qquad \vdots \qquad\qquad \vdots$$

$$a_{M_1 1} x_1 \quad + a_{M_1 2} x_2 \quad + \cdots + a_{M_1 n} x_n \quad \leq b_{M_1}$$

$$a_{M_1+1,1} x_1 \quad + a_{M_1+1,2} x_2 \quad + \cdots + a_{M_1+1,n} x_n \quad = b_{M_1+1}$$

$$\vdots \qquad\qquad \vdots \qquad\qquad\qquad \vdots \qquad\qquad \vdots \qquad\qquad \text{(C.1)}$$

$$a_{M_1+M_2,1} x_1 \quad + a_{M_1+M_2,2} x_2 \quad + \cdots + a_{M_1+M_2,n} x_n \quad = b_{M_1+M_2}$$

$$a_{M_1+M_2+1,1} x_1 \quad + a_{M_1+M_2+1,2} x_2 \quad + \cdots + a_{M_1+M_2+1,n} x_n \quad \geq b_{M_1+M_2+1}$$

$$\vdots \qquad\qquad \vdots \qquad\qquad\qquad \vdots \qquad\qquad \vdots$$

$$a_{M_1+M_2+M_3,1} x_1 = a_{M_1+M_2+M_3,2} x_2 \quad + \cdots + a_{M_1+M_2+M_3,n} x_n \quad \geq b_{M_1+M_2+M_3}$$

$$x_i \geq 0, \ 1 \leq i \leq n.$$

The number of variables n is the LINPRO calling parameter N. The first M_1 constraints are $\leq$ type, followed by M_2 constraints of $=$ type, and M_3 constraints of $\geq$ type. The calling parameters $M1$, $M2$, and $M3$ are as just described. The input parameter EPS is a tolerance bound for the magnitude of the pivot element.

Output vector X gives the optimum objective function value and the optimal solution as $X(N+1)$ and $(X(1), \ldots, X(N))$, respectively. The output flag KOD shows which one of three possibilities has occurred during the elimination process. If KOD $= 1$, the optimum exists and has been determined; if KOD $= 2$, no feasible solution exists; the constraints of the problem are contradictory. If KOD $= 3$, the objective function does not have an upper bound. That is, for an arbitrarily large number Q there exists a feasible solution such that the corresponding value of the objective function is greater than Q.

As written, LINPRO assumes that the calling program has the array declaration COMMON A(52, 151), B(50), C(50). These arrays should be assigned values according to the linear programming structure of (C.1). The common declaration was used to avoid unnecessary memory burden. The user may redimension these arrays in LINPRO in accordance with his or her needs.

───── **EXAMPLE C.1** ─────

Here we apply LINPRO to an example problem that is worked out in detail in Luenberger (1973, p. 45). As stated in Luenberger, it is a minimization problem, but any such problem can be converted to a maximization problem by simply taking the negative of the coefficients of the objective function. After such a conversion the problem is

$$\text{Maximize } -4x_1 - x_2 - x_3$$
$$\text{subject to}$$

$$2x_1 + x_2 + 2x_3 = 4$$

$$3x_1 + 3x_2 + x_3 = 3$$

and

$$x_1 \geq 0, \quad x_2 \geq 0, \quad x_3 \geq 0.$$

Here $N = 3$, $M1 = M3 = 0$, and $M2 = 2$. Somewhat arbitrarily, we took $EPS = 10^{-4}$. The calling program associated with this linear programming problem is given as Table C.2. From the output value $KOD = 1$, we see that the program "thinks" it was successful. The solution values x_1, x_2, and x_3 are given by the corresponding coordinates of the output array X, and are 0, 0.4, and 1.8, respectively, which is in accordance with Luenberger's calculations. The optimal objective function value, as given by output X(4), is -2.2.

TABLE C.2   A Calling Program for Linear Programming

```
C       PROGRAM LINIARP
C
C       *****************************************************************
C       THIS PROGRAM ASSIGNS THE VALUES REQUIRED (A,B, AND C) FOR A
C       SPECIFIC LINEAR PROGRAMMING PROBLEM. SUBROUTINE LINPRO IS
C       THEN CALLED TO SOLVE THE PROBLEM.  THE RESULTING SOLUTION
C       IS THEN PRINTED
C       CALLS:   LINPRO
C       OUTPUT:
C                KOD=THE CODE SPECIFYING THE CATEGORY OF THE PROBLEM
C                X(I)=THE LEVELS OF THE VARIABLES IN THE SOLUTION
C                     FOR I=1,2,3
C                X(4)THE OPTIMAL VALUE OF THE OBJECTIVE FUNCTION
C       *****************************************************************
C
        COMMON A(52,151),B(50),C(50)
        DIMENSION X(51)
        N=3
C
C       *** M1,M2,M3 SPECIFY THE NUMBER AND TYPE OF CONSTRAINTS   ***
C
        M1=0
        M2=2
        M3=0
        EPS=1.E-4
```

TABLE C.2 (Continued)

```
C
C     *** A IS THE MATRIX OF COEFFICIENTS,                          ***
C     *** B IS THE RIGHT HAND SIDE VECTOR, AND                      ***
C     *** C IS THE VECTOR OF OBJECTIVE FUNCTION COEFFICIENTS        ***
C
      A(1,1)=2.
      A(1,2)=1.
      A(1,3)=2.
      A(2,1)=3.
      A(2,2)=3.
      A(2,3)=1.
      B(1)=4.
      B(2)=3.
      C(1)=-4.
      C(2)=-1.
      C(3)=-1.
C
C     *** SUBROUTINE LINPRO WILL SOLVE THE PROBLEM                  ***
C
      CALL LINPRO(N,M1,M2,M3,EPS,X,KOD)
      WRITE(10,*)(X(I),I=1,N+1),'KOD=  ',KOD
      STOP
      END
```

Answers to Selected Problems

Chapter 1

1. (a) $(111001)_2 = 57$ (b) $(110.1101)_2 = 6\frac{13}{16}$

2. (a) 34.1   (b) 301.   (d) 1

3. Mantissa $= 2^{-1} + 2^{-2} + 0 + \cdots + 0 = 0.75$

exponent $= 31$

so $x* = 0.75 \cdot 2^{31} \simeq 1.61 \cdot 10^9$.

4. 2^t.

10. "Wise" procedure: Use the fact that $\sin(10) = -\sin(10 - 3\pi)$ (Why?) and evaluate the series at this smaller argument.

12. $2^{-1}2^{-(2^e-1)}$, where e is the number of bits allotted to the exponent in the floating-point segmentation of a computer word. We calculated 2^{-128} on a VAX.

13. (a) 0.0724

14. (a) $2.0286 \cdot 10^{-3}$

19. Absolute error, using the formula: area $= \frac{1}{2}(x*)^2 \tan(\theta)$, is 2.68.

Chapter 2

6. $x^{1/2} = x_0^{1/2} + \frac{1}{2}x_0^{-1/2}(x - x_0) + \frac{1}{2!}\left(\frac{1}{2}\left(-\frac{1}{2}\right)x_0^{-3/2}(x - x_0)^2\right)$,

$+ \cdots + \frac{1}{n!}\left[\left(\frac{1}{2}\right)\left(-\frac{1}{2}\right)\cdots\left(-\frac{2n-3}{2}\right)\right]x_0^{(-(2n-1)/2)}(x - x_0)^n + \cdots,$

for $|x - x_0| < x_0$.

12. **(a)** Values of $\sin(2t_i) - p(t_i)$:

t_i	(i) Taylor's Polynomial Error at t_i	(ii) Interpolation Error, Equally Spaced Data Points	(iii) Interpolation Error, Chebyshev Points
0.000E + 00	0.295E − 01	0.000E + 00	0.356E − 02
0.100E + 00	0.126E − 01	− 0.564E − 02	− 0.313E − 02
0.200E + 00	0.414E − 02	− 0.490E − 02	− 0.326E − 02
0.300E + 00	0.847E − 03	− 0.131E − 02	− 0.301E − 03
0.400E + 00	0.546E − 04	0.224E − 02	0.286E − 02
0.500E + 00	0.000E + 00	0.375E − 02	0.424E − 02
0.600E + 00	0.575E − 04	0.236E − 02	0.301E − 02
0.700E + 00	0.939E − 03	− 0.145E − 02	− 0.333E − 03
0.800E + 00	0.484E − 02	− 0.573E − 02	− 0.381E − 02
0.900E + 00	0.155E − 01	− 0.695E − 02	− 0.386E − 02
0.100E + 01	0.383E − 01	0.119E − 06	0.463E − 02

13. **(a)** $f(x) = \dfrac{6x(x+1)(x-1)}{-6} + \dfrac{4(x+2)x(x-1)}{2} + \dfrac{3(x+2)(x+1)(x-1)}{-2}$

$$+ \dfrac{3x(x+2)(x+1)}{6}$$

21. **(a)** Is a linear ($m = 1$) spline.

Chapter 3
1. **(a)** $q = 1$
2. For $f(x) = e^x$, the derivative estimate at $x = 0$ by the three-point formula given in Table 3.3. For the four-point formula, using double precision, we find

h	Four-Point Estimate	Error
0.1	0.9999131903696	8.681E − 05
0.01	0.9999999163319	8.377E − 08
0.001	0.9999999999166	8.44E − 11
0.0001	0.9999999999997	3.E − 13

7. The weight at $x = 1.0$ is 0; at 1.5 is 1.0.
8. The weight at x_2 is 0.3555. Find the others.
10. **(b)** No.
12. **(c)** $\qquad s_2 = -1.31$

error of $s_4 = 1.36$

$$s_{10} = -0.019$$

15. **(a)** Using a VAX computer in single precision, we read out from subroutine ROMB, values $E - T(1, M+1)$. Call these differences d_M. For $M = 2, 3, 4, \ldots,$

we found the ratios to be

M	$\dfrac{d_M}{d_{M-1}}$
2	0.021271
3	0.249195
4	0.249807
5	0.250012
6	0.250199
7	0.249601
8	0.253968
9	0.259259
10	0.200000
11	0.363636
12	0.500000

20. (b)

Number of Interpolation Points	Error by	
	Newton-Cotes	Chebyshev Points
3	$-1.165E-2$	$-4.29E-3$
4	$-5.245E-3$	$-7.62E-4$
5*	$-6.866E-5$	$-8.58E-6$
6*	$3.886E-5$	$-2.86E-6$

*Machine-dependent. At these accuracies, the roundoff error is of the same order as truncation.

Chapter 4
1. (a) $X(1) = 3$, $X(2) = X(3) = 0$. But remember to show your derivation.
2. For $n = 5$, on the VAX (or essentially any other computer) the residual is in the order of machine epsilon; the solution is, correctly rounded, $X(1) = 0.414$, $X(2) = 3.52$, $X(3) = -3.41$, $X(4) = 2.45$, and $X(5) = -0.742$.
9. For $n = 5$, on a VAX, the residual is in the order of 10^{-4}, and the solution vector was computed to be $X(1) = 28.04$, $X(2) = -660.7$, $X(3) = 3391.1$, $X(4) = -5888.0$, and $X(5) = 3225.0$.

12. $A^{-1} = \begin{bmatrix} 0.500 & 0.500 & 0.000 \\ 0.750 & -0.417 & -0.167 \\ -0.250 & -0.083 & 0.167 \end{bmatrix}$

Remember to show your derivation.
15. -11.0
18. (a) $X(1) = -1.0$, $X(2) = 1.0$, and $X(3) = 0$.

26.

After Iteration	X(1)	X(2)	X(3)
1	0.80000	−0.47300	0.75318
2	0.80395	−0.470557	0.75318
Answer	0.80395	−0.470557	0.75318

Chapter 5

1. $p(x)$ is increasing for the intervals:

$$-\infty < x < 2 - \frac{\sqrt{6}}{3} \approx 1.184 \quad \text{and} \quad \infty > x > 2 + \frac{\sqrt{6}}{3} \approx 2.816$$

The polynomial has three real roots.

2. The interval $(0.1, 1.1)$ contains a root. Can you find another interval?

4. (a) $(4, 5)$

7.

k	x_k	$f(x_k)$
0	3	-2
1	4	3
2	3.5	-0.625
3	3.75	0.859375

8. -3.045

9. $-0.768039, 1.678348$

10. 40 steps

13. $-0.768039, 1.678347$

14. $x^2 - 30 = 0$, $\sqrt{30} \approx 5.477226$

16. $x_0 = 3$, $x_1 = 5$, $x_2 = 4.2$.

19. 35.1283361

25. $q(z) = z^3 + (-1 + i)z^2 + z + (-1 + i)$

26. $q(z) = z^2 - 2z + 2$

29. $$\begin{pmatrix} x^{(k+1)} \\ y^{(k+1)} \end{pmatrix} = \begin{pmatrix} x^{(k)} \\ y^{(k)} \end{pmatrix} - \begin{pmatrix} -\sin(x^{(k)}) & -1 \\ 1 & -\cos(y^{(k)}) \end{pmatrix}^{-1} \begin{pmatrix} \cos(x^{(k)}) - y^{(k)} \\ x^{(k)} & -\sin(y^{(k)}) \end{pmatrix}$$

Chapter 6

1. $p_0(x) = 4$, $p_1(x) = -x + \frac{7}{2}$

2. For $N = 7$ points, using a VAX in single precision, degree $M = 0$: $a_0 = 0.533$; degree $M = 5$: $a_0 = 3.18 \cdot 10^{-5}$, $a_1 = 0.987$, $a_2 = 4.98 \cdot 10^{-2}$, $a_3 = -0.232$, $a_4 = 3.68 \cdot 10^{-2}$, $a_5 = 1.8 \cdot 10^{-5}$

3. (a) $a = \frac{5}{6}$, $b = \frac{3}{2}$

6. $p_0(x) = \dfrac{n + 1}{4n + 2}$, $\lim\limits_{x \to \infty} p_0(x) = \frac{1}{4}$

8. $p_0(x) = 4.5$, $p_1(x) = \frac{7}{2} - x$.

11. $p_0(x) = \frac{1}{2}$, which differs from the result obtained in Problem 6.

12. For $N = 7$ points, using a VAX in double precision, degree $M = 0$: $a_0 = 0.5$; degree $M = 5$: $a_0 = 3.005 \cdot 10^{-4}$, $a_1 = 0.9846$, $a_2 = 5.3 \cdot 10^{-2}$, $a_3 = -0.233$, $a_4 = 3.713 \cdot 10^{-2}$, $a_5 = 2 \cdot 10^{-17}$. (Note that the larger coefficients are not far from those of the solution to Problem 2.)

13. (a) $\varphi_0(x) = 1$, $\varphi_1(x) = x - 2$, $\varphi_2(x) = x^2 - 4x + \frac{10}{3}$
    (b) $\varphi_0(x) = 1$, $\varphi_1(x) = x - \frac{1}{2}$, $\varphi_2(x) = x^2 - x + \frac{1}{6}$

Chapter 7

1. (a) $y = 1 + \frac{1}{3}x^3$

3. $y(x) = 1 + x + x^2 + \frac{5}{6}x^3 + \frac{11}{24}x^4 + \frac{11}{40}x^5$

5. Euler solution to Problem 1(a), at time $t = 1$.

Using Step Size h	Solution Estimate $y(1)$
0.1	1.320
0.001	1.395

7. Classical Runge-Kutta solution to Problem 4(a), at time $t = 1$.

Using Step Size h	Solution Estimate $y(1)$
0.1	1.396
0.001	1.396

8. $\alpha_1 = \frac{1}{3}$, $\alpha_2 = \frac{2}{3}$, $\mu_2 = \frac{3}{4}$, $\lambda_{21} = \frac{3}{4}$, and it is unique.

14. Multidimensional classical Runge-Kutta solution to Problem 13(a), at time $t = 1$:

For Step Size h	Solution Estimate $y_1(1)$	$y_2(1)$
0.1	8.68	16.0
0.01	8.76	16.4

15. **(a)** $y_1' = y_2$ $\qquad\qquad y_1(0) = 0$

$\qquad\quad y_2' = y_3$ $\qquad\qquad y_2(0) = 0$

$\qquad\quad y_3' = x + y_1 + 2y_2 + y_3^2$ $\qquad y_3(0) = 0$

18. $y(x) = xe^x - e^x$ is the only solution with $y(0) = -1$, $y(1) = 0$. It satisfies $y(2) = e^2$, but therefore $y(2) \neq 0$.

List of Computer Programs

List of Subroutines

Index